Assessing the Accuracy of Remotely Sensed Data

Principles and Practices, Third Edition

T0314423

Assessing the Accuracy of Remotely Sensed Data

Principles and Practices, Third Edition

by

Russell G. Congalton and Kass Green

CRC Press
Taylor & Francis Group
Boca Raton London New York

CRC Press is an imprint of the
Taylor & Francis Group, an **informa** business

CRC Press
Taylor & Francis Group
6000 Broken Sound Parkway NW, Suite 300
Boca Raton, FL 33487-2742

First issued in paperback 2020

© 2019 by Taylor & Francis Group, LLC
CRC Press is an imprint of Taylor & Francis Group, an Informa business

No claim to original U.S. Government works

ISBN-13: 978-1-4987-7666-0 (hbk)
ISBN-13: 978-0-367-65667-6 (pbk)

Visit the Taylor & Francis Web site at
http://www.taylorandfrancis.com

and the CRC Press Web site at
http://www.crcpress.com

The third edition of this book is dedicated to our spouses, Jeanie Congalton and Gene Forsburg, for their constant support, encouragement, and love. It is also dedicated to a very special friendship between the authors that has stood the test of time for over 32 years and is still going strong.

Contents

Preface

Our ability to map and monitor our earth using remotely sensed data has improved rapidly over the last decade since the second edition of this book was published. These advancements include amazing leaps forward in microelectronics, digital sensor technology, object-based analysis, sensor platforms, machine learning, and cloud computing. Therefore, the principles and especially the practices for assessing the accuracy of these maps created from remotely sensed data have also continued to develop and mature. The first edition of this book, published in 1999, had only eight chapters. The second edition, published about 10 years later, expanded to 11 chapters, including the very significant addition of a new chapter on positional accuracy. This third edition contains 15 chapters, including a new chapter on better planning for an assessment, a thoroughly revised positional accuracy chapter, a new chapter on object-based accuracy assessment, two case studies (a polygon-based assessment of an object-based classification map and a more classic example), and a new summary chapter with helpful bullets emphasizing key issues. While it presents the salient principles needed to conduct a valid accuracy assessment, the strength of this book is the thorough presentation and discussion of the practical considerations that must be understood for a geospatial analyst to conduct a successful assessment. These concepts are also important for the map user to understand so as to make effective use of any map. Not every method or approach or descriptive measure suggested in the literature is covered here. Rather, this book is written for those who wish to conduct a valid and successful assessment of a map to better understand the errors, limitations, and usefulness of that map and for those who wish to be effective users of the map. As such, this book emphasizes the practical considerations, tradeoffs, and decisions that must be deliberated at the beginning of a mapping project and the awareness that must be maintained throughout the project and during the accuracy assessment. The authors have conducted many map accuracy assessments over the last 35 years and have faced every possible situation, limitation, and difficulty imaginable (and some that were unimaginable). It is our hope that this book provides you with the guidance and encouragement necessary to be successful, whether you are assessing the accuracy of a map or using the map for a specific objective and needing to understand its accuracy.

Acknowledgments

Every book is the responsibility of the authors but comes from a collection of ideas, encouragement, and suggestions of countless others. By the time a third edition of a book is published, there are many people to thank and acknowledge. The first edition of this book was inspired by Dr. John Lyon and dedicated to Dr. Roy Mead. In addition, we are especially thankful to Dr. John Jensen, Dr. Greg Biging, Dr. Tom Lillesand, Dr. Jim Smith, Mr. Ross Lunetta, Mr. Mike Renslow, Dr. George Lee, and Dr. Jim Campbell for their positive feedback and support. The second edition added a very significant chapter on positional accuracy with the help of Drs. George Lee, Greg Biging, and Dave Maune.

This chapter has been significantly revised in the third edition with significant input from Dr. Qassim Abdullah. Thanks also to the American Society for Photogrammetry and Remote Sensing (ASPRS) for allowing the publication of the new positional accuracy standards as an appendix in this edition. We would also like to thank Dr. Michael Kearsley of the National Park Service for allowing us to include the Grand Canyon National Park vegetation map accuracy assessment as a case study in this edition.

Along this 30-year adventure, many colleagues at Pacific Meridian Resources, Space Imaging Solutions, Fugro-Earthdata, Dewberry, Tukman Geospatial, and Sanborn Solutions, as well as numerous graduate students in the Department of Natural Resources and the Environment at the University of New Hampshire, helped make all the editions of this book better.

Significant contributions to this third edition were made by Mark Tukman of Tukman Geospatial, Dr. J.B. Sharma from the University of North Georgia, and Chad Lopez of Dewberry. Kamini Yadav, Heather Grybas, and Ben Fraser, graduate students at the University of New Hampshire, also contributed extensively to this edition. We also gratefully acknowledge all of our friends and colleagues in the geospatial community who inspired and encouraged us on many occasions. Finally, we would like to thank our families for the time they managed without us while we worked on this book.

Authors

Russell G. Congalton is Professor of Remote Sensing and GIS in the Department of Natural Resources and the Environment at the University of New Hampshire. He is responsible for teaching courses in geospatial analysis, including The Science of Where, Remote Sensing, Photogrammetry and Image Interpretation, Digital Image Processing, and Geographic Information Systems. Russ has authored or coauthored more than 200 papers and conference proceedings. He is the author of 12 book chapters, co-editor of a book on spatial uncertainty in natural resource databases titled *Quantifying Spatial Uncertainty in Natural Resources: Theory and Applications for GIS and Remote Sensing,* and co-author of five books, including *Imagery and GIS: Best Practices for Extracting Information from Imagery,* published by Esri in 2017. Russ served as president of the ASPRS in 2004–2005, as the National Workshop Director for ASPRS from 1997 to 2008, and as editor-in-chief of *Photogrammetric Engineering and Remote Sensing* from 2008 to 2016. He was elected a Fellow in 2007 and an Honorary Member in 2016.

Dr. Congalton received a BS (Natural Resource Management) from Rutgers University in 1979. He earned an MS (1981) and a PhD (1984) in remote sensing and forest biometrics from Virginia Tech. In addition to his academic position, Russ served as chief scientist of Pacific Meridian Resources from its founding in 1988 until 2000 and then as chief scientist of Space Imaging Solutions from 2000 to 2004. From 2004 until 2015, Russ held the position of senior technical advisor with the Solutions Group of Sanborn, the oldest mapping company in the United States. Finally, he has been the New Hampshire View Director, part of the AmericaView Consortium, since 2007 and has served as member, secretary, vice-chair, and chair of the AmericaView Board of Directors.

Kass Green's experience spans 30 years of managing and supervising GIS and remote sensing professionals for vegetation mapping, as well as leadership in GIS and remote sensing research and policy. In 1988, Ms. Green co-founded Pacific Meridian Resources, a GIS/remote sensing firm, which she grew to 75 employees in seven offices nationwide and sold to Space Imaging (now Digital Globe) in 2000. After running half of Space Imaging for 3 years, Ms. Green decided to focus her career on challenging remote sensing mapping and policy projects for public agencies, development organizations, and nongovernmental organizations. Over the last 10 years, Ms. Green has had the pleasure of using object-oriented techniques to create detailed vegetation maps of Grand Canyon National Park, the national parks of Hawaii, and Sonoma County, California from high-resolution optical imagery, lidar data, Landsat imagery, and multiple other data sets. Her work also includes international market studies and strategy papers for organizations such as the Omidyar Network, The Bill and Melinda Gates Foundation, and the UK Department of International Development (DFID).

Ms. Green is a fellow and a lifetime honorary member of ASPRS and a past president of both Management Association for Private Photogrammetric Surveyors (MAPPS) and ASPRS. She is the principal author of a recently published text for Esri Press titled *Imagery and GIS: Best Practices for Extracting Information from Imagery*. Ms. Green chairs NASA's Earth Science Applications Committee, co-founded and chaired the Department of the Interior's Landsat Advisory Group, and participated in the Landsat Science Team from 2013 to 2017 as an *ex officio* member.

1 Introduction

WHY MAP?

The earth's resources are scarce. As we continue to add more people to the earth, the scarcity of resources increases, as does their value. From land use conversion throughout the world, to fragmentation of tropical bird habitat, to polar bear habitat loss in the Arctic, to the droughts in Africa, and wars worldwide, people have significantly affected the resources and ecosystems of the world. The ever-increasing world population and need for all types of resources continue to cause the price of these resources to increase and to intensify conflicts over resource allocation.

As resources become more valuable, the need for timely and accurate information about the type, quantity, and extent of resources multiplies. Allocating and managing the earth's resources requires accurate knowledge about the distribution of resources across space and time. To efficiently plan emergency response, we need to know the location of roads relative to fire and police stations, hospitals, and emergency shelters. To improve the habitat of endangered species, we need to know what the species habitat requirements are, where that habitat exists, where the animals exist, and how changes to the habitat and surrounding environments will affect species distribution, population, and viability. To plan for future developments, we need to know where people will work, live, shop, and go to school. To grow enough food for an ever-increasing population, we need information about the spatial distribution and yields of our agricultural regions. Because each decision (including the decision to do nothing) impacts (1) the status and location of resources and (2) the relative wealth of individuals and organizations who derive value from the resources, knowing the location of resources and how they interact spatially is critical to effectively managing those resources and ourselves over time.

WHY ASSESS THE ACCURACY OF A MAP?

Decisions about resources require maps; and effective decisions require accurate maps or, at least, maps of known accuracy. For centuries, maps have provided important information concerning the distribution of resources across the earth. Maps help us to measure the extent and distribution of resources, analyze resource interactions, and identify suitable locations for specific actions (e.g., development or preservation), plan future events, and monitor change. If our decisions based on map information are to have the expected results, the accuracy of these maps must be known. Otherwise, implementing any decisions based on these maps will result in surprises, and often these surprises may be unacceptable.

For example, suppose that you wish to have a picnic in a forest on the edge of a lake. If you have a map that displays forest, crops, urban, water, and barren land cover types, you can plan the location of your picnic. If you don't know the accuracy of the map, but the map is 100% accurate, you will be able to travel to your forest lakeside location and, in fact, find yourself in a nice picnic spot. However, if the map is not spatially accurate, you may find that your picnic location falls in the middle of the lake rather than on the shore; and if the map is not labeled correctly (i.e., themati-cally accurate), you may find yourself in a city next to a fountain or in an agricultural field next to an irrigation ditch. Conversely, if you know the accuracy of the map, you can incorporate the known expectations of accuracy into your planning and create contingency plans in situations when the accuracy is low. This type of knowl-edge is critical when we move from our lighthearted picnic example to more critical decisions such as endangered species preservation, resource allocation, feeding our growing populations, peace-keeping actions, and emergency response.

There are many reasons for performing an accuracy assessment. Perhaps the sim-plest reason is curiosity—the desire to know how good a map you have made. In addition to the satisfaction gained from this knowledge, we also need or want to increase the quality of the map information by identifying and correcting the sources of errors. Third, analysts often need to compare various techniques, algorithms, ana-lysts, or interpreters to test which is best. Also, if the information derived from the remotely sensed data is to be used in some decision-making process (i.e., geographic information system [GIS] analysis), then it is critical that some measure of its qual-ity be known. Finally, it is more and more common that some measure of accuracy is included in the contract requirements of many mapping projects. Therefore, valid accuracy is not only useful but may be required.

Accuracy assessment determines the quality of a map created from remotely sensed data. Accuracy assessment can be qualitative or quantitative, expensive or inexpensive, quick or time-consuming, well designed and efficient, or haphazard. The goal of quantitative accuracy assessment is the identification and measurement of map errors so that the map can be as useful as possible to the persons using it to make decisions.

The central purpose of this book is to present the necessary theory and principles for conducting a quantitative accuracy assessment along with the practical consider-ations of how to effectively and efficiently design and implement such an assessment. Throughout the book, we emphasize that no one single recipe exists for conducting an accuracy assessment. Just as there is no one way to produce a map; there is no one way to assess the accuracy of a map. Instead, this book will teach you to consider every aspect of a mapping project and to design and implement the best possible assessment given the strengths and limitations of each mapping project you conduct, fund, or rely on. This book is not written to be an academic review of every possible idea or method ever published on map accuracy assessment. Instead, it is written for the geospatial analyst who wishes to best conduct a valid and effective assessment of their particular mapping project. As such, the considerations and limitations of such an assessment are emphasized here to best lead the analyst through the process.

TYPES OF MAP ACCURACY ASSESSMENT

There are two types of map accuracy assessment: positional and thematic. Positional accuracy deals with the accuracy of the location of map features and measures how far a spatial feature on a map is from its true or reference location on the ground (Bolstad, 2005). Thematic accuracy deals with the labels or attributes of the features of a map and measures whether the mapped feature labels are different from the true or reference feature label. For example, in the picnic example, the earth's surface was classified as forest, water, crops, urban, or barren. We are interested in both the accuracy of the location of the features, so that we can locate our picnic spot in a forest on the shore of a lake, and in the thematic accuracy, so that we truly end up in a forest and not in a city, desert, or agricultural field that was erroneously mapped as forest.

The accuracy of any map or spatial data set is a function of both positional accuracy and thematic accuracy, and this book considers both. However, because thematic accuracy is much more complex than positional accuracy, the book devotes considerably more attention to thematic accuracy assessment.

CRITICAL STEPS IN ACCURACY ASSESSMENT

As previously stated, there is no one single procedure for conducting either a positional or a thematic accuracy assessment. However, all accuracy assessments include these fundamental steps:

1. Consider the factors involved in the assessment.
2. Design the appropriate sampling approach to collect the reference data.
3. Conduct the sampling.
4. Analyze the data.
5. Report the statistics/results.

Each step must be rigorously planned and implemented. First, the accuracy assessment sampling procedures are designed, and the sample areas on the map are selected. We use sampling because time and funding limitations preclude the assessment of every spatial unit on the map. Next, information is collected from both the map and the reference data for each sample site. Thus, two types of information are collected from each sample:

- Reference accuracy assessment sample data: the position or map class label of the accuracy assessment site, which is derived from data collected that are assumed to be correct
- Map accuracy assessment sample data: the position or map class label of the accuracy assessment site derived from the map or image being assessed

Third, the map and reference information are compared with one another, the results of the comparison are analyzed for statistical significance and for reasonableness, and a report is prepared, which presents the methods and results of the assessment. In summary, effective accuracy assessment requires (1) design and implementation of unbiased sampling procedures, (2) consistent and accurate collection of sample data, and (3) rigorous comparative analysis of the sample map and reference data and reporting of the results.

Because there is no one procedure for designing and implementing an accuracy assessment, there are quite a number of important questions to ask and considerations to think about when conducting a valid assessment. This book addresses the most important ones, including:

1. Questions concerning the design of an accuracy assessment sample approach:
   - What are the map classes to be assessed, and how are they distributed across the landscape?
   - What is the appropriate sampling unit?
   - How many samples should be taken?
   - How should the samples be chosen?
2. Questions concerning how the reference data should be collected:
   - What should be the source of the reference data?
   - How should the reference data be collected?
   - When should the reference data be collected?
   - How do I ensure consistency and objectivity in my data collection?
3. Questions concerning how the analysis should be conducted:
   - What are the different analysis techniques for continuous versus discontinuous map data?
   - What is an error matrix, and how should it be used?
   - What are the statistical properties associated with the error matrix, and what analysis techniques are applicable?
   - What is fuzzy accuracy, and how can you conduct a fuzzy accuracy assessment?
   - What is object-based accuracy, and how can you conduct an object-based accuracy assessment?
   - How is an accuracy assessment conducted on change detection maps?
   - How is an accuracy assessment conducted on maps created from multiple layers of data?

ORGANIZATION OF THE BOOK

The organization of this book takes you through each of these fundamental accuracy assessment steps as follows:

- Chapter 2 begins with a review of the history and basic assumptions of map making and accuracy assessment.
- Chapter 3 is a new chapter in this edition and provides the reader with frameworks for planning out accuracy assessments. An important component of any good assessment is proper planning from the beginning of the mapping project. This chapter aids the reader in getting started on making an effective plan by identifying the components of the assessment necessary for their project. It is expected that this chapter will be reviewed by the analyst prior to beginning each and every accuracy assessment.

- Chapter 4 is a thoroughly revised presentation on positional accuracy assessment including a review of past standards and concluding with the latest and most effective *ASPRS Positional Accuracy Standards for Digital Geospatial Data* (ASPRS, 2014). This complete document is included as an appendix.
- Chapters 5 and 6 provide some of the basic methods and considerations for thematic map accuracy assessment, including the introduction of the error matrix and a thorough review of sample design considerations.
- Chapter 7 is devoted to factors that must be taken into account during the collection of reference data. The collection of sufficient and valid reference data is key to the success of any thematic accuracy assessment, and this chapter describes the many considerations, decisions, and practical details that are part of this process.
- Chapters 8 through 11 detail thematic accuracy assessment analysis, which is much more complex than positional accuracy assessment analysis and hence requires more chapters. The basic analysis techniques that can be applied to an error matrix are discussed in Chapter 8. Chapter 9 discusses the causes of differences in the error matrix, whether from map errors or from other non-error sources. Chapter 10 presents a solution to some of the non-error differences in the error matrix by suggesting the use of fuzzy accuracy assessment. Finally, Chapter 11 is a new chapter that presents the idea of an object-based accuracy assessment. Given the shift from pixel-based to object-based classification approaches, especially for high–spatial resolution imagery, it is necessary to explore accuracy assessment approaches that incorporate objects and not simply pixels in the assessment.
- Chapters 12 and 13 provide two very different case studies that give the reader some real-life examples of the complexities involved in conducting a map accuracy assessment. Neither case study is perfect, and both have flaws, considerations, and limitations that must be addressed. The goal is not to have the reader follow either case study for their assessment but rather, to use the thought process and detailed explanations provided to better understand the entire process. Chapter 12 presents a case study using a polygon-based accuracy assessment of an object-based vegetation map created for Grand Canyon National Park in 2014. It reviews all the design, data collection, and analysis methods and considerations presented in Chapters 5 through 11. Chapter 13 is a case study from the 1990s, and while dated, it presents a classic accuracy assessment that evaluates two maps—the first a polygon map created from manual photo interpretation, and the second a pixel-based map created using semi-automated image analysis. This case study contains extensive detail and explanation, beyond what would ever be published in a peer-reviewed paper. Reading through the thought process in this case study will help the reader to better understand the entire assessment.
- Chapter 14 delves into more advanced topics in accuracy assessment, including change detection accuracy assessment and multi-layer accuracy assessment.

- Chapter 15 summarizes the book by emphasizing lessons learned from the authors' implementation of dozens of accuracy assessments over the last three decades. While mapping technologies have changed significantly over the last 30 years and will continue to evolve rapidly, the basic principles of mapping and accuracy assessment endure with little change. This chapter's action statements remind and encourage the reader regarding the key components of any valid and effective accuracy assessment.

2 The History of Map Accuracy Assessment

HOW MAPS ARE MADE

Before the invention of aircraft, maps were created from human observations made on the earth's surface using survey equipment and the most basic yet sophisticated remote sensing devices—the human eyes and the analytical capabilities of the human brain. By the early sixteenth century, Portuguese navigators were able to map the coast of Africa (see Figure 2.1) by relying on measurements taken at sea from astrolabes, quadrants, cross-staffs, and other early navigation tools. During their exploration of the American Northwest, Lewis and Clark were able to produce the remarkably detailed map shown in Figure 2.2. Indian pundits secretly mapped the Himalayas to high precision in the mid-1800s by pretending to be Buddhist pilgrims (Hopkirk, 1992)—keeping count of their paces using holy beads, and concealing compasses and other instruments in their clothing and walking sticks. However, none of these maps was without error, and when observations on the earth's surface were unobtainable, map makers often interpolated between field observations with questionable results, as illustrated in one of the earliest maps of California displayed in Figure 2.3.

One of the most notorious examples of the results of reliance on an erroneous map created from field observations was the disastrous use by the Donner party in 1846 when they chose to follow the Hastings cutoff rather than the established Oregon–California trail during their immigration from the Midwest. As a result, they added hundreds of miles to their journey, as shown in Figure 2.4 (the positional accuracy was in error), were forced to cross unexpected expanses of waterless desert (the thematic accuracy was in error), and ended up attempting to cross the Sierra Nevada mountains in late fall rather than during the summer. As a result the group became stranded in 20 feet of snow just below the summit for the entire winter and lost almost half of their party to starvation, hypothermia, and cannibalism. Regardless of how the map is made, not knowing the accuracy of maps can have catastrophic results!

Today, most map makers use remote sensing* rather than field observations as the main source of spatial information. While field observations are still important, they are ancillary to the remote sensing data; providing information at sample locations instead of a total enumeration of the area to be mapped. Since the first aerial

---

* Remote sensing is defined as the collection and interpretation of information about an object from a distant vantage point. Remote sensing systems involve the measurement of electromagnetic energy reflected or emitted from an object and include instruments on balloons, aircraft, satellites, and unmanned aerial systems (UAS).

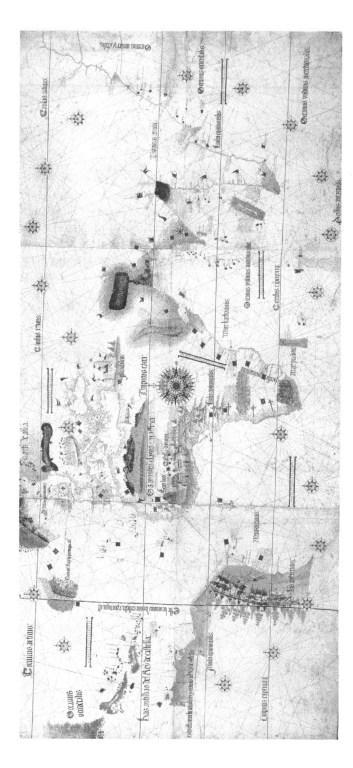

FIGURE 2.1 The Cantino world map, which is a map of the known coastlines of the world, created by sixteenth-century navigators.

FIGURE 2.2 Map of the American Northwest created by Lewis and Clark. (From Lewis, M., et al., *Maps of Lewis and Clark's Track across the Western Portion of North America*, Bradford and Inskeep, Philadelphia, 1814. With permission.)

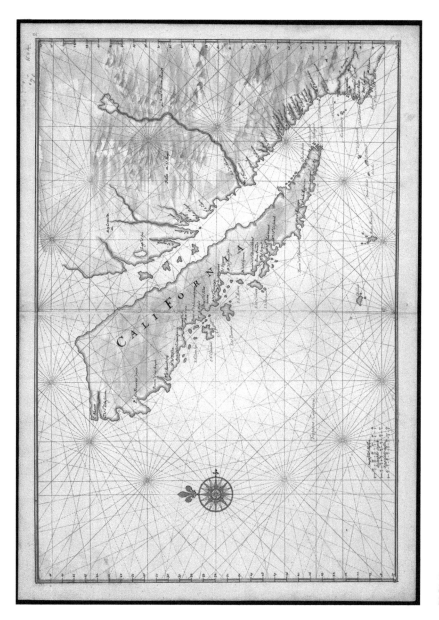

FIGURE 2.3   A seventeenth-century map of California.

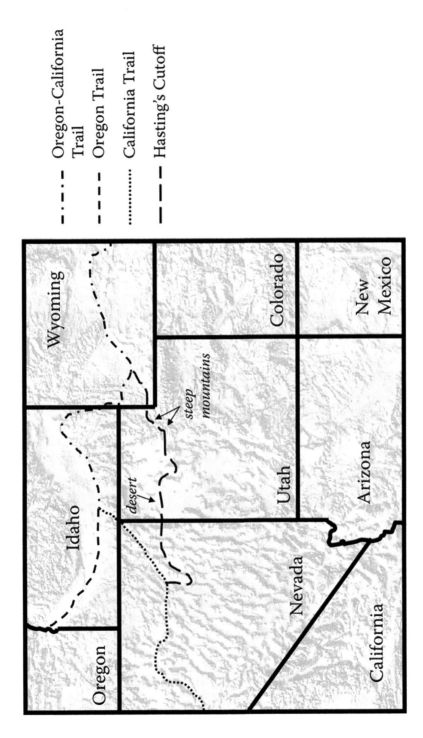

FIGURE 2.4　Hastings cutoff versus the safer California and Oregon trails used in 1846 by emigrants.

photograph was captured from a balloon in 1858, data collected using remote sensing have supplanted ground observations for map making. Satellites and aircraft offer humans a view of their surroundings that humans cannot obtain on their own. Well before the first man went aloft in a balloon in 1783, humans had been fantasizing about flight. Once humans invented successful flying machines, it was an easy step to put cameras in those flying machines so that the pilot's perspective could be shared with those on the ground.

We use remotely sensed data to make maps because it:

- is significantly less expensive and more efficient than creating maps from observations on the earth's surface
- offers a perspective from above (the "bird's eye or synoptic view"), improving our understanding of spatial relationships and the context of our observations
- permits capturing imagery and information in electromagnetic wavelengths that humans cannot sense, such as the infrared portions of the electromagnetic spectrum

Remotely sensed data are irresistible, because they provide a view that can be readily understood, are inimitably useful, and yet are impossible to obtain without the use of technology. The innovation of air and space remote sensing has fundamentally changed the way we conduct war, inventory and manage resources, perform research, and respond to disasters.

Map making with remotely sensed data requires:

1. precise linkage of the distances in the remote sensing imagery to distances on the ground so that spatial features can be accurately located
2. understanding what causes variation in the features to be mapped and understanding how the remotely sensed data and ancillary information respond to those variations so the features can be labeled

Remotely sensed data provide an excellent basis for making maps because (1) remote sensing instruments and platforms are highly calibrated and (2) a high correlation exists between variation in remotely sensed data and variation across the earth's surface

However, there is never a complete one-to-one correlation between variation in remotely sensed data and variation on the earth's surface. Aircraft movement, topography, lens distortions, clouds, shadows, and a myriad of other factors can combine to reduce the strength of the relationships between the imagery and the earth's surface. Thus, much judgment, analysis, and interpretation are required to turn remotely sensed data into a map, and as a result, errors can occur during the many steps throughout any remote sensing project. As illustrated in Figure 2.5, the possible sources of error are multiple and compounding. Error can derive from the acquisition of imagery, from its rectification and classification, from its presentation as a map, and from the application of the map in a decision-making process. And, of course, error can also occur in the accuracy assessment itself.

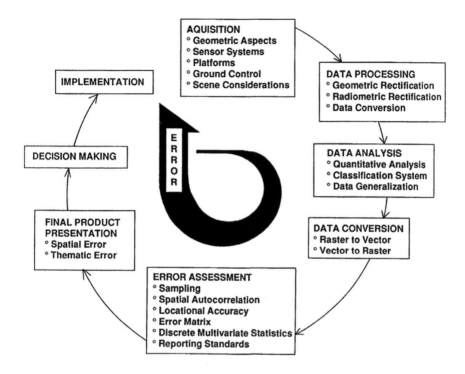

FIGURE 2.5   Sources of error in remotely sensed data. (From Lunetta et al., *Photogrammetric Engineering and Remote Sensing*, 57, 6, 677, 1991. With permission.)

Accuracy assessment estimates, identifies, and characterizes the impact that arises from all of the sources of error.

HISTORY OF ACCURACY ASSESSMENT

The widespread acceptance and use of remotely sensed data have been and will continue to be dependent on the quality of the map information derived from them. As we learned in the previous section, the history of using remotely sensed data for mapping and monitoring the earth is a relatively short one. Aerial photography (analog or film-based remote sensing) has been used as an effective mapping tool only since the 1930s. Digital image scanners and cameras on satellites and aircraft have even a shorter history, beginning in only the mid-1970s. The use of high-resolution digital imagery has exploded only since the turn of the century. The following two sections briefly review the history of positional and thematic accuracy assessment of maps created from remotely sensed data.

POSITIONAL ACCURACY ASSESSMENT

Photogrammetry—the science of determining the physical dimensions of objects from measurements on aerial photographs or imagery—was first implemented in 1849 using terrestrial photographs taken on the earth's surface (McGlone, 2004).

Aerial photogrammetry, which uses images taken from aerial or satellite platforms, followed soon after the first photographs were taken from aircraft. The adoption of the use of aerial photographs to create maps exploded with:

- the need to rebuild Europe following World War I
- the development of roll film by George Eastman (the founder of Kodak)
- the reduction of camera lens distortion
- improvements in camera bodies, including increased sturdiness and permanent mounting of lenses, techniques for holding the film flat, and inclusion of a mechanism for aligning the camera axis
- employment of fiducial marks for the definition of the image plane
- the development of analytical photogrammetry equations
- the invention of the stereo plotter (Ferris State University, 2007).

From the very first days of aerial photogrammetry, positional accuracy has been assessed by comparing the coordinates of sample points on a map against the coordinates of the same points derived from a ground survey or some other independent source that has been deemed to be more accurate than the map. In the early twentieth century, mapping scientists focused on map production and attempted to characterize each different contributor to positional error. Now, positional error assessment is more user focused, emphasizing the estimation of the total error, regardless of the source.

In 1937, the American Society of Photogrammetry (now the American Society for Photogrammetry and Remote Sensing [ASPRS]) established a committee to draft spatial accuracy standards for maps made from remotely sensed data. Soon after, the U.S. Bureau of the Budget published the United States National Map Accuracy Standards (NMAS) in 1941. The first version of the National Map Accuracy Standards was published in 1947 (U.S. Bureau of the Budget, 1947) and is included here:

1. Horizontal accuracy. For maps on publication scales larger than 1:20,000, not more than 10 percent of the points tested shall be in error by more than 1/30 inch, measured on the publication scale; for maps on publication scales of 1:20,000 or smaller, 1/50 inch. These limits of accuracy shall apply to positions of well-defined points only. Well-defined points are those that are easily visible or recoverable on the ground, such as the following: monuments or markers, such as bench marks, property boundary monuments; intersections of roads and railroads; corners of large buildings or structures (or center points of small buildings). In general, what is well defined will also be determined by what is plottable on the scale of the map within 1/100 inch. Thus, while the intersection of two roads or property lines meeting at right angles would come within a sensible interpretation, identification of the intersection of such lines meeting at an acute angle would not be practicable within 1/100 inch. Similarly, features not identifiable upon the ground within close limits are not to be considered as test points within the limits quoted, even though their positions may be scaled closely upon the map. This class would cover timber lines and soil boundaries.

2. Vertical accuracy, as applied to contour maps on all publication scales, shall be such that not more than 10 percent of the elevations tested shall be in error by more than one-half the contour interval. In checking elevations taken from the map, the apparent vertical error may be decreased by assuming a horizontal displacement within the permissible horizontal error for a map of that scale.

3. The accuracy of any map may be tested by comparing the positions of points whose locations or elevations are shown upon it with corresponding positions as determined by surveys of a higher accuracy. Tests shall be made by the producing agency, which shall also determine which of its maps are to be tested, and the extent of such testing.

4. Published maps meeting these accuracy requirements shall note this fact in their legends, as follows: "This map complies with National Map Accuracy Standards."

5. Published maps whose errors exceed those aforestated shall omit from their legends all mention of standard accuracy.

6. When a published map is a considerable enlargement of a map drawing (manuscript) or of a published map, that fact shall be stated in the legend. For example, "This map is an enlargement of a 1:20,000-scale map drawing," or "This map is an enlargement of a 1:24,000-scale published map."

7. To facilitate ready interchange and use of basic information for map construction among all Federal mapmaking agencies, manuscript maps and published maps, wherever economically feasible and consistent with the use to which the map is to be put, shall conform to latitude and longitude boundaries, being 15 minutes of latitude and longitude, or 7.5 minutes, or 3.75 minutes in size.

The establishment of the standards was a critical step in implementing consistency in positional accuracy across the United States. However, NMAS focuses on errors measured at the map instead of the ground scale, which became problematic over the years as maps migrated from paper to digital formats that can be printed at variable map scales. Additionally, the standards state the requirements for spatial accuracy but only briefly discuss the procedures for collecting samples (reference data) to determine whether or not those standards have been met. Thus, while the accuracy percentage was standardized, the procedures for measuring accuracy were not.

In the 1960s, a precursor to the present day National GeoSpatial-Intelligence Agency (NGA), the Aeronautical Chart and Information Center, printed a report entitled *Principles of Error Theory and Cartographic Applications* (Greenwalt and Schultz, 1962, 1968), which meticulously provided the statistical foundation for estimating the distribution of positional map error from a sample of reference points. The basic concepts of the report derive from the probability theories developed in the 1800s to predict the probable distribution of artillery shells fired at a target. The report develops positional accuracy standards based on the probable distribution of errors. It became and has remained the foundation for all other publications that stipulate the calculation of map error from a set of sample points (ASPRS, 1990; DMA,

1991; FGDC, 1998; MPLMIC [Minnesota Planning Land Management Information Center], 1999; Bolstad, 2005; Maune, 2007; ASPRS, 2014). However, unlike later publications, the report focused only on how to calculate error and did not address how the sample points should be chosen or measured.

In the late 1970s, the ASPRS Specifications and Standards Committee began a review of the 1947 standards with the goal of updating them to include standards for both hardcopy and digital maps. The result was the 1990 publication of the ASPRS Interim Accuracy Standards for Large-Scale Maps (ASPRS, 1990), which stipulated that accuracy be reported at ground scale rather than map scale, thereby allowing the consideration of digital as well as hardcopy maps. The standards established the maximum root mean squared error (RMSE) (measured at ground distances) permissible for map scales from 1:50 to 1:20,000. Finally, they provided guidance on how accuracy sample points should be identified, measured, and distributed across the map and how these reference points should be collected.

Soon after the release of the ASPRS Standards, the Ad Hoc Map Accuracy Standards Working Group of the Subcommittee on Base Cartographic Data of the Federal Geographic Data Committee (FGDC) produced the U.S. National Cartographic Standards for Spatial Accuracy (NCSSA) (FGDC, 1998) to create positional accuracy standards for medium- and small-scale maps.

Following public review, the NCSSA was significantly modified so as to adopt positional accuracy assessment procedures in lieu of accuracy assessment standards. The result was the 1998 publication of the FDGC National Standard for Spatial Data Accuracy (NSSDA) (FGDC, 1998), which relies heavily on the ASPRS standards and "implements a statistical and testing methodology for estimating the positional accuracy of points on maps and in digital geospatial data, with respect to georeferenced ground positions of higher accuracy." The Standard explicitly does not establish threshold standards (as did the NMAS and ASPRS) but encourages map users to establish and publish their standards, which it was recognized would vary depending on the user's requirements.

Relying on Greenwalt and Schultz (1962, 1968), the NSSDA specifies that positional accuracy be characterized using RSME, requires that accuracy be reported in ground distance units at the "95% confidence level," and provides guidance on how samples are to be selected. For 20 years, NSSDA was the accepted standard on positional accuracy assessment. It was often used in conjunction with the ASPRS large-scale map standards with NSSDA providing standardized processes for assessing positional accuracy and the ASPRS standards setting the maximum errors allowable for different map scales.

Over the last 10 years, several new guidelines have been established for assessing the accuracy of digital elevation data. All call for the stratification of positional accuracy assessment samples into land cover types. Most suggest that accuracy be reported at the "95th percentile error" in addition to the NSSDA statistic.

Finally, in 2014, the ASPRS developed the *ASPRS Positional Accuracy Standards for Digital Data*. This document provides the most comprehensive discussion of positional accuracy that has been developed to date and establishes standards for maps of different quality and scale. Key components of this standard, along with a review of previous standards, are presented in Chapter 4 of this book.

THEMATIC ACCURACY ASSESSMENT

Unlike positional accuracy, there is no government or professional society standard for assessing and reporting thematic accuracy. This omission is partially due to the inherent complexity of thematic accuracy but primarily to the fact that when maps were made from aerial photographs, thematic accuracy was generally assumed to be at acceptable levels. It was the development and use of digital remote sensing devices that had the most profound impact on thematic accuracy assessment of maps created from all remotely sensed data.

Spurr, in his excellent book entitled *Aerial Photographs in Forestry* (1948), presents the early prevailing opinion about assessing the accuracy of photo interpretation. He states:

> Once the map has been prepared from the photographs, it must be checked on the ground. If preliminary reconnaissance has been carried out, and a map prepared carefully from good quality photographs, ground checking may be confined to those stands whose classification could not be agreed upon in the office, and to those stands passed through in route to these doubtful stands.

In other words, a qualitative visual check to see if the map looks right has traditionally been the recommended course of action for assessing photo interpretation.

However, in the 1950s, some researchers saw the need for quantitative assessment of photo interpretation to promote their discipline as a science (Sammi, 1950; Katz, 1952; Young, 1955; Colwell, 1955). In a panel discussion entitled "Reliability of Measured Values" held at the 18th Annual Meeting of the American Society of Photogrammetry, Mr. Amrom Katz (1952), the panel chair, made a very compelling plea for the use of statistics in photogrammetry. Other panel discussions were held and talks were presented that culminated with a paper by Young and Stoeckler (1956). In this paper, these authors actually proposed techniques for a quantitative evaluation of photo interpretation, including the use of an error matrix to compare field and photo classifications, and a discussion of the boundary error problem.

Unfortunately, these techniques never received widespread attention or acceptance. The *Manual of Photo Interpretation* published by the American Society of Photogrammetry (ASPRS, 1960) does mention the need to train and test photo interpreters. However, it contains no description of the quantitative techniques proposed by a brave few in the 1950s.

There is no doubt that photo interpretation has become a time-honored skill, and the prevailing opinion for decades was that a quantitative thematic accuracy assessment was unnecessary. Some of the old-time photo interpreters remember those times when quantitative assessment was an issue. In fact, they mostly agree with the need to perform such an assessment and are usually the first to point out the limitations of photo interpretation. However, it was mostly agreed that the results of any photo interpretation grouped areas that were similar, and that there was more variation between these polygons or vegetation types or forest stands than between them. Hence, with this goal achieved, no quantitative assessment was necessary. Therefore, the quantitative assessment of photo interpretation is typically not a requirement of any project.

Rather, the assumption that the map was correct, or at least good enough, prevailed. Then, along came digital remote sensing, and some of these fundamental assumptions about photo interpretation needed to be further scrutinized and adapted.

As in the early days of aerial photography, the launch of Landsat 1 in 1972 resulted in a great burst of exuberant effort as researchers and scientists charged ahead trying to develop the field of digital remote sensing. In those early days, much progress was made, and there wasn't much time to sit back and evaluate how they were doing. This "can do" mentality is common in many developing technologies. The geographic information systems (GIS) community has experienced a similar development pattern. However, as a technology matures, more effort is dedicated to data quality and error/accuracy issues. By the early 1980s, some researchers began to consider and realistically evaluate where they were going and, to some extent, how they were doing with respect to the quality of maps derived from digital remotely sensed data.

The history of assessing the thematic accuracy of maps derived from remotely sensed data is relatively brief, beginning around 1975. Researchers, notably Hord and Brooner (1976), van Genderen and Lock (1977), and Ginevan (1979), proposed criteria and basic techniques for testing overall map accuracy. In the early 1980s, more in-depth studies were conducted and new techniques proposed (Aronoff, 1982; Rosenfield et al., 1982; Congalton and Mead, 1983; Congalton et al., 1983; Aronoff, 1985). Finally, from the late 1980s up to the present time, a great deal of work has been conducted on thematic accuracy assessment. More and more researchers, scientists, and users are discovering the need to adequately assess the thematic accuracy of maps created from remotely sensed data, such that accuracy assessment has become a key component of most mapping projects.

The history of digital accuracy assessment can be effectively divided into four parts or epochs. Initially, no real accuracy assessment was performed; rather, an "it looks good" mentality prevailed. This approach is typical of a new, emerging technology in which everything is changing so quickly that there is no time to sit back and assess how well you are doing. Despite the maturing of the technology over the last 45 years, some remote sensing analysts and map users are still too dependent on this mentality.

The second epoch is called the age of *non-site-specific assessment*. During this period, total acreages by map class were compared between reference estimates and the map without regard for location. It did not matter whether you knew where it was; only the total amounts were compared. While total acreage is useful, it is far more important to know where a specific land cover or vegetation type exists. Therefore, this second epoch was relatively short-lived and quickly led to the age of site-specific assessments.

In a site-specific assessment, actual locations on the ground are compared with the same locations on the map, and a measure of overall accuracy (i.e., percent correct) is presented. This method far exceeded the non-site-specific assessment but lacked information about individual land cover/vegetation categories. Only overall map accuracy was assessed. Site-specific assessment techniques were the dominant method until the late 1980s.

Finally, the fourth and current age of accuracy assessment could be called *the age of the error matrix*. An error matrix compares information from reference sites with information on a map for a number of sample areas. The matrix is a square array of

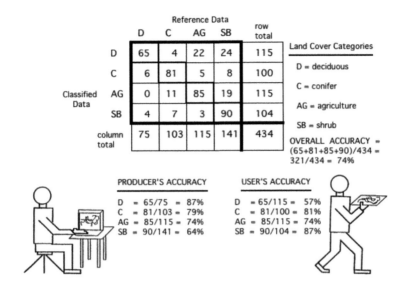

FIGURE 2.6   Example error matrix.

numbers set out in rows and columns, which express the labels of samples assigned to a particular category in one classification relative to the labels of samples assigned to a particular category in another classification (Figure 2.6). One of the classifications, usually the columns, is assumed to be correct and is termed the *reference data*. The rows usually are used to display the map labels or classified data generated from the remotely sensed image. Thus, two labels from each sample are compared with one another:

- Reference data labels: the class label of the accuracy assessment site derived from data collected, which is assumed to be correct
- Classified data or map labels: the class label of the accuracy assessment site derived from the map

Error matrices are very effective representations of map accuracy, because the individual accuracies of each map category are plainly described along with both the errors of inclusion (commission errors) and the errors of exclusion (omission errors) present in the map. A commission error occurs when an area is included in an incorrect category. An omission error occurs when an area is excluded from the category to which it belongs. Every error on the map is an omission from the correct category and a commission to an incorrect category.

In addition to clearly showing errors of omission and commission, the error matrix can be used to compute overall accuracy, producer's accuracy, and user's accuracy, which were introduced to the remote sensing community by Story and Congalton (1986). Overall accuracy is simply the sum of the major diagonal (i.e., the correctly classified sample units) divided by the total number sample units in the error matrix.

This value is the most commonly reported accuracy assessment statistic and was part of the older, site-specific assessment. Producer's and user's accuracies are ways of representing individual category accuracies instead of just the overall classification accuracy (see Chapter 5 for more details on the error matrix).

Proper use of the error matrix includes correctly sampling the map and rigorously analyzing the matrix results. The techniques and considerations for generating and analyzing an error matrix are main themes of this book.

3 Planning for Conducting an Accuracy Assessment

Chapter 1 of this book presented an overall introduction to the concept of accuracy assessment and provided a list of questions that might be asked when conducting such an assessment. Chapter 2 provided a review of the history of accuracy assessment. In this chapter, it is time to dig in a little more deeply and study the entire process more closely, so that the analyst has a thorough overview to plan their assessment. The overarching goal of this book is to provide the necessary considerations and knowledge for the remote sensing analyst to effectively conduct a full and complete quantitative accuracy assessment of their own or another's map. However, there are many ways to consider the validity of a map. Some of these are qualitative, while others are quantitative. Finally, some look at map quality from an error budgeting approach, while others provide a more visual comparison. It is important for the analyst to be aware of all the possible approaches for evaluating their map when planning just how to conduct their assessment.

It is also important to realize that the map assessment process does not follow a simple recipe but rather, depends on many decisions along the way to achieve success. Therefore, it is of key importance that the analyst spends adequate time thinking about and planning the accuracy assessment before beginning the mapping process. Commonly, the analyst jumps into the process too quickly, resulting in the discovery of an issue or complication too late in the assessment to be able to account for it properly. Failure to properly plan the assessment from the start of the project can lead to significant problems, including large cost overruns, lack of sufficient reference data, inappropriate sampling strategies, and flawed analysis. This chapter provides an overview of all the possible assessment methods that could be used and then guides the analyst through planning the most effective accuracy assessment possible.

WHAT TYPE OF ASSESSMENT?

There are two general types of map assessments: qualitative and quantitative (Figure 3.1). Performing a quantitative assessment is typically the preferred approach. However, it can be argued that if the map fails to meet the requirements imposed in a qualitative assessment, then it should be fixed/corrected before the effort is spent to conduct a quantitative assessment. Therefore, this chapter presents a discussion of the qualitative assessment methods first, then follows with some methods that bridge the gap between qualitative and quantitative approaches, and concludes with an overview of the quantitative methods.

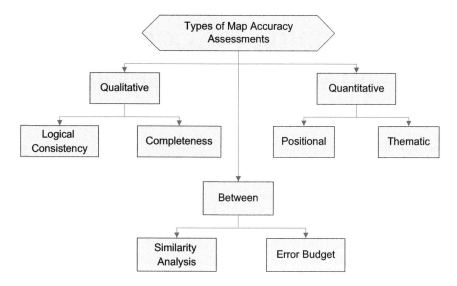

FIGURE 3.1   Flowchart documenting the various types of accuracy assessment that can be used to evaluate a map.

QUALITATIVE ACCURACY ASSESSMENT

Qualitative accuracy assessment ultimately comes down to whether or not the map "looks good" to the map producers and users. When a map is evaluated, it is a necessary, but not sufficient, condition that the map appears to be satisfactory. If the map contains obvious errors, then these should be analyzed and corrected before any further assessment is conducted. The conditions for defining that the map "looks good" can be divided into two categories (Bolstad, 2016): (1) logical consistency and (2) completeness. Logical consistency tests whether the map makes sense or whether situations represented on the map are inherently flawed. For example, is there a lake occurring on a slope on the side of a mountain? Do urban development pixels appear in the center of a corn field? Or do street signs occur in the middle of the street? All of these examples demonstrate an inconsistency in the map that needs to be corrected.

Completeness assesses whether objects are missing from the map that clearly should be there. A poor map will be obviously incomplete to even the most casual user. For example, a map of Grand Canyon National Park would be incomplete if it omitted the Colorado River. However, a more subtle situation can occur when the map user examines areas that he or she knows very well (e.g., the area around their home or their favorite hiking spot). If these areas are wrong or missing, then the map will also be considered incomplete even if the rest of the map is satisfactory. These issues are much harder to correct than the obvious misses.

BETWEEN QUALITATIVE AND QUANTITATIVE ASSESSMENT

In addition to the qualitative assessment methods discussed, there are two additional approaches that can be used to help visualize the error in maps generated from remotely sensed data. The first approach is to conduct a similarity analysis, and the second is to create an error budget. Both methods provide additional insight into the quality of the map.

Similarity Analysis

A similarity analysis can be performed when another map exists of the same area as the map that is to be assessed. The two maps are registered to each other and then overlaid to compare, pixel by pixel, or polygon by polygon, using simple geographic information system (GIS) functions. The result of this overlay is called a *difference map* and shows the similarity and differences between the two maps. If the existing map is deemed correct, then this analysis will show actual agreement and error, and an error matrix can be generated. However, in the vast majority of cases, the existing map cannot be assumed to be correct and is not a valid tool for assessing the new map, because changes may have occurred between the dates of the two maps, the maps may rely on different classification schemes, or the methods used to create the two maps may be dissimilar. However, an older map, or a map created using a slightly different classification scheme or a different image source or classification methodology, might exist for all or part of the area. Such a map would allow a similarity analysis to be conducted as long as these differences are recognized and considered. Figure 3.2 shows a difference map created to compare two different maps of crop extent of the United States. One map was created by Northern Arizona University (Massey et al., 2018) and is compared with the United States Department

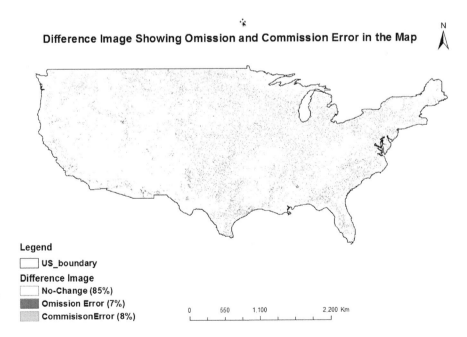

FIGURE 3.2 Difference map comparing one map with another and showing agreement (similarity) and disagreement for the cropped areas of the United States.

		CDL			
		Crop	No-Crop	Total	User Accuracy
Map	Crop	2,078,684,902	411,167,925	2,489,852,827	83.49%
	No-Crop	362,134,634	2,147,483,648	2,509,618,282	85.57%
Total		2,440,819,536	2,558,651,573	4,999,471,109	
Producer Accuracy		85.16%	83.93%		84.53%

FIGURE 3.3   Similarity matrix generated from the difference map shown in Figure 3.2.

of Agriculture (USDA) Cropland Data Layer (CDL) generated yearly (Cropscape at https://nassgeodata.gmu.edu/CropScape). The CDL data were collapsed to a crop extent map (i.e., crop vs. no crop) and overlaid on the Northern Arizona University map. The white areas represent where the two maps agree, while the black areas represent where they are different (i.e., disagree). The patterns of the black areas reveal information about how the differences are distributed throughout the study area, which can be used to review and edit the map where warranted. If an old map exists of an area for which you are creating a new map, you can be sure that someone, at some time, will compare the old map with the new map. It is better for the map producer to perform the similarity analysis first, before the new map is published or a quantitative accuracy assessment is performed, than to have easily corrected errors discovered by map users after publication.

The difference map can also be represented quantitatively as a similarity matrix where each pixel is represented in a contingency table (like the error matrix introduced in Chapter 2). Figure 3.3 shows the results of this type of analysis.

Error Budget

An error budget approach is a very different way of considering map error from any other method discussed in this book. While it is obvious that error can enter a mapping project from a great variety of sources, little research has been done to evaluate these errors and prioritize methods for dealing with them. Building on the sources of error discussed in Chapter 2 (see Figure 2.5), Congalton and Brennan (1999) and Congalton (2009) proposed an error budget analysis method that not only clearly lists the possible sources of error in any mapping project but also suggests a method for evaluating the (1) error contribution, (2) implementation difficulty, and (3) implementation priority for each error. Table 3.1 shows an example of this error budgeting approach. This table has five columns, which are generated as follows. The first step is to list the sources of error in this mapping project (i.e., place in the first column of the table). Each list will be slightly different depending on the mapping methods used, but there will be some error sources that are common to every project. The list found in Table 3.1 is not meant to be exhaustive but rather, to aid the analyst in reflecting about what errors are involved in their specific project. Once a list of potential sources of error is generated, then a qualitative evaluation of the potential of that source to contribute error (high, medium, or low) to the project must be determined and entered into the second column. This determination is made by analyst experience combined with what can be gleaned from the remote sensing literature.

TABLE 3.1
Example Error Budget Analysis for a Map Generated from Remotely Sensed Data

Source of Error	Error Contribution Potential	Implementation Difficulty	Implementation Priority	Error Assessment Technique
Systematic Error				
Sensor	Low	5	21	Calibration and analysis
Natural Error				
Atmosphere	Medium	3	20	Analysis and correction
Pre-processing Errors				
Geometric registration	Low	2	19	Positional accuracy assessment
Image masking	Medium	3	18	Single-date error matrix
Derivative Data Errors				
Band ratios	Low	1	12	Data exploration
Indices (i.e., NDVI)	Low	1	14	Data exploration
Principal components analysis	Low	1	15	Data exploration
Tassled-cap analysis	Low	1	16	Data exploration
Other geospatial data	Medium	3	17	Data exploration
Classification Errors				
Classification scheme	Medium	2	6	Single-date error matrix
Training data collection	Medium	3	7	Single-date error matrix
Classification algorithm	Medium	3	8	Single-date error matrix
Post-processing Errors				
Data conversion	High	2	13	Single-date error matrix
Accuracy Assessment Errors				
Labeling errors	High	1	2	QC/QA procedures
Sample unit	Medium	1	5	Single-date error matrix
Sample size	Low	2	10	Single-date error matrix
Sampling scheme	Medium	3	11	Single-date error matrix
Spatial autocorrelation	Low	1	4	Geostatistical analysis
Positional accuracy	Medium	3	3	RMSE/NSSDA
Final Product Errors				
Decision-making	Medium	2	1	Sensitivity analysis
Implementation	Medium	2	9	Sensitivity analysis

Error Contribution Potential—ranked from low to medium to high.
Implementation Difficulty—ranked from 1: not very difficult to 5: extremely difficult.
Implementation Priority—ranked from 1 to n, showing the order in which to implement improvements.
NDVI, Normalized Difference Vegetation Index; NSSDA, National Standard for Spatial Data Accuracy; QA, quality assurance; QC, quality control; RMSE, root mean squared error.

Next, the implementation difficulty of dealing with or mitigating a specific error is ranked from 1 to 5, with 1 being not difficult and 5 being very difficult. Therefore, by evaluating the combination of error potential with the implementation difficulty, an implementation priority can be generated in Column 4. This value is the order in which the error sources should be addressed to gain the maximum benefit (i.e., to minimize the error in the project). In other words, those sources of error that are largest but are the least difficult to correct should be dealt with first. Finally, the last column in Table 3.1 shows the method or technique that can be used to evaluate the error. Altogether, creating a table such as this one provides the analyst with an excellent mechanism for thinking through the errors that may occur in their mapping project. Even if the analyst struggles with ranking the error contribution potential or the implementation difficulty or cannot exactly determine the error assessment technique, there is great value in working on an error budget, as doing so will encourage the analyst to really consider the error sources in their project.

QUANTITATIVE ACCURACY ASSESSMENT

Qualitative accuracy assessment, as presented in the previous sections, can be a valuable beginning to evaluating a map. However, it is just the beginning. The rest of this book is dedicated to presenting the considerations, methodologies, and practicals for conducting quantitative map accuracy assessment. As introduced in Chapter 1, there are two types of quantitative map assessments that can be performed: positional accuracy assessment and thematic accuracy assessment. Positional accuracy evaluates whether objects on the map are in the correct location, while thematic accuracy determines whether the objects have been correctly labeled. These two assessments are inextricably related. If the object is in the wrong place, then it is possible that this will cause not only a positional error but also a thematic error. The converse is also true. Therefore, in many assessments, it is necessary to conduct both a positional and a thematic validation.

Regardless of whether the analyst is conducting a positional or thematic assessment or both, the same general process holds, including the following steps:

1. Consider the factors involved in the assessment.
2. Design the appropriate sampling approach to collect the reference data.
3. Conduct the sampling.
4. Analyze the data.
5. Report the statistics/results.

Positional Accuracy Assessment

There are a great many similarities between conducting a positional assessment and a thematic assessment. There are also some important differences. In general, a positional accuracy assessment is simpler to perform and has fewer considerations. However, it is still critical to effectively and efficiently plan a positional accuracy assessment.

Figure 3.4 shows a flowchart of the process of conducting a positional accuracy assessment. There are three overarching considerations here: the sources of error, the

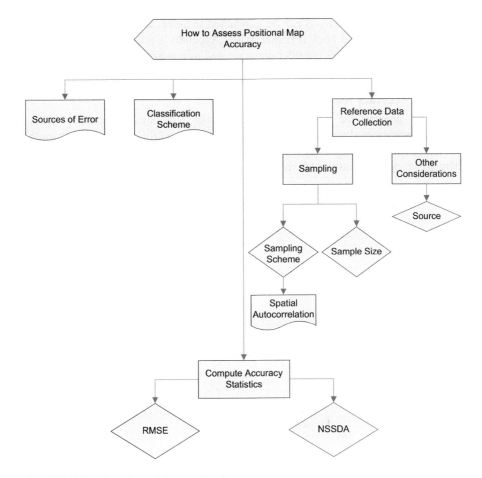

FIGURE 3.4 Flowchart of the positional accuracy assessment process.

classification scheme, and the reference data. Careful consideration of the sources of error is vital to the effective and efficient planning of the assessment. This process could be as simple as making a list of the major sources of error and deciding how best to control them or as complex as a full error budgeting approach as described earlier. The classification scheme in positional accuracy assessment has become increasingly important recently. Historically, no classification scheme was used, and samples were taken throughout the map. Recently, it has been recognized that the positional accuracy of a sample depends on the land cover type. Therefore, the classification scheme now plays an important role in positional assessment.

The final consideration, reference data collection, is the most complex to plan. There are many statistical considerations revolving around the collection of the reference data that are used to compare with the map. Issues related to the sampling scheme selected, the number of samples needed, and spatial autocorrelation are important to consider. In addition, the source of the reference data must be determined. Once all these factors are considered and properly controlled, then the appropriate statistical analysis can be conducted to assess the positional accuracy. The next chapter in this book (Chapter 4) and Appendix 1 provide the analyst with the concepts and practical considerations needed to conduct a positional accuracy assessment.

Thematic Accuracy Assessment

Figure 3.5 presents the flowchart of the thematic accuracy assessment process. A quick comparison shows that many of the components of this assessment are quite similar to the positional accuracy assessment. However, as shown in Figure 2.5, there are many more sources of error that come into play in a thematic assessment. In

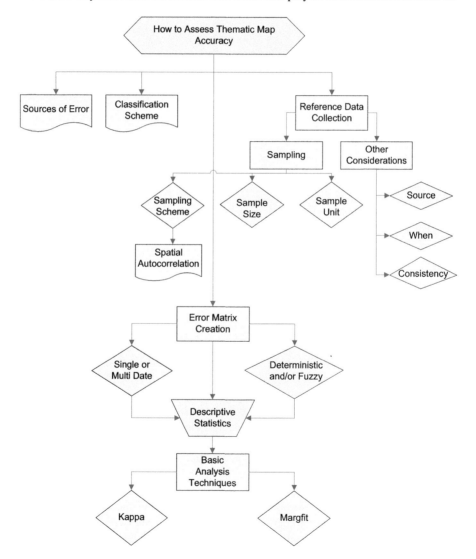

FIGURE 3.5   Flowchart of the thematic accuracy assessment process.

addition, the classification scheme (as discussed in Chapter 6) is often more complex than those typically used in positional assessment. Again, there are also statistical considerations that revolve around the collection of the reference data. However, there are more factors to consider here, including not only the source of the reference data but also when they are collected, as well as ensuring the consistency or objectivity of the collection. The sampling issues related to collecting the reference data for thematic accuracy assessment are also more complex than for a positional assessment. They involve not only the sampling scheme, including spatial autocorrelation and the number of samples, but also the consideration of the sample unit (pixel, cluster of pixels, or polygon). Finally, the analysis of the thematic map is also more complex, involving the consideration of not only the appropriate error matrix to generate (single or multi-date; deterministic and/or fuzzy) but also the appropriate descriptive statistics and basic analysis techniques used once the matrix has been generated. The majority of the rest of this book is devoted to the concepts and practical considerations necessary to conduct a quantitative thematic accuracy assessment.

A careful study of these flow charts will provide an effective list of considerations for the analyst and lead them to the more detailed descriptions of these considerations in the chapters that follow. Having a solid, overall understanding of the entire flow of the assessment process from the beginning of the mapping provides the best assurance that the project will be successful.

WHY DO THE ASSESSMENT?

In addition to understanding the basic assessment considerations for assessing map accuracy, it is important for the analyst to know why they are doing the assessment. As introduced in Chapter 1, there are many reasons for conducting an accuracy assessment. Today, many mapping contracts require that an assessment be conducted and some minimum overall accuracy or other metric be achieved. While this reason is perhaps the least intellectually satisfying, it does provide sufficient motivation for planning out the process from the very beginning of the project, so that these minimum metrics are achieved while keeping within the project budget. A poorly planned accuracy assessment that then does not produce the information needed to effectively evaluate the map is one of the major reasons for failure to meet the project timeline and budget.

In other cases, it is perhaps only necessary to know the overall accuracy of the map. This requirement simplifies the assessment process and should reduce the cost and effort involved. There is no reason to spend the time and effort to conduct a full assessment if only the overall map accuracy is required. However, in most assessments, the accuracies of individual map categories are required (i.e., a full assessment). In this situation, effective sampling strategies for collecting the appropriate and complete reference data must be determined.

Finally, the results of the assessment may be used for comparison with other maps to determine which is better. Perhaps a new classification algorithm has been developed, and the analyst wishes to know whether it performs better than the current method. Or, perhaps a prospective client is looking to hire some additional mapping firms to expand its capabilities and wants to see how these new firms compare with current ones. In these cases, it is vital that full consideration of the assessment process be performed to produce the most effective and efficient assessment possible.

HOW WILL THE ASSESSMENT BE USED?

Another factor that influences the assessment is how it will be used. Again, there are a large number of possibilities here, with some crossover between reasons why the assessment was done. In most cases, a map is generated for a specific purpose, and the assessment is performed to assess the map accordingly. Therefore, if only an overall accuracy is required, then that type of assessment can be performed. If the map generated is to be used as input to some ecological model, then the value of the map for this purpose should be assessed. However, given the time, effort, and value of many maps generated today, the majority of maps will be used for other reasons beyond their original intent. Therefore, in such cases, there is great benefit to conducting the most complete and generic assessment possible so as to make the map widely acceptable to all.

WHAT SPECIAL CONSIDERATIONS EXIST?

Lastly, there are always some special considerations to think about when assessing the map accuracy. These might include specific knowledge about the map user (i.e., the entity the map is made for). For example, if the map was made for a natural resource entity (e.g., Forest Service, National Park Service, etc.), then paying particular attention to the map categories involving natural resources could be more important. Conversely, if the client was more interested in urban or developed areas, then the focus might be on buildings and infrastructure.

Additionally, any special interests of the organization requesting the map should be considered, so that the assessment is well planned and responsive to the needs of the users. In some situations, political considerations may be more determinative of map acceptance than the quantitative accuracy of a map, and those political considerations need to be addressed. If people are strongly vested in a particular result, it is very difficult to change their minds if the assessment proves differently.

Involvement in the creation of the map can also be of critical importance. For example, during one of our first mapping projects, the organization funding the map required that the eventual map users be excluded from the process of creating the map. As a result, the users felt alienated from the project and suspicious of the final map, regardless of the quantitative accuracy results. More importantly, the map suffered from not being reviewed by the eventual users. People who live in the area being mapped will always know more about the local landscape than will a map maker from outside the area. The more local users are involved, the higher the acceptance of the map, regardless of the quantitative accuracy results.

CONCLUSIONS

This chapter has described many ways to think about assessing map accuracy. Some of these methods are qualitative and easy to implement. Others are quantitative and require more consideration and planning to be valid, effective, and efficient. The remainder of the book presents the details necessary to conduct a quantitative map accuracy assessment based on the overview presented here. The analyst should carefully review Chapters 1 through 3 to fully appreciate the many ways to conduct an assessment and to carefully consider all the issues that are involved in producing a valid and effective assessment. Then, the next chapter (Chapter 4) is dedicated to positional accuracy assessment. The remaining chapters are dedicated to all the considerations and methods for conducting a thematic accuracy assessment. Finally, Appendix 1 includes the latest ASPRS positional accuracy standards.

4 Positional Accuracy

INTRODUCTION

As we learned in Chapter 1, accuracy assessment is characterized by two measures: positional and thematic accuracy. For example, it is possible to be in the correct location and mislabel (incorrectly measure or classify) the attribute (i.e., theme). It is also possible to correctly label the attribute but be in the wrong location. In either case, error is introduced into the map or spatial data set. These two factors are not independent of each other, and great care needs to be taken not only to assess each of these factors but also to control them to minimize the errors.

This chapter reviews the concepts of positional accuracy. The first section reviews the basic concepts of positional accuracy assessment. Section 2 briefly reviews historical positional accuracy assessment standards or guidelines and examines the most recently compiled and widely accepted accuracy assessment standard, which incorporates concepts from earlier standards and guidelines developed over the last 70+ years. Developed by the American Society of Photogrammetry (ASPRS) and Remote Sensing, this standard is titled *ASPRS Positional Accuracy Standards for Digital Geospatial Data* (ASPRS, 2014). It can be found at http://www.asprs. org/a/society/committees/standards/Positional_Accuracy_Standards.pdf and is included as Appendix 1. The standard is comprehensive and addresses a wide variety of issues in positional accuracy assessment. The most important, but not all, of the concepts reviewed in the ASPRS 2014 standards are examined in this chapter.

The next section reviews positional accuracy sample design and collection within the overall framework set forth in the ASPRS 2014 standard. The fourth section introduces basic statistical concepts and uses them to explain how to analyze the accuracy assessment sample data to estimate positional accuracy under the APSRS 2014 standards.

A major goal of this chapter is to bring clarity to the language and equations of positional accuracy assessment. Since the development of the first standards in 1942, each new standard or guideline has introduced new concepts and interpreted old concepts in new ways. As a result, the language of positional accuracy assessment is often confusing, and the equations that are used to compute accuracy assessment standards sometimes require assumptions that may not be appropriate. Therefore, additional care and consideration are needed to ensure that your positional assessment is appropriate and valid.

WHAT IS POSITIONAL ACCURACY?

Positional accuracy assesses the correctness of the coordinates of an image or a map. All locations on maps and geo-referenced images are expressed in x and y coordinates for horizontal location. Many data sets also include elevations, which are represented by the letter z. Positional accuracy assessment employs sampling to estimate the discrepancy between the coordinates or elevations of features in a map or image and their actual or "true" location on the earth's surface. The actual position is determined from a source of higher accuracy than the instruments used to create the map or geo-referenced image—often a survey. Positional accuracy can refer to either horizontal (planimetric) and/or vertical (elevational) accuracy, and this chapter discusses both.

The *Glossary of the Mapping Sciences* (ASPRS and ASCE, 1994) defines positional accuracy as "the degree of compliance with which the coordinates of points determined from a map agree with the coordinates determined by survey or other independent means accepted as accurate." The ASPRS 2014 standards are slightly different and define accuracy as "The closeness of an estimated value (for example, measured or computed) to a standard or accepted (true) value of a particular quantity" and positional accuracy as "The accuracy of the position of features, including horizontal and vertical positions, with respect to horizontal and vertical datums."

Several factors can affect the positional accuracy of a map or geo-referenced image. For example, the sensor lens may be distorted, or the aircraft carrying the sensor may suddenly tilt or yaw, changing the relationship of the sensor's image plane to the ground. However, the most important cause of positional error arises from the impact of topography on remotely sensed imagery. Because the sensor image plane is flat, and the earth has relief such as hills and ravines, the scale of the remotely sensed imagery relative to the earth varies with topographic changes, requiring that some sort of adjustment be made to "terrain-correct" the image. This correction is a complex process, which has been highly prone to error in the past but is becoming less problematic with technological advances and the creation of increasingly accurate digital elevation models for vast areas of the earth.

For example, Figure 4.1 presents an illustration of an inaccurate road layer displayed over the top of an ortho-corrected digital image. The reference data that have been "accepted as accurate" are survey points indicated by a colored circle on the figure. As you can see, the road layer does not align with the points in many places (i.e., there are positional errors). In general, the roads are shifted to the north and west of their "true" location, as determined by the survey points. While we can clearly see that the position of the road is inaccurate, we need to use quantitative accuracy assessment to estimate the magnitude and direction of the errors.

In statistics and accuracy assessment, there are two terms that are commonly used and confused and which need clarification. Accuracy and precision are often thought of as synonymous but actually have very different meanings. Accuracy refers to the bias of an estimator. It measures how close an estimated or calculated value is to its true value. Precision refers to the variability in an estimator. It quantifies how repeated measures of the same estimator will vary. Inaccurate measurements can be very precise, and accurate measurements can be imprecise. Therefore, we need to measure both the accuracy and the precision of maps and imagery. Figure 4.2 illustrates the concepts of accuracy and precision with an example of multiple measurements made of one location.

In positional accuracy assessment, we are interested in characterizing the accuracy and precision of a geospatial data set. We take samples to estimate the bias

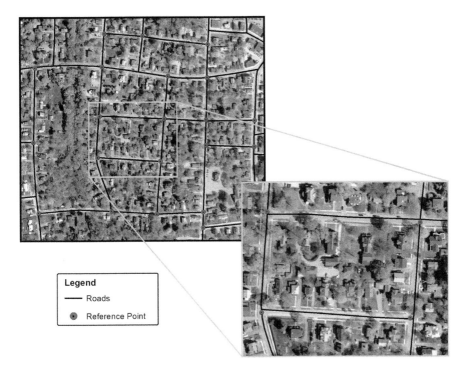

FIGURE 4.1 Illustration of positional errors in a road map (in black) compared with the survey (reference) points, which are accepted as accurate.

and precision of the modeled or estimated values. We also strive to ensure that our measurements of the location of each sample's reference and geospatial data set are themselves accurate, and we must take enough samples, so that our estimate of the bias (if it exists) is precise.

The steps for positional accuracy assessment are the same as for thematic accuracy assessment and require:

1. Designing a sample—how many samples will be collected and where they will be placed
2. Collecting reference data
3. Comparing the reference data with the map or image data
4. Analyzing the differences between the reference data and the map or image data
5. Reporting results

WHAT ARE THE COMMON STANDARDS FOR POSITIONAL ACCURACY?

The second edition of this book (Congalton and Green, 2009) reviewed past positional accuracy assessment standards and guidelines, and noted the deficiencies in each.

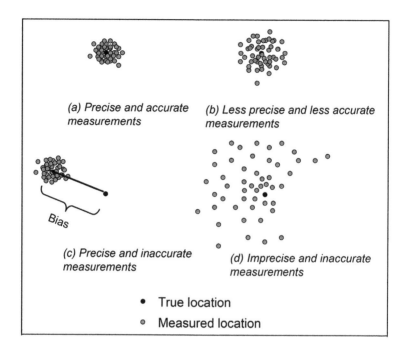

FIGURE 4.2 Comparison of precision versus accuracy.

More recently, Annex A of the *ASPRS Positional Accuracy Standards for Digital Geospatial Data* (ASPRS, 2014) summarized past standards and guidelines. They are as follows:

1. *National Map Accuracy Standards* (NMAS) (U.S. Bureau of the Budget, 1947) https://nationalmap.gov/standards/pdf/NMAS647.PDF. Using the "percentile method", NMAS states that no more than 10 percent of the samples may exceed the maximum error allowed. Maximum errors are stipulated for scales below and above 1:20,000. NMAS relies on map versus ground units and provides no guidance for creating statistically valid bounds on the estimated error.
2. *Principles of Error Theory and Cartographic Applications* (Greenwalt and Schultz, 1962, 1968), which is cited by all subsequent standards: www.fgdc.gov/standards/projects/accuracy/part3/tr96. The report interprets NMAS's (1947) "10 percent of the points taken" to limit the size of errors to that "which 90% of the well-defined points will not exceed" (Greenwalt and Schultz, 1962, 1968). The report uses probability theory to develop equations for calculating one-dimensional elevation (z) "map accuracy standard" (MAS) and two-dimensional (x and y) "circular map accuracy standard" (CMAS) statistics by assuming that map errors are normally distributed.[*] MAS is the estimated interval around the mean vertical error, and CMAS is

---

[*] We will examine the implications of the assumption of normality later in this chapter.

the estimated interval around the horizontal mean error within which 90% of the error is predicted to occur. The report does not specify 90% as the only probability level to be employed. Instead, it shows how to estimate the distribution of errors under various probability levels and provides tables for converting from one probability level to another.

3. *ASPRS 1990 Accuracy Standards for Large-Scale Maps* (ASPRS, 1990) www.asprs.org/a/society/committees/standards/1990_jul_1068-1070.pdf. The ASPRS 1990 standards stipulate a maximum distance which errors may not exceed for scales from 1:50 to 1:20,000. The standard states that the mean error estimated from the samples may not exceed the stipulated maximum distance. Most importantly, the ASPRS standards migrate the units of measurement of error from map units to ground units. The ASPRS standards also restate the Greenwalt and Schultz CMAS equations but do not imply that the equations should necessarily be used.

4. *National Standard for Spatial Data Accuracy* (NSSDA) (FGDC, 1998) www.fgdc.gov/standards/projects/accuracy/part3/chapter3. NSSDA explicitly rejects setting a maximum allowable error at any scale and suggests instead that the maximal allowable error threshold be determined as needed. Accuracy is to be reported "in ground distances at the 95% confidence level," which is interpreted as allowing "one point to fail the threshold given in the product specifications" when a sample of 20 points is used. Until the development of the ASPRS 2014 standards, it was not unusual for positional accuracy projects to use the equations of NSSDA to calculate accuracy statistics and also to require that those statistics not exceed the distances established in the ASPRS 1990 standards. NSSDA also incorporates the approach of Greenwalt and Schultz by referencing its equations and defining accuracy as a measure of the maximum error expected at a specific probability level.

5. Between 1998 and 2010, FEMA published several versions of their *Guidelines and Specifications for Flood Hazard Mapping Partners* (FEMA, 1998 -2010), which increased the statistical rigor of positional accuracy assessment for flood hazard mapping by requiring that a minimum of 20 samples be collected and accuracy reported for each major vegetation type, of which there may be a minimum of three, resulting in a minimum of 60 total sites sampled. These guidelines are applicable only to flood hazard mapping.

6. *Guidelines for Digital Elevation Data* (National Digital Elevation Program [NDEP], 2004). Like NSSDA, the NDEP guidelines do not establish accuracy thresholds. They call for vertical accuracy to be reported in three different ways depending on the ground cover of the area being mapped or imaged:

   a. "Fundamental vertical accuracy" (FVA) is computed only from samples measured in open terrain and relies on the NSSDA equations for calculating accuracy.

   b. "Supplemental vertical accuracy" (SVA) is measured from samples taken in non-open terrain cover types and is determined using the "95th percentile error" method, which is defined as the "absolute value in a dataset of errors. It is determined by dividing the distribution of the individual sample errors in the dataset into one hundred groups of equal frequency." By definition, 95% of the sampled errors will be less than the 95th percentile value.

   c. "Consolidated vertical accuracy" (CVA) is a combination of the samples from both open terrain and other ground cover classes and is reported as a 95th percentile error.

7. *ASPRS Guidelines: Vertical Accuracy Reporting for Lidar Data* (ASPRS, 2010) https://nationalmap.gov/standards/pdf/NDEP_Elevation_Guidelines_ Ver1_10May2004.pdf. The *ASPRS Guidelines for Reporting Vertical Accuracy of Lidar Data* (ASPRS, 2004) ratify the NDEP guidance to stratify the landscape into different land cover classes and report FVA, SVA, and CVA. These guidelines also do not establish accuracy thresholds.

8. *Guidelines and Specifications for Flood Hazard Mapping Partners* (FEMA, 2003) www.fema.gov/media-library-data/1388520285939-754d a930e9d1d081955e4cce0b279ccd/Guidelines_and_Specifications_for_Fl ood_Hazard_Mapping_Partners_Volume_1-_Flood_Studies_and_Mapp ing_(Apr_2003).pdf. FEMA's *Guidelines and Specifications for Flood Hazard Mapping Partners* (FEMA, 2003) increased the minimum number of samples required by stipulating that a minimum of 20 samples be collected for each major vegetation type, of which there may be a minimum of three, resulting in a minimum of 60 total sites sampled.

9. *United States Geological Survey (USGS) Lidar Base Specification Version 1.2* (USGS, 2014), which specifies the base requirements for source lidar data collected under the national interagency 3D Elevation Program (3DEP). These specifications were updated by the USGS in February, 2018 to Version 1.3: https://pubs.usgs.gov/tm/11b4/pdf/tm11-B4.pdf. They incorporate recent technical advances resulting in improved lidar resolution and accuracy, the ASPRS (2014) positional accuracy standards, new industry standards, new lidar applications, and the need for interoperable data. The new version explicitly requires that "horizontal accuracy of each lidar project shall be reported using the form specified by the ASPRS (2014)" and "Absolute vertical accuracy of the lidar data and the derived DEM will be assessed and reported in accordance with ASPRS (2014)".

Development of the ASPRS 2014 standard was needed because:

- Several of the old standards relied on measuring error in hardcopy maps, which is no longer suitable in today's digital era.
- Many new technologies (e.g., digital cameras and scanners, lidar, and phodar) were not in existence when the old standards and guidelines were developed, requiring the expansion of accuracy assessment methods to account for the improved accuracies resulting from these technological advances (Congalton and Green, 2009; ASPRS, 2014).

- None of the standards or guidelines comprehensively specified positional accuracy assessment design, equations, and accuracy thresholds for all applications and scales.

The ASPRS 2014 standard systematically incorporates the relevant concepts of the standards or guidelines that preceded it. It is an exhaustive document and rightfully has become the de facto standard worldwide for positional accuracy assessment. However, even this new standard incorporates assumptions from past standards, which need elucidation and will be examined in later sections of this chapter. Perhaps the most significant contribution of the ASPRS 2014 standards can be found in their Annex B, which presents examples of accuracy measures for digital orthoimagery and planimetric data at various ground sample distances along with their equivalent map scales in relationship to the legacy standards of ASPRS 1990 and NMAS of 1947.

HOW SHOULD POSITIONAL ACCURACY ASSESSMENT BE DESIGNED AND THE SAMPLES BE SELECTED?

Positional accuracy assessment requires the selection of samples to estimate the statistical parameters of the population of errors (e_i) occurring in the spatial data being assessed. Estimates are made of the mean (μ), the standard deviation (σ), the standard error (σ_μ), and the root mean square error (RMSE) to characterize the distribution of the population of errors and the reliability of estimators. The mean (μ) is the expected value of a random variable. The variance measures how much the variables of a population deviate from the population mean. The standard deviation (σ) is the square root of the population variance. The standard error (σ_μ) is the square root of the variance of the estimate of the mean. It measures how estimates of the population mean will deviate from the true mean and is central in the creation of a confidence interval around an estimate of the mean. The RMSE measures the difference between a predicted or estimated value versus the true value. RMSE is the most commonly used measure of spatial accuracy. Equations for each of these parameters are presented in the following section "How Is Positional Accuracy Analyzed?"

Estimating error parameters requires the comparison of coordinates and/or elevations of identical sample locations from:

- the spatial data set to be assessed (map or imagery)
- the reference data, which must be an "independent source of higher accuracy" (FGDC, 1998; ASPRS, 2014).

We rely on samples, because measuring every point in the geospatial data set being assessed would be prohibitively expensive and time consuming. Instead, sampling can provide highly reliable estimates of the error population's parameters.

The standards in use today outline several requirements that govern positional accuracy sampling design and collection. They are as follows:

Data Independence. As noted earlier, both the NSSDA and ASPRS require that the reference data must be an "independent source of higher accuracy" (FGDC, 1998; ASPRS, 2014). To ensure the objectivity and rigor of the assessment, it is critically important that the reference data be independent from the data being assessed. In other words, the reference data cannot have been relied on during the creation of the map or image being tested. Thus, control points or digital elevation models (DEMs) used to create the spatial products being tested are unsuitable sources of reference data. A different RMSE value is often calculated as part of the spatial data set registration process. We will call this $RMSE_{reg}$. The calculation of $RMSE_{reg}$ during the registration process is a test of the goodness of fit of the registered data set to its control points. Because of its lack of independence from the data set being assessed, $RMSE_{reg}$ is not a valid measure of positional accuracy. Independent positional accuracy assessment requires the collection of a separate and independent set of test sample points that were not used as control points in the registration process.

Source of Reference Data. The source of the reference points depends on a number of factors. In some cases, a map of larger scale than the map or image being assessed may provide sufficiently detailed reference coordinates. This is especially true if the map/image to be tested is of small scale and covers a large area. In other cases, such as engineering site drawings, much more precision is required for the reference data points, which may require a field survey or the use of a high-precision GPS. While NSSDA (FGDC, 1998) stipulates that the reference source data "be of the highest accuracy feasible and practicable," the ASPRS 2014 standard requires that the accuracy assessment reference data must be at least "three times more accurate that the required accuracy of the geospatial data set being tested."

Number of Samples. While NSSDA (FGDC, 1998) suggested that a minimum of 20 samples be used to estimate positional accuracy, later guidelines and the ASPRS 2014 standard stipulate a minimum of 20 samples and suggest many more samples, especially for large-area projects. The ASPRS 2014 standard states: "Whereas 100 or more is a desirable number of checkpoints, that number of checkpoints may be impractical and unaffordable for many projects, especially small area projects." Accordingly, Table C.1 in the ASPRS 2014 standards (see Appendix 1) lists the number of samples required for horizontal and vertical accuracy assessment by ranges of project size up to 2500 km^2. For areas over 2500 km^2, the standard calls for an additional five vertical samples for each additional 500 km^2. For the horizontal accuracy assessment of areas over 2500 km^2, the standard states that "clients should determine the number of additional horizontal checkpoints, if any, based on criteria such as resolution of imagery and extent of urbanization."

Earlier guidelines have also stipulated a minimum number of samples by general land cover type (FEMA, 2003). The ASPRS 2014 standards do not require a minimum number of samples by land cover but rather, recognize "that some project areas are primarily non-vegetation, whereas other areas are primarily vegetated. For these reasons, the distribution of checkpoints can vary based on the general proportion of vegetation and non-vegetation area in the project. Checkpoints should be distributed generally proportionally among the various vegetated land cover types in the project."

Identification of Samples. ASPRS (2014) specifies that the samples for assessing horizontal accuracy must be "well-defined points" that are "easily visible or identifiable on the independent source of higher accuracy, and on the product itself." Conversely, samples for assessing vertical accuracy do not require clearly identifiable features, but they should be collected on flat or uniformly sloped open terrain of less than 10% slope. Areas with abrupt elevation changes should be avoided. The purpose of these requirements is to minimize the impact of differences caused by interpolation in the data set being assessed.

Distribution of Samples. The ASPRS (2014) standard affords a large amount of discretion in the placement of the accuracy assessment samples. Because some project areas are primarily non-vegetated, whereas other areas are primarily vegetated, and others are mixed, the distribution of checkpoints is expected to fluctuate based on the proportion of vegetated and non-vegetated areas in a project. For largely unvegetated and rectangular areas, the ASPRS (2014) standard calls for the same distribution of samples as suggested first in the ASPRS (1990) standards and adopted in the NSSDA (FGDC, 1998). To implement that sample distribution, first, the map or image is divided into quadrants. Next, a minimum of 20% of the sample points are allocated to each quadrant. To ensure adequate spacing between the sample points, no two points should be closer than $d/10$ distance from each other, where d is the diagonal dimension of the map or image as illustrated in Figure 4.3. This spacing will minimize spatial autocorrelation (a topic that will be discussed in detail in later chapters). This systematic sample distribution requires assuming that the sample distribution is not correlated with map or image error, which is a reasonable assumption, because most positional error is correlated with topography, and topography is rarely distributed on a grid pattern.

Overall, the distribution of positional accuracy samples requires the simultaneous consideration of several factors. Often, there is a trade-off between well-distributed, accessible, and easily identifiable sample points. It is not uncommon for some of the desired sample points to fall on private land, which may be inaccessible if a ground survey is being used to collect the reference data. Often, easily identifiable points are concentrated in small areas and are not evenly distributed throughout the map. Care must be taken to obtain the best possible combination of good test points that are appropriately distributed throughout the map or image being assessed. Most importantly, the client requesting the map or imagery should be involved in designing the distribution of the accuracy assessment samples. As stated in the ASPRS 2014 standards, "The general location and distribution of checkpoints should be discussed between and agreed upon by the vendor and the customer as part of the project plan."

HOW IS POSITIONAL ACCURACY ANALYZED?

Analyzing positional accuracy involves using sample data to estimate the fit of the spatial data layer (map or image) being assessed to the reference layer, which is assumed to be correct. The APSRS (2014) standards recommend two different types of analysis depending on whether or not the errors are believed to be normally distributed. When the errors are believed to be normally distributed, the RMSE and the error population's estimated mean, standard deviation, and standard error are

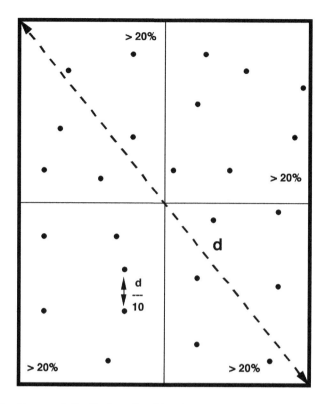

FIGURE 4.3 Suggested distribution of positional accuracy assessment sample locations.

calculated and used to report accuracy. When the errors are assumed not to be normally distributed, the percentile rank below or above a given maximum value is commonly used to assess accuracy. Because much confusion exists between the commonly used accuracy standards, we will begin this section with a review of the basic statistics and then move on to the specific equations for depicting positional accuracy.

REVIEW OF BASIC STATISTICS

The concepts in this section may be found in any standard statistics textbook. Documents directly relied on for this text include *Principles of Error Theory and Cartographic Applications* (Greenwalt and Schultz, 1962, 1968), *Biostatistical Analysis* (Zar, 1974), and *Analysis and Adjustment of Survey Measurements* (Mikhail and Gracie, 1981).

This section first provides the equations for calculating and estimating the parameters of a population of values. Next, it discusses the assumptions and equations required to estimate the dispersal of values around the mean. Finally, it provides the equations for calculating a confidence interval around the estimate of the mean.

Parameters and Statistics

The arithmetic mean (μ) of a population of random variables (X_i) is the expected value of any random variable and is calculated by:

$$\mu_{X_i} = \sum_{i}^{N} X_i / N \qquad (4.1)$$

where:
   X_i= the value of the ith individual in the population
   N= the total number of individuals in the population

The mean is estimated from a sample by the variable $\overline{X}$ and is calculated by

$$\overline{X} = \sum_{i}^{n} x_i / n \qquad (4.2)$$

where:
   x_i= the value of the ith sample unit chosen from the population
   n= the total number of sample units chosen

The standard deviation (σ) is the square root of the population variance, which measures how much the variables of a population deviate from their expected value (i.e., the population mean). The standard deviation is calculated by:

$$\sigma = \sqrt{\sum_{i}^{N} (X_i - \mu)^2 / (N - 1)} \qquad (4.3)$$

where X_i, μ, and N are defined as earlier.

The standard deviation is estimated from a sample by the variable S, and is calculated by

$$S = \sqrt{\sum_{i}^{n} (x_i - \overline{X})^2 / (n - 1)} \qquad (4.4)$$

where X_i, μ, and N are defined as earlier.

Another key parameter in statistics is the standard error ($\sigma_{\overline{X}}$), which helps us to characterize the spread in the distribution of the possible means that could be derived from a single *sample* of a population (rather than the entire population itself). According to the Central Limit Theorem, the standard error, which is the square root of the variance of the population of estimated means, is a valuable parameter, because it allows us to estimate our confidence in our estimate of the mean. There is a population of possible estimated means (instead of just one), because there are many possible values of $\overline{X}$, each resulting from a different selection of samples of size n from the population. The standard error is calculated by:

$$\sigma_{\overline{X}} = \sigma / \sqrt{n} \qquad (4.5)$$

where σ and n are defined as earlier.

The standard error is estimated from a sample by the variable $S_{\bar{X}}$ and is calculated by:

$$S_{\bar{X}} = S / \sqrt{n} \tag{4.6}$$

where S and n are defined as earlier.

The final and most important statistic in positional accuracy assessment is the RMSE, which is the square root of the mean squared differences between the sample map points and the reference points. The equation for calculating RMSE in mapping applications is:

$$\text{RMSE} = \sqrt{\sum_i^n (e_i)^2 / n} \tag{4.7}$$

where:

$$e_i = e_{ri} - e_{mi} \tag{4.8}$$

e_{ri} equals the reference location or elevation at the ith sample point
e_{mi} equals the map or image location or elevation at the ith sample point
n is the number of samples

ESTIMATING THE DISPERSAL OF VARIABLES

If the errors are normally distributed about the mean, as depicted in Figure 4.4, then the normal or Gaussian distribution can be used to approximate the distribution of population variables. Additionally, the standard normal distribution can be used to estimate an interval of X_i at specified probabilities within which the mean of the population (μ) will fall. To do so, the distribution of the population variables must be standardized by transforming the scale of the standard normal distribution to the scale of the population being studied.

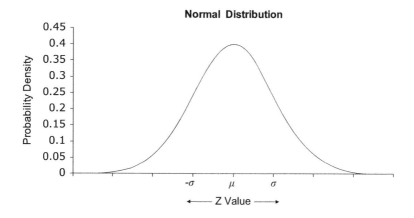

FIGURE 4.4 The shape of the normal distribution.

All normal distributions are shaped like the curve in Figure 4.4, with the area underneath the curve equal to 1. The *standard* normal distribution represents the distribution of the standard normal variable (Z_i) and is unique because it has a mean of 0 and a standard deviation of 1, as illustrated in Figure 4.5.

The standard normal variable, Z_i, is defined as:

$$Z_i = (X_i - \mu)/\sigma \qquad (4.9)$$

where:

Z_i is the value from the x-axis of the standard normal distribution at the *i*th probability level

X_i is the corresponding value from the x-axis of the population of interest, and

μ and σ are defined as earlier

By using algebra, we can transform the x-axis scale of the normal distribution to that of our population by solving for values of X_i such that:

$$Z_i * \sigma = (X_i - \mu) \qquad (4.10)$$

and

$$X = (Z * \sigma) + \mu \qquad (4.11)$$

With this formula, we could transform every Z_i value of the standard normal distribution into a X_i value of our population. More commonly, the transformation is used to calculate an interval at a specified probability level within which values of X_i will occur such that $X_i < \mu < X_1$, or using Equation 4.11, the interval becomes:

$$\left[\mu - Z_i * \sigma, \mu + Z_i * \sigma \right] \qquad (4.12)$$

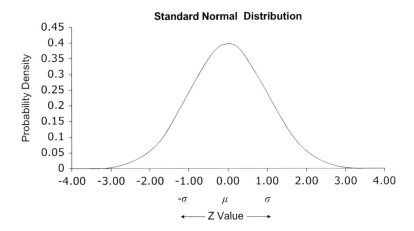

FIGURE 4.5   The standard normal distribution.

Transforming the values of the normal standard distribution into that of our population of interest requires that the distributions of variables of the normal distribution and of the population of interest be almost identical. This is not a big leap of faith, as the normal distribution characterizes a multitude of natural phenomena ranging from organism population dynamics to human polling behavior. However, it is always important to fully understand whether or not the population you are studying is actually normally distributed or not. Equation 4.12 expresses the dispersal around the mean of the variable X_i at the stipulated probability level *if, and only if, the population of X_i's is normally distributed*. Figure 4.6 illustrates the portions of the normal distribution and the corresponding Z_i values that match various levels of probability.

To summarize, determining the interval at a specific probability within which the mean (μ) of our population of interest will fall requires simply:

1. looking up the Z_i value for the specified probability level in a standard normal table (which may be found in the back of any statistics text or by searching on the Internet)
2. multiplying the Z_i value times the standard deviation (σ) of the population of interest
3. adding and subtracting the resulting $Z_i * \sigma$ value from the mean (μ)

For example, the interval within which 90% of the values of a normally distributed population with a mean (μ) of 20 and a standard deviation (σ) of 4 can be determined by:

1. Looking up the Z_i value for 90% probability in a Z table or from Figure 4.6. At 90% probability, Z_i is equal to 1.645.
2. Calculating $Z_i * \sigma$ by multiplying 1.645 times the standard deviation of 4, which equals 6.58.
3. Adding and subtracting 6.58 from the mean to determine the interval at 90% probability $= 20 - 6.58$, $20 + 6.58$.

This results in the interval ranging from 13.42 to 26.58.

Therefore, we know that 90% of the values of our population will fall within a range between 13.42 and 26.58. Figure 4.7 shows how the x-axis scale of the standard normal distribution transforms to that of our example.

Usually, we do not know the true mean and the standard deviation of the population. However, because $\overline{X}$ and S are unbiased estimators of μ and σ, we can use the sample estimates of the mean ($\overline{X}$) and the standard deviation (S) to calculate the interval, which becomes:

$$\overline{X} - Z_i * S < \mu < \overline{X} + Z_i * S \qquad (4.13)$$

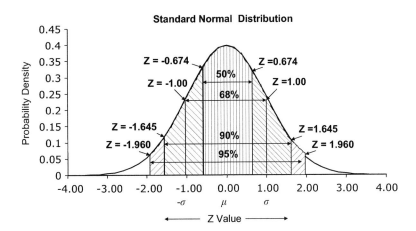

FIGURE 4.6 Probability areas and corresponding Z_i values of the standard normal distribution.

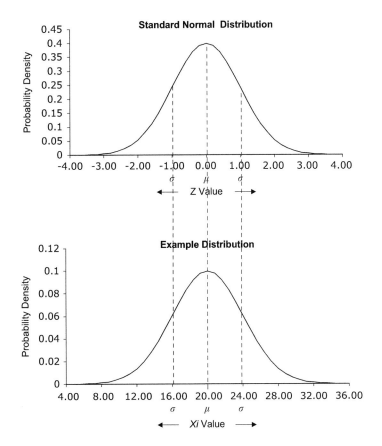

FIGURE 4.7 Transformation of the standard normal distribution x-axis scale to the x-axis scale of the example.

ESTIMATING A CONFIDENCE INTERVAL AROUND THE ESTIMATE OF THE MEAN

Often, we want to understand how reliable our estimate of the mean is. To do so requires using the sample data to develop a "confidence interval" around the estimated mean. Estimating the confidence interval once again employs the standard normal variable (Z_i), which for the population of sample means is defined as:

$$Z_i = (\overline{X}_i - \mu) / \sigma_{\overline{X}} \tag{4.14}$$

and can be estimated by:

$$Z_i = (\overline{X}_i - \overline{X}) / S_{\overline{X}} \tag{4.15}$$

where:

   $\overline{X}_i$ is the value from the population of estimated means that corresponds to the Z_i value from the normal distribution and

   μ, σ, $\overline{X}$, and $S_{\overline{X}}$ are defined as earlier.

The confidence interval on the estimate of the mean is calculated as:

$$\overline{X} - Z_i(S_{\overline{X}}) < \mu < \overline{X} + Z_i(S_{\overline{X}}) \tag{4.16}$$

when sample sizes are large and by:

$$\overline{X} - t_i(S_{\overline{X}}) < \mu < \overline{X} + t_i(S_{\overline{X}}) \tag{4.17}$$

when sample sizes are small, where t_i is the value from the x-axis of the Student's t distribution* at the ith probability level.

There is a subtle but very important distinction between estimating the *dispersal* interval of population values around the mean at a specified probability, as calculated in Equation 4.13, and the calculation of a *confidence interval* around the estimate of the mean, as shown in Equations 4.16 and 4.17. The former expresses the dispersal of a population of values around the mean at specified probabilities. The latter expresses the reliability of the estimate of the mean at specified probabilities.

An interesting aspect of the population of sample means is that it will be normally distributed even when the underlying population of variables is not. This important concept, which is derived from the Central Limit Theorem, tells us that when the sample size is large enough, and the samples are chosen without bias, then the distribution of the population of means *will* be normally distributed even when the population distribution from which the samples were chosen to estimate the means *is not* normally distributed. The Central Limit Theorem

---

* The Student's t distribution should be used instead of the Z distribution when sample sizes are below 30. As with the Z distribution, the values of the Student's t distribution can be found in the back of any statistics text or on the Internet.

permits us to state our confidence in our estimate of $\overline{X}$ regardless of the distribution of X_i, which allows us to rely on, rather than assuming, the fact that the shape of the standard normal distribution is the shape of the distribution of sample means.

STATISTICS IN POSITIONAL ACCURACY ASSESSMENT

All positional accuracy measures are estimated by comparing reference coordinates or elevations with the map or image coordinates or elevations of the data set being assessed at each sample location. This section reviews the equations used to report positional accuracy if the errors are assumed to be normally distributed. First, one-dimensional vertical accuracy assessment is discussed. Next, two-dimensional horizontal accuracy assessment is reviewed. The following section reviews the techniques which the ASPRS standard stipulated should be used if the errors are assumed not to be normally distributed.

If the Errors Are Assumed to Be Normally Distributed

Vertical Accuracy

The ASPRS 2014 standards call for the mean vertical positional error (μ_v)* to be calculated as the simple mean of the error values calculated by:

$$\overline{e_v} = 1/n \sum_{i=1}^{n} e_{vi} \tag{4.18}$$

However, using the simple mean can result in errors cancelling one another out. An alternative equation for estimating the mean magnitude of error, suggested by Greenwalt and Schultz (1968), is to calculate the mean of the absolute values of the errors as:

$$\overline{|e_v|} = 1/n \sum_{i=1}^{n} |e_{vi}| \tag{4.19}$$

The ASPRS 2014 standards acknowledge the advantage of using absolute values "as it is the magnitude of the errors, not the sign that is of concern."

However, if our goal is to understand the distribution of errors around the mean, we need to use the arithmetic mean as stipulated in Equation 4.18. The standard deviation (σ_v) of the population of vertical errors is estimated by:

$$S_v = \sqrt{\sum_i^n \left(e_{vi} - \overline{e_v}\right)^2 / (n-1)} \tag{4.20}$$

---

* Some mapping texts use the subscript z to denote vertical error. Because this text (and most statistics texts) uses the variable Z_i to denote the standard normal variable, we use the subscript v to denote vertical error.

and the standard error of estimates of $\overline{e_v}$ is estimated by:

$$S_{\overline{e_v}} = S_v / \sqrt{n} \qquad (4.21)$$

Assuming that the vertical errors are normally distributed, the estimated interval of errors at a specific probability can be expressed as:

$$\overline{e} \pm Z_i(S_v) \qquad (4.22)$$

At a 95% probability level, the equation becomes:

$$\overline{e_v} \pm 1.96\,(S_v). \qquad (4.23)$$

If $\overline{e_v}$ equals zero, then the factor $\pm Z_i(S_v)$ will express the interval of error at the probability level specified by the Z_i variable, and the interval at 95% will equal $\pm 1.96(S_v)$. A 90% interval, with $\overline{e_v}$ equal to zero, will be $1.645(S_v)$.

The *Principles of Error Theory and Cartographic Application* report was the first report to propose the use of the $Z_i(S_v)$ interval as a standard in estimating positional accuracy (Greenwalt and Schultz, 1962, 1968). The report relies on estimating the interval $Z_i(S_v)$ at various probability levels, where it is referred to as the *probable error* (PE) at 50% and the *map accuracy standard* (MAS) at 90% (Greenwalt and Schulz, 1962, 1968). Figure 4.8 illustrates the portions of the normal distribution that correspond to the PE at 50%, the MAS at 90%, and the NSSDA standard at 95%.

The Greenwalt and Schultz (1962, 1968) and subsequent FGDC (1998) and ASPRS (2014) equations estimate the interval of errors around the mean error at different probability levels. Derived from the military science of ballistics, the equations result in an estimate of the probable dispersal of error around the mean error $(\overline{e_v})$ at specified probabilities.*

Note that the $Z_i(S_v)$ interval is not a confidence interval around the estimate of $\overline{e_v}$, nor is it the range of expected errors at a given probability. Rather, it is an estimate of the maximum interval of error that will exist at a specified probability assuming that the errors are normally distributed and the mean error is equal to zero. Unfortunately, spatial errors are often biased and inter-related, bringing the assumption of normality into question.

---

* Greenwalt and Schultz (1962, 1968) define the vertical map accuracy standard as "the size of error which 90% of the elevations will not exceed." However, the interval $Z_i(S_z)$ meets this definition *only* when $\overline{e_v}$ equals zero. The estimated size of elevation errors which will not be exceeded at a probability level specified by Z_i is $\overline{e_v} \pm Z_i(S_v)$.

To measure the reliability of (or our confidence in) the estimate of $\overline{e_v}$, we can calculate a confidence interval around $\overline{e_v}$ by converting the general confidence interval equations (Equations 4.16 and 4.17):

$$\overline{X} - Z_i(S_{\overline{X}}) < \mu < \overline{X} + Z_i(S_{\overline{X}})$$

to our mapping application terminology such that:

$$\overline{e_v} - Z_i\left(S_{\overline{e_v}}\right) < \mu < \overline{e_v} + Z_i\left(S_{\overline{e_v}}\right) \qquad (4.24)$$

for large sample sizes and:

$$\overline{e_v} - t_i\left(S_{\overline{e_v}}\right) < \mu < \overline{e_v} + t_i\left(S_{\overline{e_v}}\right) \qquad (4.25)$$

for small sample sizes, where all variables are defined as earlier.

In most situations, if we have more than 30 samples, at a 95% confidence level, the equation becomes:

$$\overline{e_v} - 1.96\left(S_{\overline{e_v}}\right) < \mu < \overline{e_v} + 1.96\left(S_{\overline{e_v}}\right). \qquad (4.26)$$

That means that we are 95% certain that the interval contains the true, but unknown, population average error.

Table 4.1 displays the map and reference elevations for a hypothetical digital elevation data set. The errors at each sample point are calculated as well as the

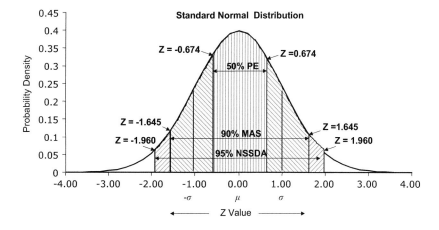

FIGURE 4.8 Areas and Z_i values of the standard normal distribution corresponding to the probability levels of various map accuracy standards.

TABLE 4.1

Vertical Accuracy Example

Point ID	v_{ri} Reference	v_{mi} Map	Error $= e_{vi} =$ Reference − Map $(v_{ri} - v_{mi}) = e_{vi}$	Error Squared $(v_{ri} - v_{mi})^2 = e_{vi}^2$	Absolute Error	(Absolute e_{vi} − Absolute Mean)^2	$(e_{vi}$ − Arithmetic Mean)^2
1202	2362.2075	2361.3100	0.8975	0.8055	0.8975	0.4519	0.6710
1230	2421.5855	2420.9000	0.6855	0.4699	0.6855	0.2118	0.3686
1229	2701.6110	2701.1700	0.4410	0.1945	0.4410	0.0465	0.1315
125	705.3117	705.0190	0.2927	0.0857	0.2927	0.0045	0.0459
316	1009.2344	1009.0300	0.2044	0.0418	0.2044	0.0004	0.0159
369	920.0574	919.8740	0.1834	0.0336	0.1834	0.0018	0.0110
292	586.3659	586.2400	0.1259	0.0159	0.1259	0.0099	0.0023
143	761.4684	761.3910	0.0774	0.0060	0.0774	0.0219	0.0000
132	712.1791	712.1320	0.0471	0.0022	0.0471	0.0317	0.0010
1005	1190.4284	1190.4000	0.0284	0.0008	0.0284	0.0387	0.0025
274	809.0433	809.0500	−0.0067	0.0000	0.0067	0.0478	0.0072
112	387.2611	387.2960	−0.0349	0.0012	0.0349	0.0362	0.0128
339	965.6910	965.7480	−0.0570	0.0032	0.0570	0.0283	0.0183
130	1059.1342	1059.2300	−0.0958	0.0092	0.0958	0.0168	0.0303
113	428.7700	428.9630	−0.1930	0.0372	0.1930	0.0010	0.0736
122	1012.0117	1012.3100	−0.2983	0.0890	0.2983	0.0053	0.1419
136	308.7100	309.0110	−0.3010	0.0906	0.3010	0.0057	0.1439
104	529.4721	529.8260	−0.3539	0.1252	0.3539	0.0166	0.1868
101	427.1653	427.5840	−0.4187	0.1753	0.4187	0.0374	0.2471
1221	2690.1380	2689.5200	0.6180	0.3819	0.6180	0.1543	0.2912
129	483.4317	483.0480	0.3837	0.1472	0.3837	0.0251	0.0932

(Continued)

TABLE 4.1 (CONTINUED)
Vertical Accuracy Example

Point ID	v_{ri} Reference	v_{mi} Map	Error $= e_{vi} =$ Reference $-$ Map $(v_{ri} - v_{mi}) = e_{vi}$	Error Squared $(v_{ri} - v_{mi})^2 = e_{vi}^2$	Absolute Error	(Absolute $e_{vi} -$ Absolute Mean)^2	$(e_{vi} -$ Arithmetic Mean)^2
128	492.7014	492.5810	0.1204	0.0145	0.1204	0.0110	0.0018
114	799.9452	799.8560	0.0892	0.0080	0.0892	0.0185	0.0001
367	1273.0857	1273.0300	0.0557	0.0031	0.0557	0.0287	0.0005
108	1235.0128	1235.0300	−0.0172	0.0003	0.0172	0.0433	0.0091
325	1040.9078	1040.9700	−0.0622	0.0039	0.0622	0.0266	0.0198
250	211.4375	211.5230	−0.0855	0.0073	0.0855	0.0195	0.0268
1010	1189.4876	1189.6200	−0.1324	0.0175	0.1324	0.0086	0.0444
Sum			2.19	2.77	6.31	1.35	2.60

estimated $RMSE_v$, $\left|\overline{e_v}\right|$, S_v, $S_{\overline{e}}$, NSSDA accuracy statistic, MAS, and a 95% confidence interval around the estimate of $\overline{e_v}$. All these equations for these calculations are shown in Table 4.2.

ASPRS (2014) Standards

The ASPRS (2014) standards direct that the NSSDA equations be used to report vertical accuracy. The NSSDA (FGDC, 1998) requires that accuracy be reported at the 95% level, which is defined by NSSDA as meaning "that 95% of the positions in the data set will have an error with respect to true ground positions that is equal to or smaller than the reported accuracy." NSSDA references the Greenwalt and Schultz (1962, 1968) equations for calculation of the 95% level, but substitutes $RMSE_v$ for the

---

TABLE 4.2
Vertical Accuracy Example Equations and Statistics

Definition	Equation	Value
Estimated root mean square error of the population of vertical errors	$RSME_v = \sqrt{\sum_i^n (e_{vi})^2 / n}$	0.315
Estimated absolute mean of the population of vertical errors	$\left\|\overline{e_v}\right\| = \sum_1^n \left\|e_{vi}\right\| / n$	0.225
Estimated variance of the population of vertical errors	$S_{\|v\|}^2 = \sum_1^n \left(\left\|e_{vi}\right\| - \left\|\overline{e_v}\right\|\right)^2 / (n-1)$	0.050
Estimated standard deviation of the population of vertical errors	$S_{\|v\|} = \sqrt{\sum_i^n \left(\left\|e_{vi}\right\| - \left\|\overline{e_v}\right\|\right)^2 / (n-1)}$	0.224
Estimated standard error of the population of mean errors	$S_{\|e_v\|} = S_{\|v\|} / \sqrt{n}$	0.042
Estimated arithmetic mean of the population of vertical errors	$\overline{e_v} = \sum_1^n e_{vi} / n$	0.078
Estimated standard deviation of the population of arithmetic errors	$S_v = \sqrt{\sum_i^n (e_{vi} - \overline{e_v})^2 / (n-1)}$	0.310
Estimated standard error of the population of mean arithmetic errors	$S_{\overline{e_v}} = S_v / \sqrt{n}$	0.059
Greenwalt & Schultz MAS standard normal interval of ev_i at 90% probability	$1.645 * S_v$	0.510
Greenwalt & Schultz standard normal interval of ev_i at 95% probability	$1.96 * S_v$	0.608
NSSDA statistic	$1.96 * RMSE_v$	0.617
Test on whether the mean arithmetic error is significantly different from zero at the 95% confidence level.	$\overline{e_v} \pm 1.96 * S_{\overline{e_v}}$	$0.078 \pm 1.96*0.059$

Which results in an interval from –0.038 to 0.194. The assumption that the mean error could equal zero is correct at the 95% confidence level, which means that RMSE can be substituted for S_v and the equation for the NSSDA statistic is correct.

---

estimated standard deviation (S_v) in the equations. The resulting NSSDA equation for calculating the NSSDA vertical accuracy statistic is*:

$$\text{NSSDA vertical accuracy} = 1.96\left(\text{RMSE}_v\right) \qquad (4.27)$$

rather than the Greenwalt and Schultz (1962, 1968) equation, which is:

$$\text{Vertical accuracy} = 1.96\left(S_v\right) \qquad (4.28)$$

Looking at the following equations for the RMSE and the standard deviation, we can see that the two statistics are not the same and are equal to one another only when the mean error is equal to zero and the sample size is large, so that the denominator (i.e., $(n-1)$) in the standard deviation equation approaches n, which is the denominator in the RMSE equation.

$$\text{RMSE} = \sqrt{\sum_i^n (e_{vi})^2 / n}$$

$$S_v = \sqrt{\sum_i^n (e_{vi} - \overline{e_v})^2 / (n-1)}$$

If the mean is equal to zero, then the interval becomes the Greenwalt and Schultz (1962, 1968) statistic of $Z_i\,(S_v)$. However, assuming that the mean error is equal to zero seems to defeat the purpose of positional accuracy assessment, which is to use a sample to understand the size and distribution of positional errors. While the NSSDA standard is applied ubiquitously, it is valid only when $\text{RMSE}_v = S_v$. If S_v is less than RMSE_v, then the NSSDA statistic will overestimate the error interval, and if S_v is greater than RMSE_v, the NSSDA statistic will underestimate the error interval. Therefore, we suggest that this assumption not be made and that the estimate of the mean error be calculated and the interval of probable errors be estimated as specified in Equations 4.24 and 4.25 and repeated here:

$$\overline{e_v} - Z_i\left(S_{\overline{e_v}}\right) < \mu < \overline{e_v} + Z_i\left(S_{\overline{e_v}}\right) \qquad (4.24)$$

for large sample sizes and

$$\overline{e_v} - t_i\left(S_{\overline{e_v}}\right) < \mu < \overline{e_v} + t_i\left(S_{\overline{e_v}}\right) \qquad (4.25)$$

for small sample sizes.

---

* 1.96 is the standard normal distribution Z statistic (the value from the x-axis of the normal distribution) for a probability of 95%.

Horizontal Accuracy

Horizontal accuracy is more complex than vertical accuracy, because the error is distributed in two dimensions (both the x and y dimensions), requiring the calculation of the radial error and reliance on the bivariate normal distribution to estimate probabilities. To calculate the horizontal root mean square error ($RMSE_h$),* first, the x coordinate from the reference data is recorded, followed by the x coordinate from the spatial data set being assessed. Then, the difference between the two locations is computed, followed by a squaring of this difference. The same process is used for the y coordinate. Each test point then has an associated error distance, e_i, defined by the following equation:

$$e_h = \sqrt{\left(x_{ri} - x_{mi}\right)^2 + \left(y_{ri} - y_{mi}\right)^2} \tag{4.29}$$

and

$$e_h^2 = (x_{ri} - x_{mi})^2 + (y_{ri} - y_{mi})^2 \tag{4.30}$$

where:

x_r and y_r are the reference coordinates, and
x_m and y_m are the map or image coordinates for the ith sample point in the spatial data set being assessed.

The equation for $RMSE_h$ is calculated from the errors of the individual test sample points using the following equation:

$$RMSE_h = \sqrt{\sum_i^n ((x_{ri} - x_{mi})^2 + (y_{ri} - y_{mi})^2)/n}$$
$$= \sqrt{(RMSE_x^2 + RMSE_y^2)/n} \tag{4.31}$$

or

$$RMSE_h = \sqrt{\dfrac{\sum_i^n e_{h_i}^2}{n}} \tag{4.32}$$

where:

e_{h_i} is defined in Equation 4.32
n is the number of test sample points

As in vertical accuracy, the ASPRS 2014 standards call for the mean horizontal positional error (μ_h) to be calculated as the simple mean of the error values calculated by

---

* Greenwalt and Schultz (1962, 1968) refer to horizontal error as *circular error*, which they designate with the subscript *c*. NSSDA (FGDC, 1998) refers to horizontal error as *radial*, designated by the subscript *r*. Because the errors are usually elliptical rather than circular, and because we have already designated the subscript *r* to indicate a reference value of an accuracy assessment sample, this text uses the subscript *h* to designate horizontal error.

$$\overline{e_h} = 1/n \sum_{i=1}^{n} e_i \qquad (4.33)$$

However, using the simple mean can result in errors cancelling one another out. An alternative equation is that suggested by Greenwalt and Schultz (1968), which is to calculate the absolute mean of the errors as:

$$\left|\overline{e_h}\right| = 1/n \sum_{1}^{n} \left|e_{hi}\right| \qquad (4.34)$$

Once the mean has been estimated, the standard deviation (S_h) of the population of horizontal errors can also be estimated from the samples by using the Greenwalt and Schultz (1962, 1968) equation:

$$S_h = \left(S_x + S_y\right)/2 \qquad (4.35)$$

where:

$$S_x = \sqrt{\sum_{i}^{n} ((x_{ri} - x_{mi}) - \overline{e_x})^2 / (n-1)} \qquad (4.36)$$

and

$$S_y = \sqrt{\sum_{i}^{n} ((y_{ri} - y_{mi}) - \overline{e_y})^2 / (n-1)} \qquad (4.37)$$

The estimated standard error of the population of horizontal errors is:

$$S_{\overline{e_h}} = S_h / \sqrt{n} \qquad (4.38)$$

Assuming that the errors are normally distributed, the estimated interval of errors at a specified probability can be expressed as:

$$\overline{e_h} \pm Z_i(S_h) \qquad (4.39)$$

If $\overline{e_h}$ is equal to zero, then the error interval becomes the Greenwalt and Schultz (1962, 1968) specified $Z_i(S)$.

The confidence interval around the estimate of the mean horizontal error can be calculated as follows:

$$\overline{e_h} - Z_i(S_{\overline{e_h}}) < \mu < \overline{e_h} + Z_i(S_{\overline{e_h}}) \qquad (4.40)$$

for large sample sizes and

$$\overline{e_h} - t_i\left(S_{\overline{e_h}}\right) < \mu < \overline{e_h} + t_i\left(S_{\overline{e_h}}\right) \tag{4.41}$$

for small sample sizes.

Because horizontal error is measured in two dimensions, the *bivariate* standard normal distribution must be used to characterize the distribution of errors. Figure 4.9 provides a three-dimensional illustration of the bivariate normal distribution. Figure 4.10 is an overhead view of the bivariate standard normal probability distribution with the commonly used map standards (circular error probable [CEP] at 50%, the circular map accuracy standard [CMAS] at 90%, and NSSDA at 95%) delineated.

Relying on the bivariate standard normal distribution to characterize the distribution of horizontal errors requires that we assume that the horizontal errors are distributed in a circle with S_x equal to S_y. We can test for circularity by calculating the ratio of the S_{min} to S_{max} (where S_{min} is the lower of S_x or S_y, and S_{max} is the larger of S_x or S_y). Figure 4.11 shows how differences in S_x and S_y affect the shape of the distribution of errors. If the ratio of S_{min} to S_{max} is 0.2 or above, Greenwalt and Schultz (1962, 1968) state that the circular distribution can be assumed. NSSDA again substitutes RMSE for the S and requires that the ratio of $RMSE_{min}$ to $RMSE_{max}$ be above 0.6.

As with vertical accuracy, ASPRS 2014 relies on the $Z_i(S_h)$ as the statistic to estimate horizontal accuracy. The statistic estimates the maximum interval of error on either side of $\overline{e_h}$ that will exist at a specified probability. The bivariate standard normal distribution Z_i statistic at 95% probability is 2.4477 (Greenwalt and Schultz, 1962, 1968), and the resulting interval of errors at 95% probability is:

$$= 2.4477\left(\left(S_x + S_y\right)/2\right) \tag{4.42}$$

$$= 2.4477 S_h \tag{4.43}$$

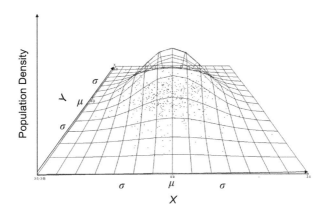

FIGURE 4.9   Three-dimensional representation of the standard normal bivariate distribution.

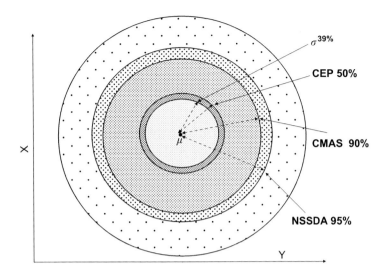

FIGURE 4.10 Two-dimensional representation of the normal standard bivariate or circular distribution with the probabilities of common horizontal map standards. (From Greenwalt, C. and Schultz, M., United States Air Force, Aeronautical Chart and Information Center, ACIC Technical Report Number 96, St. Louis, Missouri, 1962 and 1968.)

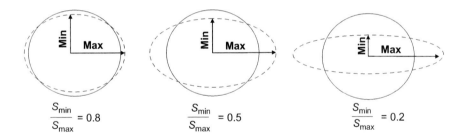

FIGURE 4.11 Comparison of circular and elliptical distributions for various ratios of S_{min}/S_{max} (From DMA (Defense Mapping Agency), Defense Mapping Agency Technical Report 8400.1, Fairfax, VA, 1991.)

The interval of errors within which 95% of the errors will occur (assuming the errors are normally distributed) is:

$$\left[\overline{e_h} - 2.447 * S_h, \overline{e_h} + 2.447 * S_h \right] \tag{4.44}$$

If $\overline{e_h}$ is equal to zero, the estimated interval reduces to the Greenwalt and Schultz (1962, 1968) and ASPRS (1989) accuracy statistic of $2.447(S_h)$.

Because the distribution of $\overline{e_h}$ is one-dimensional (even though the distribution of errors is two-dimensional), a confidence interval on $\overline{e_h}$ at the 95% level is expressed by:

$$\overline{e_h} - 1.96 S_{\overline{e_h}} < \mu < \overline{e_h} + 1.96 S_{\overline{e_h}} \tag{4.45}$$

for large samples

$$\overline{e}_h - t_{95\%,\,n\text{-}1\text{ degrees of freedom}} S_{\overline{e}_h} < \mu < \overline{e}_h + t_{95\%,\,n\text{-}1\text{ degrees of freedom}} S_{\overline{e}_h} \tag{4.46}$$

for small samples.

ASPRS 2014 Standards

As with vertical accuracy, the ASPRS (2014) standards direct that the NSSDA equations be used to report horizontal accuracy. And as with the vertical accuracy equations, the NSSDA accuracy statistic incorrectly applies the $RMSE_h$ rather than S_h to calculate the NSSDA accuracy statistic under two different conditions described in the following: when the $RMSE_y$ and $RMSE_x$ are equal and when they are not equal.

When the Errors are Circular

NSSDA defines errors as circular if $RMSE_y = RMSE_x$ (rather than when $S_x = S_y$). Under NSSDA, if $RMSE_y = RMSE_x$, then:

$$RMSE_h = \sqrt{(2RMSE_x)^2} = \sqrt{(2RMSE_y)^2}$$

$$= 1.4142 * RMSE_x = 1.4142 * RMSE_y \tag{4.47}$$

Applying the circular error normal distribution Z statistic at 95% probability of 2.4477 results in NSSDA Horizontal Accuracy $= 2.4477 * RMSE_h$

or

$$= 2.4477 * RMSE_h / 1.4142$$

$$= 1.7308 * RMSE_h \tag{4.48}$$

Most organizations use this simplified equation regardless of whether the errors are distributed circularly or not.[*] However, as with elevational accuracy, the NSSDA horizontal accuracy value assumes that the standard deviation is equal to RMSE, which requires the mean error to be zero and the sample size to be large. The population parameter that should be used to determine the interval of error at a specific probability level is the standard deviation of the horizontal errors (S_h) and not $RMSE_h$ (Ager, 2004).

When the Errors Are Not Circular

If $RMES_y \neq RMES_x$, then the NSSDA stipulates that the NSSDA *accuracy* statistic is:

$$= 2.4477\big((RMSE_x + RMSE_y)/2\big) \tag{4.49}$$

---

[*] Circularity is defined by Greenwalt and Schultz as S_{min}/S_{max} greater than or equal to 0.2. However, NSSDA restricts application of the circular distribution to those situations where S_{min}/S_{max} is between 0.6 and 1.0.

Tables 4.3 and 4.4 present sample reference and map coordinates from our earlier example and calculate the $RMSE_h$, S_h, S_{RMSE_h}, and $Zi*S$ at 95% probability, as well as the CMAS accuracy interval at 90%, the NSSDA statistic, and a 95% confidence interval around $RMSE_h$.

If the Errors Are Assumed Not to Be Normally Distributed

The ASPRS 2014 standards distinguish between vegetated vertical accuracy (VVA) and non-vegetated vertical accuracy (NVA), because they assume that non-vegetated terrain errors typically follow a normal distribution suitable for RMSE statistical analysis, whereas vegetated terrain does "not necessarily follow a normal distribution and ... the 95th percentile value more fairly estimates vertical accuracy at a 95% confidence level." No citations are given in the ASPRS 2014 standards as to why it might be valid to assume that errors in vegetated terrain do not follow a normal distribution. The ASPRS 2014 equations for determination of the 95th percentile are given as:

$$R = ((P/100)*(n-1)+1)$$

where:

R   is the rank of the observation that contains the Pth percentile
P   is the proportion (of 100) at which the percentile is desired (e.g., 95 for 95th percentile)
n   is the number of observations in the sample data set

Once the rank of the observation is determined, the percentile (Q_p) can then be interpolated from the upper and lower observations using the following equation:

$$Q_p = (A[n_w]+(n_d*(A[n_w+1]-A[n_w])))$$

where:

Q_p is the Pth percentile
         the value at rank n, A is an array of the absolute values of the samples, indexed in ascending order from 1 to N
$A[i]$ is the sample value of array A at index i (e.g., n_w or n_d)
i must be an integer between 1 and N
n is the rank of the observation that contains the Pth percentile
n_w is the whole number component of n (e.g., 3 of 3.14)
n_d is the decimal component of n (e.g., 0.14 of 3.14)

While this procedure will determine the range within which 95% of the sampled error values fall, it is not a measure of "confidence," as it does not rely on a model of the distribution of the errors.

HOW IS POSITIONAL ACCURACY REPORTED?

The ASPRS 2014 standards are very specific about the reporting of positional accuracy and differentiate clearly between reporting on actual accuracy assessment

TABLE 4.3
Horizontal Accuracy Example

Point ID	x_{ri} reference	x_{mi} map	Error in x Dimension = Reference − Map $(x_{ri} - x_{mi}) = e_{xi}$	Absolute Error in the x Dimension $|e_{xi}|$	Error in x Dimension Squared $(x_{ri} - x_{mi})^2 = e_{xi}^2$	Square of Absolute Error Minus Absolute Mean Error $\left(|e_{xi}| - |\bar{e}_x|\right)^2$	$(e_{xi} -$ Arithmetic Mean$)^2$	y_{ri} Reference	y_{mi} Map	Error in y Dimension = Reference − Map $(y_{ri} - y_{mi}) = e_{yi}$	Absolute Error in the y Dimension $|e_{yi}|$	Error in y Dimension Squared $(y_{ri} - y_{mi})^2 = e_{yi}^2$	Square of Absolute Error Minus Absolute Mean Error $\left(|e_{yi}| - |\bar{e}_y|\right)^2$	$(e_{yi} -$ Arithmetic Mean$)^2$	Sum of Squared Errors $e_{xi}^2 + e_{yi}^2$
107	6463928.4275	6463928.2891	0.1384	0.1384	0.0192	0.0004	0.0551	1740487.9905	1740488.2089	−0.2184	0.2184	0.0477	0.0079	0.0490	0.0669
108	6478942.9446	6478942.9707	−0.0261	0.0261	0.0007	0.0177	0.0049	1757945.7986	1757945.5996	0.1991	0.1991	0.0396	0.0118	0.0384	0.0403
110	6498179.1383	6498179.2172	−0.0789	0.0789	0.0062	0.0065	0.0003	1736983.2778	1736983.7799	−0.5021	0.5021	0.2521	0.0378	0.2551	0.2583
111	6500864.5792	6500866.2526	−1.6734	1.6734	2.8004	2.2927	2.4876	1758833.2498	1758830.8834	2.3664	2.3664	5.5999	4.2387	5.5856	8.4003
116	6527762.0733	6527762.1410	−0.0677	0.0677	0.0046	0.0084	0.0008	1731210.4027	1731210.7259	−0.3232	0.3232	0.1045	0.0002	0.1064	0.1091
117	6539890.0536	6539890.2650	−0.2113	0.2113	0.0447	0.0027	0.0133	1755842.1176	1755841.9103	0.2073	0.2073	0.0430	0.0101	0.0417	0.0876
122	6452053.8265	6452053.8601	−0.0336	0.0336	0.0011	0.0158	0.0039	1728034.3838	1728034.6916	−0.3078	0.3078	0.0948	0.0000	0.0966	0.0959
123	6435447.0261	6435446.7694	0.2567	0.2567	0.0659	0.0095	0.1245	1737489.6870	1737489.9830	−0.2960	0.2960	0.0876	0.0001	0.0894	0.1535
124	6445012.8528	6445012.7143	0.1385	0.1385	0.0192	0.0004	0.0551	1757524.8057	1757524.7919	0.0138	0.0138	0.0002	0.0863	0.0001	0.0194
206	6523662.6628	6523662.7526	−0.0898	0.0898	0.0081	0.0048	0.0000	1753217.8809	1753218.0854	−0.2045	0.2045	0.0418	0.0106	0.0431	0.0499
216	6503988.9073	6503989.0881	−0.1808	0.1808	0.0327	0.0005	0.0072	1728652.7232	1728653.2982	−0.5750	0.5750	0.3306	0.0715	0.3341	0.3633
222	6497217.5322	6497217.6331	−0.1009	0.1009	0.0102	0.0034	0.0000	1751316.3332	1751316.3331	0.0001	0.0001	0.0000	0.0946	0.0000	0.0102
227	6532154.2998	6532154.2726	0.0272	0.0272	0.0007	0.0174	0.0152	1740450.9200	1740451.2630	−0.3430	0.3430	0.1177	0.0013	0.1198	0.1184
228	6514726.6170	6514726.6231	−0.0061	0.0061	0.0000	0.0235	0.0081	1748724.1696	1748724.4427	−0.2731	0.2731	0.0746	0.0012	0.0762	0.0746
229	6480333.2958	6480333.3200	−0.0242	0.0242	0.0006	0.0182	0.0052	1742388.0615	1742388.2686	−0.2071	0.2071	0.0429	0.0101	0.0442	0.0435
283	6510536.2705	6510536.4059	−0.1354	0.1354	0.0183	0.0006	0.0015	1757706.5081	1757706.5519	−0.0438	0.0438	0.0019	0.0696	0.0022	0.0202
200	6509030.6018	6509030.5422	0.0596	0.0596	0.0036	0.0099	0.0243	1746587.3294	1746587.4407	−0.1113	0.1113	0.0124	0.0385	0.0131	0.0159
112	6502026.5461	6502026.5552	−0.0091	0.0091	0.0001	0.0225	0.0076	1779378.9142	1779378.7511	0.1631	0.1631	0.0266	0.0209	0.0256	0.0267
232	6509030.6018	6509030.5422	0.0596	0.0596	0.0036	0.0099	0.0243	1793670.4405	1793670.5122	−0.0717	0.0717	0.0051	0.0556	0.0056	0.0087
125	6436524.8263	6436525.0783	−0.2520	0.2520	0.0635	0.0086	0.0243	1782491.7176	1782490.8878	0.8298	0.8298	0.6886	0.2727	0.6836	0.7521
126	6464717.9797	6464718.1392	−0.1595	0.1595	0.0254	0.0000	0.0040	1778968.4568	1778968.1097	0.3472	0.3472	0.1205	0.0016	0.1184	0.1459

(Continued)

TABLE 4.3 (CONTINUED)
Horizontal Accuracy Example

Point ID	x_{ri} reference	x_{mi} map	Error in x Dimension = Reference − Map $(x_{ri}-x_{mi})=e_{xi}$	Absolute Error in the x Dimension $|e_{xi}|$	Error in x Dimension Squared $(x_{ri}-x_{mi})^2 = e_{xi}^2$	Square of Absolute Error Minus Absolute Mean Error $\left(|e_{xi}|-\overline{|e_x|}\right)^2$	$(e_{xi}-\text{Arithmetic Mean})^2$	y_{ri} Reference	y_{mi} Map	Error in y Dimension = Reference − Map $(y_{ri}-y_{mi})=e_{yi}$	Absolute Error in the y Dimension $|e_{yi}|$	Error in y Dimension Squared $(y_{ri}-y_{mi})^2 = e_{yi}^2$	Square of Absolute Error Minus Absolute Mean Error $\left(|e_{yi}|-\overline{|e_y|}\right)^2$	$(e_{yi}-\text{Arithmetic Mean})^2$	Sum of Squared Errors $e_{xi}^2 + e_{yi}^2$
128	6536017.6536	6536017.6690	−0.0154	0.0154	0.0002	0.0207	0.0065	1791341.5236	1791341.5819	−0.0583	0.0583	0.0034	0.0621	0.0038	0.0036
207	6523447.4186	6523447.4146	0.0040	0.0040	0.0000	0.0241	0.0100	1781813.7690	1781813.5640	0.2050	0.2050	0.0420	0.0105	0.0408	0.0420
208	6458661.4231	6458661.4229	0.0002	0.0002	0.0000	0.0253	0.0093	1763512.1326	1763512.0318	0.1008	0.1008	0.0102	0.0428	0.0096	0.0102
210	6432704.3136	6432704.3469	−0.0333	0.0333	0.0011	0.0159	0.0040	1797681.4156	1797681.5817	−0.1661	0.1661	0.0276	0.0200	0.0286	0.0287
214	6524150.2574	6524150.1541	0.1033	0.1033	0.0107	0.0031	0.0398	1766691.1359	1766691.0337	0.1022	0.1022	0.0105	0.0422	0.0098	0.0211
221	6490159.7535	6490159.5953	0.1582	0.1582	0.0250	0.0000	0.0647	1774521.0769	1774520.9524	0.1245	0.1245	0.0155	0.0335	0.0148	0.0405
223	6464915.1344	6464915.4456	−0.3112	0.3112	0.0968	0.0231	0.0462	1795190.2232	1795190.4426	−0.2194	0.2194	0.0481	0.0078	0.0495	0.1450
224	6446211.1711	6446211.4834	−0.3123	0.3123	0.0975	0.0234	0.0467	1776288.7842	1776289.2253	−0.4411	0.4411	0.1946	0.0178	0.1973	0.2921
226	6513283.3804	6513283.4917	−0.1113	0.1113	0.0124	0.0023	0.0002	1771237.4139	1771237.6203	−0.2064	0.2064	0.0426	0.0102	0.0439	0.0550
Sum				4.7780	3.3724	2.6114	3.0947				9.2277	8.1266	5.2882	8.1263	11.4990

TABLE 4.4
Horizontal Accuracy Example Equations and Statistics

Definitions	x Dimension Equations	x Dimension Values
Estimated root mean square of the population of errors	$RMSE_x = \sqrt{\sum_i^n (e_{xi})^2 / n}$	0.3353
Estimated absolute mean of the population of errors	$\overline{\|e_x\|} = 1/n \sum_1^n \|e_{xi}\|$	0.1593
Estimated variance of the population of errors	$S_{\|e_x\|}^2 = \sum_i^n \left(\|e_{xi}\| - \overline{\|e_x\|}\right)^2 / (n-1)$	0.0900
Estimated standard deviation of the population of errors	$S_{\|e_x\|} = \sqrt{\sum_i^n \left(\|e_{xi}\| - \overline{\|e_x\|}\right)^2 / (n-1)}$	0.3001
Estimated standard error of the population of mean errors	$S_{\overline{\|e_x\|}} = \sqrt{S_{\|e_x\|}^2 / n}$	0.0548
Estimated arithmetic mean of the population of errors	$\overline{e_x} = \sum_1^n e_{xi} / n$	−0.0962
Estimated standard deviation of the population of arithmetic errors	$S_{e_x} = \sqrt{\sum_i^n (e_{xi} - \overline{e_x})^2 / (n-1)}$	0.3267
Estimated standard error of the population of mean arithmetic errors	$S_{\overline{e_x}} = S_{e_x} / \sqrt{n}$	0.0596
Greenwalt and Schultz CMAS standard normal (Z) interval of the population of errors at 90% probability	$1.645 * S_{e_x}$	0.5374
Greenwalt and Schultz standard normal (Z) interval of the population of errors at 95% probability	$1.96 * S_{e_x}$	0.6403
NSSDA statistic	$1.96 * RMSE_x$	0.6572
Test on whether the mean error is significantly different from zero at the 95% confidence level	$\overline{e_x} \pm 1.96 * S_{\overline{e_x}}$	−0.0962 ± 0.640

This results in an interval from −0.2131 to 0.0207. The assumption that the mean error in the x dimension could equal zero is correct at the 95% confidence level, which means that RMSE can be substituted for S_e, and the equation for the NSSDA statistic is also correct.

Definitions	y Dimension Equations	y Dimension Values
Estimated root mean square of the population of errors	$RMSE_y = \sqrt{\sum_i^n (e_{yi})^2 / n}$	0.5205
Estimated absolute mean of the population of errors	$\overline{\|e_y\|} = 1/n \sum_1^n \|e_{yi}\|$	0.3076
Estimated variance of the population of errors	$S_{\|e_y\|}^2 = \sum_i^n \left(\|e_{yi}\| - \overline{\|e_y\|}\right)^2 / (n-1)$	0.1824
Estimated standard deviation of the population of errors	$S_{\|e_y\|} = \sqrt{\sum_i^n \left(\|e_{yi}\| - \overline{\|e_y\|}\right)^2 / (n-1)}$	0.4270

(Continued)

TABLE 4.4 (CONTINUED)
Horizontal Accuracy Example Equations and Statistics

Definitions	y Dimension Equations	y Dimension Values
Estimated standard error of the population of mean errors	$S_{\overline{\|e_y\|}} = \sqrt{S^2_{\|e_y\|}/n}$	0.0780
Estimated arithmetic mean of the population of errors	$\overline{e_y} = \sum_1^n e_{yi}/n$	0.0030
Estimated standard deviation of the population of arithmetic errors	$S_{e_y} = \sqrt{\sum_i^n (e_{yi}-\overline{e_y})^2/(n-1)}$	0.5294
Estimated standard error of the population of mean arithmetic errors	$S_{\overline{e_y}} = S_{e_y}/\sqrt{n}$	0.097
Greenwalt and Schultz CMAS standard normal (Z) interval of the population of errors at 90% probability	$1.645 * S_{e_y}$	0.8708
Greenwalt and Schultz standard normal (Z) interval of the population of errors at 95% probability	$1.96 * S_{e_y}$	1.0375
NSSDA statistic	$1.96 * RSME_y$	1.0201
Test on whether the mean error is significantly different from zero at the 95% confidence level.	$\overline{e_y} \pm 1.96 * S_{\overline{e_y}}$	0.003 ± 0.097

This results in an interval from −0.1864 to 0.1925. The assumption that the mean error in the y dimension could equal zero is correct at the 95% confidence level, which means that RMSE can be substituted for S_e, and the equation for the NSSDA statistic is correct.

Definitions	Circular Equations	
Estimated root mean square of the population of errors	$RMSE_h = \sqrt{\sum_i^n (e_{hi})^2/n}$	0.6191
Estimate absolute mean of the population of errors	$\overline{\|e_h\|} = 1/n \sum_1^n \|e_{hi}\|$	0.4669
Estimated standard deviation of the population of errors	$S\|e_h\| = (S\|e_x\| + S\|e_y\|)/2$	0.3636
Estimated standard error of the population of mean errors	$S_{\overline{\|e_h\|}} = S_{\|e_h\|}/\sqrt{n}$	0.0664
Estimated arithmetic mean of the population of errors	$\overline{e_h} = \sum_1^n e_{hi}/n$	−0.0932
Estimated standard deviation of the population of arithmetic errors	$S_{e_h} = (S_{e_x}+S_{e_y})/2$	0.4280
Estimated standard error of the population of mean arithmetic errors	$S_{\overline{e_h}} = S_{e_h}/\sqrt{n}$	0.0781
Greenwalt and Schultz CMAS standard normal (Z) interval of the population of errors at 90% probability	$2.146 * S_{e_h}$	0.9185

(Continued)

TABLE 4.4 (CONTINUED)

Horizontal Accuracy Example Equations and Statistics

Definitions	Circular Equations	
Greenwalt and Schultz standard normal (Z) interval of the population of errors at 95% probability	$2.4477 * S_{e_h}$	1.0476
Test for circularity (Greenwalt and Schultz). Because the ratio is over 0.2, it can be assumed that the distribution of errors is circular.	$S_{min}/S_{max} = 0.3267/0.5294$	0.6171
Test for circularity (NSSDA). Because the ratio is over 0.6, it can be assumed that the distribution of errors is circular.	$RMSE_{min}/RMSE_{max} = 0.3353/0.5205$	0.6442
$NSSDA_{circular}$ statistic	$1.7308 * RMSE_h$	1.0716
$NSSDA_{elliptical}$ statistic	$2.4477 * .5 * (RMSE_x + RMSE_y)$	1.0473
Test on whether the mean error is significantly different from zero at the 95% confidence level.	$\overline{e_h} \pm 1.96 * S_{\overline{e_h}}$	−0.0932 ± 0.1532

This results in an interval from −0.2464 to 0.06. The assumption that the mean error could equal zero is correct at the 95% confidence level, which means that RMSE can be substituted for S_e, and the equation for the NSSDA statistic is correct.

results and the statement of an accuracy assessment goal under which the product was produced. Reporting is specified as follows:

The horizontal accuracy of digital orthoimagery, planimetric data, and elevation data sets are to be documented in the metadata in one of the following manners:

1. "This data set was tested to meet ASPRS Positional Accuracy Standards for Digital Geospatial Data (2014) for a _____ (cm) $RMSE_x/RMSE_y$ Horizontal Accuracy Class. Actual positional accuracy was found to be $RMSE_x =$ _____ (cm) and $RMSE_y =$ _____ cm which equates to Positional Horizontal Accuracy = +/− _____ at 95% confidence level." or,
2. "This data set was produced to meet ASPRS Positional Accuracy Standards for Digital Geospatial Data (2014) for a _____ (cm) $RMSE_x/RMSE_y$ Horizontal Accuracy Class which equates to Positional Horizontal Accuracy = +/− _____ cm at a 95% confidence level."

The vertical accuracy of elevation data sets is to be documented in the metadata in one of the following manners:

1. "This data set was tested to meet ASPRS Positional Accuracy Standards for Digital Geospatial Data (2014) for a _____ (cm) $RMSE_z$ Vertical Accuracy Class. Actual NVA accuracy was found to be $RMSE_z =$ _____ cm, equating

to +/– _____ cm at 95% confidence level. Actual VVA accuracy was found
to be +/– _____ cm at the 95th percentile." or,

2. "This data set was produced to meet ASPRS Positional Accuracy Standards
for Digital Geospatial Data (2014) for a _____ cm $RMSE_z$ Vertical Accuracy
Class equating to NVA=+/– _____ cm at 95% confidence level and
VVA=+/– _____ cm at the 95th percentile."

SUMMARY

The ASPRS 2014 standards are a robust body of work that represents a huge improvement in the practice of positional accuracy assessment. They clearly adapt old standards to new technologies and exhaustively review a variety of accuracy measures.
Most importantly, the ASPRS 1990 accuracy classes have been improved and refined
to account for the higher accuracies currently obtainable from the latest technologies. Annex B of these standards presents clear and concise tables that relate current
orthoimagery pixel size to suggested maximum $RMSE_x$ and $RMSE_y$ for different
quality classes.

The standards also persist in the substitution of RMSE for the standard deviation in the calculation of the NSSDA statistic, resulting in the assumption that the
mean error is zero—a curious assumption for a standard that seeks to understand the
magnitude and distribution of errors. We suggest instead that practitioners calculate
whether the mean error is statistically different from zero at the 95% confidence
level, and if it is, use the standard deviation instead of the RMSE in the calculation
of the NSSDA statistic. We also suggest that more research be conducted into understanding what the typical distribution of positional errors actually looks like.

5 Thematic Map Accuracy Basics

The major focus of this book is quantitative thematic map accuracy assessment. The previous chapter presented a thorough overview and the computations for performing a quantitative positional accuracy assessment, including a discussion of the standard measures for reporting it. This chapter introduces the universally accepted measure for representing thematic map accuracy: the error matrix. The chapter also documents the evolution of thematic map accuracy assessment, beginning with a discussion of early quantitative, non-site-specific assessments. Next, quantitative site-specific assessment techniques employing the error matrix are presented, followed by the mathematical representation of the error matrix.

NON-SITE-SPECIFIC ASSESSMENTS

In a quantitative non-site-specific accuracy assessment, only the total areas for each class mapped are computed without regard to the location of these areas. In other words, a comparison is performed between the number of acres or hectares of each class on the map generated from remotely sensed data and the number of acres or hectares of each class on the reference data. In this type of assessment, the errors of omission and commission can compensate for each other (i.e., cancel each other out), and therefore, the overall area totals by map class may compare favorably even while the map contains substantial error. The significant issue with a non-site-specific assessment is that nothing is known about any specific locations on the map or how it agrees or disagrees spatially with the reference data.

A simple example quickly demonstrates the shortcomings of the non-site-specific approach. Figure 5.1 shows the distribution of the forest class on both a reference data set and two different classifications generated from remotely sensed data. Classification #1 was generated using one type of classification algorithm (e.g., supervised, unsupervised, or non-parametric, etc.), while Classification #2 employed a different algorithm. In this example, only the forest class is being evaluated. The reference data show a total of 2435 acres of forest, while Classification #1 shows 2322 acres, and Classification #2 shows 2635 acres. In a non-site-specific assessment, you would conclude that Classification #1 is the better map for the forest class, because the total number of forest acres for Classification #1 more closely agrees with the number of acres of forest on the reference image (2435 acres – 2322 acres = 113 acres difference for Classification #1, while Classification #2 differs by 200 acres). However, a visual comparison (see Figure 5.2) between the forest polygons on Classification #1 and the reference data demonstrates little locational correspondence. Classification #2, despite being judged inferior by the non-site-specific assessment, appears to spatially agree better with the reference data forest polygons

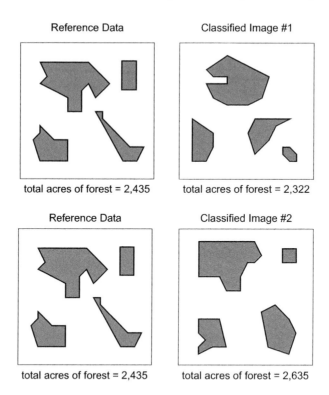

FIGURE 5.1   Example of non-site-specific accuracy assessment.

(see Figure 5.2). This example shows how the use of non-site-specific accuracy assessment can be quite misleading. In the example shown here, the non-site-specific assessment actually recommends the use of the inferior classification algorithm.

SITE-SPECIFIC ASSESSMENTS

Given the obvious limitations of a non-site-specific accuracy assessment, there is a need to know how the map generated from the remotely sensed data compares with the reference data spatially (i.e., on a locational basis). Therefore, the use of site-specific assessments was instituted. Initially, a single value representing the accuracy of the entire classification (i.e., overall accuracy) was employed. This computation was performed by comparing a sample of locations on the map with the same locations on the reference data and keeping track of the number of times there was agreement. Dividing the number that are correct by the total number of samples results in a measure called *overall accuracy*.

In the past, an overall accuracy level of 85% was often adopted as representing the cutoff between acceptable and unacceptable results. This standard was first proposed in Anderson et al. (1976) despite the lack of any research being performed to establish this standard. Obviously, the accuracy of a map depends on a great many factors, including the amount of effort, the level of landscape detail (i.e., classification

Classified Image #1 on top of the Reference Data

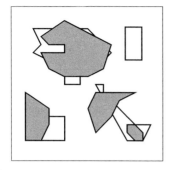

While the total acres of forest in the reference data (2,435) and the total acres of forest in the classified image #1 (2,322) is only 5% different, the spatial correspondence between the two data sets is low. There is low agreement between the actual location of the forested areas in the Reference Data and the Map.

Classified Image #2 on top of the Reference Data

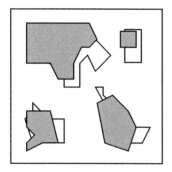

While the total acres of forest in the reference data (2,435) and the total acres of forest in the classified image #2 (2,635) is 8% different, the spatial correspondence between the two data sets is higer. There is greater agreement between the actual location of the forested areas in the Reference Data and the Map.

FIGURE 5.2 An example demonstrating the spatial correspondence that is lacking in a non-site-specific accuracy assessment.

scheme) being mapped, and the variability of the classes to be mapped. In some applications, an overall accuracy of 85% is more than sufficient; in other cases, it would not be accurate enough; and in others, such an accuracy would be way too expensive to achieve.

While having a single number to measure overall thematic map accuracy was an improvement over the non-site-specific assessment method, it soon became obvious that this single number was not sufficient to represent all the accuracy information contained within the thematic map. The need to evaluate individual map classes within the classification scheme was recognized, and so began the use of the error matrix to represent map accuracy.

THE ERROR MATRIX

As previously introduced, an error matrix is a square array of numbers set out in rows and columns, which express the number of sample units assigned to a particular class in one classification relative to the number of sample units assigned to a particular class in another classification (Table 5.1). In most cases, one of the classifications is

TABLE 5.1

Example Error Matrix (same as presented in Figure 2.6)

		Reference Data				Row Total
		D	C	AG	SB	
	D	65	4	22	24	115
	C	6	81	5	8	100
Classified Data	AG	0	11	85	19	115
	SB	4	7	3	90	104
Column Total		75	103	115	141	434

Land Cover Categories

D = deciduous
C = conifer
AG = agriculture
SB = shrub

OVERALL ACCURACY =
(65 + 81 + 85 + 90)/434 =
321/434 = 74%

PRODUCER'S ACCURACY	USER'S ACCURACY
D = 65/75 = 87%	D = 65/115 = 57%
C = 81/103 = 79%	C = 81/100 = 81%
AG = 85/115 = 74%	AG = 85/115 = 74%
SB = 90/141 = 64%	SB = 90/104 = 87%

considered to be correct (i.e., called the *reference data*) and may be generated from higher–spatial resolution imagery, ground observation, or ground measurement. The columns usually represent the reference data, while the rows indicate the classification generated from the remotely sensed data (i.e., the map to be assessed). It should be noted that the reference data have often been referred to as *ground truth*. While it is true that the reference data are assumed to be more correct than the map they are being used to assess, it is by no means true that these data are perfect, without error, or represent "the truth." Therefore, the term *ground truth* is inappropriate and in some cases, very misleading. Throughout this book, the authors use the term *reference data* to identify the sample data, which are assumed to be correct and are compared with the map sample data to generate the error matrix. We strongly urge that the term *ground truth* be abandoned by the geospatial analysis community.

An error matrix is a very effective way to represent thematic map accuracy, because it provides a clear way of deriving the individual accuracies of each class along with both the errors of inclusion (commission errors) and errors of exclusion (omission errors) present in the classification. Commission error is simply defined as including an area into a class when it doesn't belong to that class. Omission error occurs when an area is excluded from the class in which it truly does belong. Just as a coin has two sides, heads and tails, each and every error has two components: an omission from the correct class and a commission to a wrong class. For example, in the error matrix in Table 5.1, there are four sample areas that were classified as deciduous when the reference data show that they are actually conifer. Therefore, four areas are in error, as they were omitted from the correct coniferous class and committed to the incorrect deciduous class.

Overall, Producer's, and User's Accuracies

In addition to clearly showing errors of omission and commission, the error matrix can be used to compute other accuracy measures such as overall accuracy, producer's accuracy, and user's accuracy (Story and Congalton, 1986). As shown in Table 5.1, overall accuracy is simply the sum of the major diagonal (i.e., the correctly classified sample units) divided by the total number of sample units in the entire error matrix. In this example, it is 321/434, which equals 74%. This value is the most commonly reported accuracy assessment statistic and is probably most familiar to the reader. However, just presenting the overall accuracy is not enough. It is important to present the entire matrix, so that other accuracy measures can be computed as needed, and any confusion between map classes is clearly presented and understood.

Producer's and user's accuracies (Story and Congalton, 1986) can be computed to determine individual class accuracies in addition to computing the overall classification accuracy for the entire matrix. The producer of the map may want to know how well they mapped a certain map class, the producer's accuracy. This value is computed by dividing the value from the major diagonal (the agreement) for that class by the total number of samples in that map class as indicated by the sum of the reference data for that class. Looking at Table 5.1 shows that the map producer called 81 sample areas conifer, while the reference data indicate that there were a total of 103 conifer sample areas. Twenty-two areas were omitted from the conifer class, and of these, four were committed to deciduous, 11 were committed to agriculture, and seven were committed to shrub. So, 81/103 samples were correctly classified as conifer, for a conifer producer's accuracy of 79%. However, this is only half the story. If you now view the map from the user's perspective, you see once again that 81 samples were classified as conifer on the map that were actually conifer, but in addition, the map shows six samples as conifer that were actually deciduous, five samples as conifer that were actually agriculture, and eight samples as conifer that were actually shrub. The map, therefore, shows 100 samples as conifer, but only 81 are actually conifer. There was commission error of 19 samples into the conifer class that were not conifer. The conifer user's accuracy is then computed by dividing the major diagonal value for the conifer class by the total number of samples mapped as conifer, 81/100 = 81%. In evaluating the accuracy of an individual map class, it is important to consider both the producer's and the user's accuracies.

Before error matrices were the standard accuracy reporting mechanism, it was common to report the overall accuracy and only either the producer's or the user's accuracy. Sometimes, only the higher of the producer's and the user's accuracy was selected to be reported, resulting in misleading information about the map accuracy (i.e., only reporting the highest accuracies for each map class). A quick example will demonstrate the need to publish the entire matrix so that all three accuracy measures can be computed.

As previously demonstrated, the error matrix shown in Table 5.1 reveals an overall map accuracy of 74%. This value tells about how accurate the map is, in general, but not about any of the individual map classes. However, suppose we are most interested in the ability to classify deciduous forests, so we calculate a producer's accuracy for this class. This calculation is performed by dividing the total number of correct sample units in the deciduous class (i.e., 65) by the total number of decidu-ous sample units indicated by the reference data (i.e., 75 or the column total). This division results in a producer's accuracy of 87%, which is quite good. If we stopped here, one might conclude that although this classification appears to be average over-all (i.e., 74%), it is quite adequate for the deciduous class. Making such a conclusion could be a very serious mistake. A quick calculation of the user's accuracy, computed by dividing the total number of correct sample units in the deciduous class (i.e., 65) by the total number of sample units classified as deciduous (i.e., 115 or the row total) reveals a value of 57%. In other words, although 87% of the deciduous areas have been correctly identified as deciduous, only 57% of the areas called deciduous on the map are actually deciduous on the ground/reference data. The high producer's accuracy occurs because too much of the map is classified as deciduous (i.e., there is a large amount of commission error in the deciduous class). A more careful look at the error matrix reveals significant confusion in discriminating deciduous from bar-ren and shrub. Therefore, although the producer of this map can claim that 87% of the time, an area that was deciduous on the ground was identified as such on the map, a user of this map will find that only 57% of the time that the map says an area is deciduous will it actually be deciduous on the ground. Careful study and analysis of the error matrix can be quite fruitful in understanding the thematic errors in a map.

Mathematical Representation of the Error Matrix

We have presented an explanation of the error matrix in descriptive terms, including an example (Table 5.1). This section presents the error matrix in mathematical terms necessary to compute the descriptive statistics, including overall, producer's, and user's accuracies, and to perform the analysis techniques described in Chapter 8.

Assume that n samples are distributed into k^2 cells, where each sample is assigned to one of k classes in the map (usually the rows) and, independently, to one of the same k classes in the reference data set (usually the columns). Let n_{ij} denote the number of samples classified into class i ($i = 1, 2, \ldots, k$) in the map and class j ($j = 1, 2, \ldots, k$) in the reference data set (Table 5.2).
Let:

$$n_{i+} = \sum_{j=1}^{k} n_{ij}$$

be the number of samples classified into class i in the remotely sensed classification, and

$$n_{+j} = \sum_{i=1}^{k} n_{ij}$$

be the number of samples classified into class j in the reference data set. The plus sign in the notation here simply indicates all the values in that row or column. For

TABLE 5.2

Mathematical Example of an Error Matrix

		j = Columns (Reference)			Row Total
		1	**2**	**k**	n_{i+}
i = Rows (Classification)	**1**	n_{11}	n_{12}	n_{1k}	n_{1+}
	2	n_{21}	n_{22}	n_{2k}	n_{2+}
	k	n_{k1}	n_{k2}	n_{kk}	n_{k+}
Column Total n_{+j}		n_{+1}	n_{+2}	n_{+k}	n

example, as shown in Table 5.2, if $i = 1$, then the total number of samples in class 1 is the sum of all the samples in row 1; if $i = k$, then the total number of samples in class k is the sum of all the samples in row k.

The overall accuracy between the remotely sensed classification (i.e., the map) and the reference data can then be computed as follows:

$$\text{overall accuracy} = \frac{\sum_{i=1}^{k} n_{ii}}{n}$$

Producer's accuracy can be computed by:

$$\text{producer's accuracy}_j = \frac{n_{jj}}{n_{+j}}$$

and the user's accuracy can be computed by:

$$\text{user's accuracy}_i = \frac{n_{ii}}{n_{i+}}$$

Finally, let p_{ij} denote the proportion of samples in the i, jth cell, corresponding to n_{ij}. In other words, $p_{ij} = n_{ij}/n$.

Then, let p_{i+} and p_{+j} be defined by:

$$p_{i+} = \sum_{j=1}^{k} p_{ij}$$

and

$$p_{+j} = \sum_{i=1}^{k} p_{ij}.$$

It may take a little practice to get used to using this mathematical representation of the error matrix. Actually, looking at an error matrix the very first time can take some effort. However, given the importance of the error matrix in thematic accuracy assessment and the need for the mathematical representation for the descriptive statistics (overall, producer's, and user's accuracies) presented in this chapter and some of the analysis techniques presented in Chapter 8, the reader is encouraged to spend a little time with these equations until they feel comfortable that they understand the matrix and how it can be analyzed. Many examples will be provided throughout the book as well as some case studies to aid every reader in becoming an error matrix expert.

6 Thematic Map Accuracy Assessment Considerations

Now that we understand that thematic map accuracy is typically represented using an error matrix, it is important to know how to correctly generate and populate the matrix. As shown in the thematic map accuracy flow chart introduced in Chapter 3 (Figure 3.5) and presented in Figure 6.1, there are many issues to consider when conducting a thematic map accuracy assessment. Two very important considerations are (1) the sources of error, which were introduced in Chapter 2 (Figure 2.5) and discussed in detail using an error budget approach in Chapter 3, and (2) the selection of a classification scheme as presented in this chapter. The considerations related to sampling for the collection of the reference data are also discussed in this chapter. These considerations include the sample unit, the sample size, and the sampling strategy along with the practical application of spatial autocorrelation. Non-sampling considerations regarding the collection of the reference data are covered in Chapter 7.

Assessing the thematic accuracy of maps or other spatial data requires sampling, because it is not economically feasible or time effective to visit every place on the ground. The class labels of sample areas on the map are compared with the class labels of the same areas in the reference data set to generate the error matrix. Determining the appropriate sampling design to collect these sample areas requires knowledge of the distribution of the thematic map classes across the landscape; determination of the types and number of samples to be taken; and the choice of a sampling scheme for selecting the samples. The design of an effective and efficient sample to collect valid map and reference data is one of the most challenging and important components of any accuracy assessment, because the design will determine both the cost and the statistical rigor of the assessment.

Thematic map accuracy assessment assumes that the information contained within the error matrix is a true characterization of the map being assessed. Thus, an improperly designed sample will produce misleading accuracy results. It is, therefore, absolutely critical that the sampling process be valid and efficient. Several considerations are of key importance when designing an accuracy assessment sample that is truly representative of the map, including:

1. What are the thematic map classes to be assessed, and how are they distributed across the landscape?
2. What is the appropriate sample unit?
3. How many samples should be taken?
4. How should the samples be chosen, and how does spatial autocorrelation impact this decision?

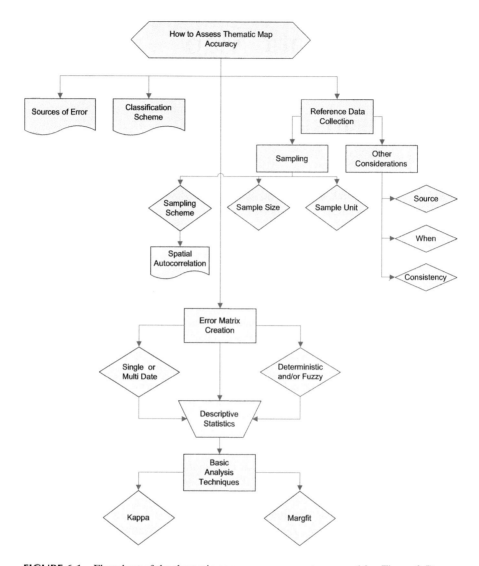

FIGURE 6.1 Flowchart of the thematic accuracy assessment process (also Figure 3.5).

While seemingly straightforward, each of these steps has many potential pitfalls. Failure to consider even one of them can lead to serious shortcomings in the assessment process. This chapter carefully considers each one of these factors and informs the reader about the principles and practices needed to make the right choices for their given mapping project.

WHAT ARE THE THEMATIC MAP CLASSES TO BE ASSESSED?

How we sample the map for accuracy will partially be driven by the number of map classes and their distribution across the landscape. This distribution will, in turn, be a function of how we have chosen to categorize the features of the earth being mapped—referred to as the *classification scheme*. Once we know the classification scheme, we can then learn more about how the map classes are distributed. Important considerations are the discrete nature of map information and the spatial inter-relationship or autocorrelation of that information. Assumptions made about the distribution of map categories will affect both how we select accuracy assessment samples and the outcome of the analysis.

THE CLASSIFICATION SCHEME

Maps categorize the earth's surface. For example, road maps tell us the type of road, its name, and its location. Land cover maps typically enumerate the types, mix, and density of vegetation covering the earth (e.g., trees, shrubs, grass). Land use maps characterize how land is used by humans (e.g., urban, agriculture, forest management).

Thematic map categories are specified by the project's classification scheme. Classification schemes are a means of organizing spatial information in an orderly and logical way (e.g., Cowardin et al., 1979). Classification schemes are fundamental to any mapping project, because they create order out of chaos and reduce the total number of classes (i.e., distinct groupings) considered to some reasonable amount. The classification scheme makes it possible for the map producer to consistently characterize landscape features and for the map user to readily recognize them. Without a classification scheme, no mapping is truly possible, nor is management of the resources portrayed on the map. The detail of the scheme is driven by (1) the anticipated uses of the map information and (2) the features of the earth that can be discerned with the remotely sensed data (e.g., aerial or satellite imagery) being used to create the map. If a rigorous classification scheme is not developed before mapping begins, then any subsequent accuracy assessment of the map will be meaningless, because it will be impossible to definitively label the accuracy assessment samples in the reference data.

A strong and effective classification scheme has five common components. These are: (1) the scheme must have a set of rules/definitions included with each map class/ label, (2) the scheme is mutually exclusive, (3) the scheme is totally exhaustive, (4) the scheme is hierarchical, and (5) a minimum mapping unit (mmu) is specified. All classification schemes have a set of *map classes/labels* (e.g., urban residential, deciduous forest, palustrine emergent wetland, etc.); however, to be effective, the scheme must also have a set of *rules* or definitions used for clearly assigning these labels (e.g., a "deciduous forest must have at least 75 percent crown closure in deciduous trees and the trees must be at least 5 meters tall"). Without a clear set of rules, the assignment of labels to classes can be arbitrary and lack consistency. For example, everyone has their own idea about what constitutes a forest, and therefore, these many definitions could result in very different maps of forest distribution. Consider a situation where one agency has a very liberal definition and defines a forest as an area where at least 10% of the ground area is covered by trees, while another agency is more conservative and uses a slightly different definition in which forest exists only if greater than 25% of the ground area is covered by trees. If analysts from each of these agencies were together in a specific plot of land, they would label the area differently based on their agency's definitions of a forest, and both of the labels would be correct, but the maps would be different. Without class definitions expressed as quantifiable rules, there can be little agreement on how the area on the ground or the image should be labeled, and therefore, any assessment of such a map would be problematic.

There are a number of methods that can be used to elucidate the map labels. They can be as simple as providing detailed definitions of each label (i.e., map class). Another method is to create a dichotomous key in which a binary choice (yes or no) is provided to lead the analyst/user to determine the appropriate map label. While developing such a key can be challenging, especially for a complex classification scheme, the benefits are tremendous, in that the process of labeling any area becomes very objective and consistent and can be done by almost anyone given some training. Finally, full descriptions, including example pictures and other graphics, can be provided to clearly demonstrate each map class. The more complex the classification scheme, the more detailed are the rules for defining each class. However, even the most simple classification scheme will suffer if adequate rules are not provided for accurately labeling the map classes.

The level of detail (i.e., the number and complexity of the map classes) in the scheme strongly influences the time and effort needed to make the map and to conduct the accuracy assessment. The more complex the scheme, the more expensive the map and its assessment. Because the classification scheme is so important, no work should begin on a mapping project until the scheme has been thoroughly reviewed and as many issues and problems as possible identified and solved. This key point cannot be emphasized enough here. Failure for all involved in the mapping project to agree on and thoroughly understand the classification scheme can easily ruin a project. Great care must be taken to ensure that all stakeholders have reviewed and accepted the scheme, or there is a risk of repeating a great deal of work later on in the project. In addition to being composed of labels with a corresponding set of rules, a classification scheme should also be *mutually exclusive* and *totally exhaustive*. Mutual exclusivity requires that each mapped area fall into one, and only one, map class or class. The best way to guarantee exclusivity is to have a clear and concise set of rules for each map class. For example, classification scheme rules would need to clearly distinguish between forest and water (seemingly simple), so that a mangrove swamp cannot receive both a forest and a water label.

A totally exhaustive classification scheme results in every area on the mapped landscape receiving a map label; no area can be left unlabeled. One way to ensure that the scheme is totally exhaustive is to have a class labeled as "other" or "unclassified." However, if a large portion of the map is labeled as "other," then it may be necessary to rethink the classification scheme used in the project, as the "other" class is typically not very useful.

If possible, it is also advantageous to use a classification scheme that is *hierarchical*. A hierarchical classification scheme typically has many levels, beginning with the most general categories at Level 1 and growing in detail when moving to higher levels. For example, Level 1 classes might include forest, developed, water, and so on, while the Level 2 forest classes might be divided into coniferous and hardwood (Figure 6.2). In hierarchical systems, specific categories within the classification scheme can be collapsed to form more general categories. This ability is especially important when it is discovered that certain map categories cannot be reliably mapped. For example, it may be impossible to separate interior live oak from canyon live oak in California's oak woodlands (these two oak types are almost indistinguishable on the ground). Therefore, these two categories may have to be collapsed to form a live oak class that can be reliably mapped.

Finally, the classification scheme must specify the mmu for each class being mapped. The mmu is the smallest area of that class to be delineated on the map. Figure 6.3 illustrates this concept. In this example, the rule for mapping a forest is specified as

an area of 1 acre or more where more than 30% of the ground, as seen from above the tree canopy, is covered by the foliage of hardwood or conifer trees.

The mmu for forests is 1 acre. Areas covered with 30% tree foliage but smaller than the 1 acre mmu will not be labeled as forests. Additionally, areas larger than 1 acre but containing less than 30% tree foliage cover will also not be labeled as forests. Reference data sample units must be at least as large as the stated mmu used

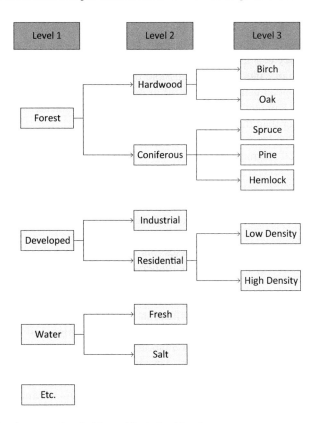

FIGURE 6.2   An example of a hierarchical classification structure.

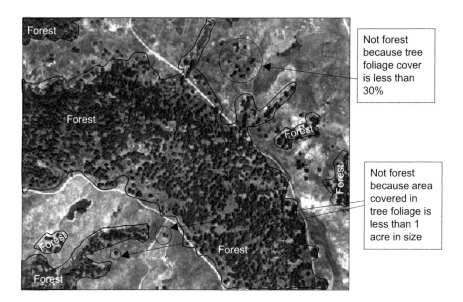

Not forest because tree foliage cover is less than 30%

Not forest because area covered in tree foliage is less than 1 acre in size

FIGURE 6.3 An example of the impact that the classification scheme rules and minimum mapping unit have on labeling areas as "forest."

to create the map from the remotely sensed data. For example, it is not possible to assess the accuracy of a Landsat 30 meter × 30 meter pixel with a single 1/20 hectare ground inventory plot; nor is it possible to assess the accuracy of an AVHRR 1.1 km × 1.1 km pixel using a 30 m × 30 m pixel. For this reason, often previously collected ground samples are unusable for accuracy assessment, because the area sampled is smaller than the map mmu.

Figure 6.4 provides a dichotomous key for a simple, yet robust, classification scheme for a fire/fuel mapping project. Notice how the scheme specifies an mmu and is:

- totally exhaustive—every piece of the landscape will be labeled.
- mutually exclusive—no one piece of the landscape can receive more than one label.
- hierarchical—detailed fuel classes can be lumped into the more general groups of non-fuel, grass, shrub, timber slash, and timber litter.

It is critical to note that accuracy assessment reference data must be collected and labeled using the identical classification scheme as was used to generate the map. This observation may seem obvious until you are tempted to use an existing map to assess the accuracy of a new map. Rarely will any two maps be created using the same classification scheme. Any differences between the classification scheme of the map and the classification scheme of the reference data may cause discrepancies between map and reference accuracy assessment sample unit labels. The result will be an assessment of classification scheme differences and not of map accuracy.

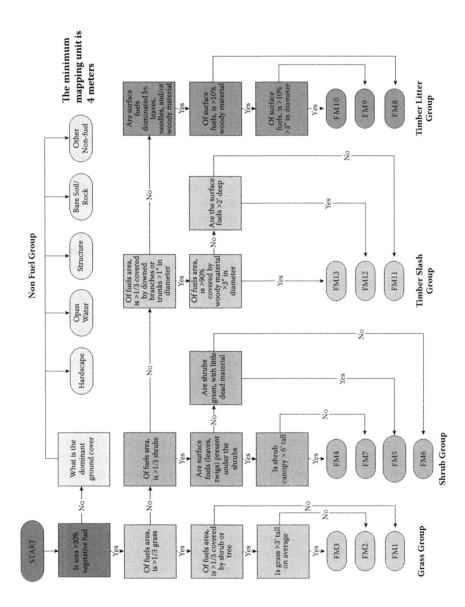

FIGURE 6.4 An example of a dichotomous key for a wildland fuel classification scheme.

Sometimes, it is possible to create a crosswalk (i.e., translation) between one classification and another, which attempts to make the two schemes equivalent. Care should be taken in these situations, as it is rare for the crosswalk to provide a perfect translation. Instead, compromises are made to try to make one scheme equivalent to the other. It must be realized that any compromises could introduce error into the assessment process that is not actually map error but rather, error created by using a classification scheme that was not the same as the one used to make the map.

OTHER DATA CONSIDERATIONS

Continuous versus Non-Continuous Data

Most statistical analysis assumes that the population to be sampled is continuous and normally distributed, and that samples will be independent. Yet we know that classification systems, for all their power in organizing chaos, take a continuous landscape and divide it into often arbitrarily discrete classes. For example, tree crown closure rarely exists in discrete classes. Yet when we make a map of crown closure, we impose discrete crown closure classes across the landscape. We may create a crown closure map with four classes: Class 1 from 0–10% crown closure, Class 2 from 11–50% crown closure, Class 3 from 51–75% crown closure, and Class 4 from 76–100% crown closure. Given this boundary between two crown closure classes at 75%, one can expect to find confusion between a forest stand with a crown closure of 73% that belongs in Class 3 and a stand of 77% that belongs in Class 4 (see Chapter 10 for a discussion on fuzzy accuracy assessment). In addition, categories tend to be related spatially, resulting in autocorrelation (discussed next in this chapter). In most situations, some balance between what is statistically valid and what is practically obtainable is desired. Therefore, knowledge of these statistical considerations is a must.

It is valuable to note that most analysts who have completed a beginning statistics course are familiar with sampling and analysis techniques for continuous, normally distributed data. It is these techniques, such as analysis of variance (ANOVA) and linear regression, that are most familiar to the reader. Only in more advanced statistics courses do techniques involving non-normal theory statistical methods get discussed. Therefore, dealing with these issues is less familiar to most readers. Those who wish to delve more deeply into some of the subjects presented in this book would do well to consider additional study of more advanced statistical methods.

Thematic map information is discrete, not continuous, and frequently not normally distributed. Therefore, normal theory statistical techniques that assume a continuous normal distribution may be inappropriate for map accuracy assessment. It is important to consider how the data are distributed and what assumptions are being made about these data before performing any statistical analysis. Sometimes, there is little that can be done about the artificial delineations in the classification scheme; other times, the scheme can be modified to better represent natural breaks. Care and thought must go into this process to achieve the best and most appropriate analysis possible.

Spatial Autocorrelation

Spatial autocorrelation occurs when the presence, absence, or degree of a certain characteristic affects the presence, absence, or degree of that same characteristic in neighboring units (Cliff and Ord, 1973), thereby violating the assumption of sample independence. This condition is particularly important in accuracy assessment if an error in a certain location can be found to positively or negatively influence errors in surrounding locations (Campbell, 1981). Clearly, if spatial autocorrelation exists, the sampling must ensure that the samples are separated by enough distance to minimize this effect, or else the sampling will not adequately represent the entire map.

The existence of spatial autocorrelation is clearly illustrated in work by Congalton (1988a) on the error in land cover maps generated from Landsat MSS imagery for three areas of varying spatial diversity/complexity (i.e., an agriculture area, a rangeland, and a forested site), which showed a positive influence over 1 mile away. Figure 6.5 presents the results of this analysis. Each map, called a *difference map* (as introduced in Chapter 3), is a comparison between the remotely sensed classification (i.e., the map) and the reference data. In this case, the reference data covered the entire study area and were not a sample but, rather, a total enumeration. The black areas represent the error, those places where the map and the reference label disagree. The white areas represent the agreement.

The pattern of differences between the land cover map and the reference labels is readily explainable in an agricultural environment, where field sizes are large, and typical misclassification would result in an error in labeling the entire field. In the agricultural difference map in Figure 6.5, the fields are center-pivot irrigation, circular fields, and examples can be seen of misclassifying entire fields. For example, a field that is mapped as corn when it is actually wheat will result in an entire field (center-pivot area in Figure 6.5) being mislabeled. Therefore, it is not surprising that the errors occur in large areas and that there is a positive autocorrelation over a large distance.

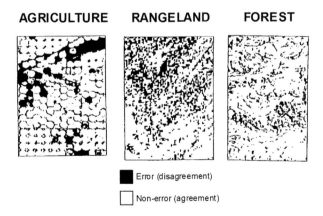

FIGURE 6.5 Difference images (7.5 minute quadrangles) showing the pattern of error for three ecosystems of varying complexity: agriculture, rangeland, and forest.

However, the results are more surprising for the rangeland and forested classes. Both classes are more spatially complex (i.e., have more fragmentation, edges, and mixtures of land cover) than the agriculture map, and therefore, one would expect them to be less spatially autocorrelated. Primarily because of rangeland fencing, the rangeland class does have some fields that are similar to agriculture, but it also reflects some of the edge effects more common to the complex forest class.

The forest class is the most spatially complex, and most map error would be expected to occur along the edges or transition zones between forest types. While viewing the forest difference map does tend to confirm these edge problems, the results of the analysis still indicate that there is strong, positive autocorrelation between errors up to 30 pixels away. In other words, if an error occurs at a given location, it is more likely to find another error even up to this rather large distance away (i.e., 30 MSS pixels or about 240 meters) than it is to find a correct classification.

The existence of spatial autocorrelation can violate the assumption of sample independence, which, in turn, can affect the sample size and especially the sampling scheme used in the accuracy assessment. This influence is especially important when collecting reference data on the ground (i.e., field sampling). If using higher–spatial resolution imagery as the source of reference data, it is easier to space the samples further apart. Spatial autocorrelation may indicate the existence of periodicity in the presence of a specific map class across the landscape, which could affect the results of any type of systematic sample if the systematic sample design repeats that same periodicity. For example, maple trees need ample water and, in arid landscapes, are usually located along streams. A systematic sample scheme based on choosing samples near streams would repeat the periodicity of the maple forest class and would result in a biased choice of samples that would oversample maple forests and undersample other map classes.

In addition, autocorrelation may affect the size and number of samples used in a cluster sampling approach, because each sample unit may be contributing not new, independent information but rather, redundant information. Therefore, it would not be effective to collect information for many sample units in a large cluster, since the contribution of each new sample unit in the cluster could very quickly be reduced to very little because of this lack of independence. However, cluster sampling is a very cost-effective method, especially in the field, when the cost of traveling from one sample location to another can be very high. Even when the accuracy assessment samples are taken in the office from aerial imagery, cluster sampling can create savings in set-up time for each image. Therefore, it is important to consider spatial autocorrelation and balance the impact of having spatially autocorrelated samples against the efficiencies of cluster sampling. This can be done by limiting the number of samples taken in the cluster to between two and four, making sure that each sample unit in the cluster is taken in a different thematic class, and spreading the samples as far apart as possible. Without understanding these considerations, so that the statistical validity can be effectively balanced with practical application, the accuracy assessment process will not be as efficient as it should be.

WHAT IS THE APPROPRIATE SAMPLE UNIT?

Sample units are the portions of the map that are selected for accuracy assessment. There are three possible choices for the sampling unit: (1) a single pixel; (2) a cluster of pixels (often a 3×3 pixel square); and (3) a polygon.

Single Pixel

Historically, a large number of accuracy assessments have been conducted using a single pixel as the sampling unit. However, a single pixel is a very poor choice for the sampling unit for many reasons:

- First, a pixel is an arbitrary rectangular delineation of the landscape that may have little relation to the actual delineation of land cover or land use type. It can be a single land cover or vegetation class (i.e., a pure pixel), or more often than not, it can be a mixture of land cover or vegetation classes.
- Second, even with the best geocoding and terrain correction procedures, it is not possible to exactly align one pixel on a map to the exact same area in the reference data. Therefore, there is no way to guarantee that the location of the reference pixel is identical to the location of the map pixel. Similarly, unless survey-grade global positioning systems (GPS) are used (not efficient for most reference data collection), the accuracy of the position of the ground-collected reference data has a typical error of 3–10 meters depending on a variety of conditions. Therefore, positional accuracy becomes a large issue, and the thematic accuracy of the map is affected because of positional error.
- Finally, few classification schemes specify a unit as small as a pixel as the mmu. If the mmu is larger than a single pixel, then a single pixel is likely to represent a component of the map class and not the class itself, especially in highly heterogeneous areas such as sparse forest lands or suburban neighborhoods. For example, consider a sparse forest with brush and grass in the understory. If the minimum mapping unit is 5 acres, then a single Landsat pixel (at 900 sq. meters or 0.22 acres) could fall onto brush, a tree, grass, or a mixed vegetation area. One pixel will not adequately characterize the vegetation within that polygon that is 5 acres or more in size.

Even with all the state-of-the-art advances in GPS, terrain correction, and geocoding, accuracy assessment sample units will still have some positional inaccuracies. It is commonly accepted that a positional accuracy of one half a pixel is sufficient for medium-resolution sensors such as Landsat Thematic Mapper, SPOT Multispectral, and Sentinel 2 imagery. As imagery increases in spatial resolution, such as that collected from digital airborne cameras, unmanned aerial systems (UASs), and high-resolution satellites, positional accuracy increases in importance and must be considered an even more critical factor in the assessment process. This factor is especially true if the sensor is not vertical but rather, collects imagery in an off-nadir condition (as most of the higher–spatial resolution satellites now do). If an image with a pixel size of 10–30 meters is registered to the ground to within half a pixel (i.e., 5–15 meters), and a GPS unit is used to locate the unit on the ground to within 3–10 meters, then it is *completely inappropriate to use a single pixel as the sampling unit for assessing the thematic accuracy of a map*. There would simply be no guarantee that the map and the reference data would be collected from the identical area, as there is no way to accurately locate the corners of a given pixel on the ground and match it to the same pixel on the image. If the positional accuracy is not up to the standard, or if GPS is not used to precisely locate the sample on the ground, then these factors increase in importance and can significantly affect the thematic accuracy assessment. This situation only increases in importance with higher–spatial resolution imagery, where the pixels may be smaller and therefore, the issue of positional error can be even greater. For example, if an analyst selected a 1 meter pixel as their sampling unit (using 1 meter imagery), the positional accuracy of the imagery was 5 pixels, and the GPS was accurate to the optimal 3 meters, it is easy to see that the chance of overlaying the ground sample exactly on the image pixel is extremely low. For all these reasons and despite far too many examples in the literature, a single pixel should never be used as the sample unit for medium to high–spatial resolution imagery.

CLUSTER OF PIXELS AS A SINGLE SAMPLE UNIT

Given the need to balance thematic accuracy with positional accuracy, a cluster of pixels, typically a 3×3 square for moderate-resolution imagery (e.g., Landsat), is often an effective choice for the sample unit. Choosing a cluster of pixels as a single sample unit minimizes registration problems, because it is easier to locate on the reference data or in the field. However, a cluster of pixels (especially a 3×3 window) may still be an arbitrary delineation of the landscape, resulting in the sample unit encompassing more than one map class. To avoid this problem, it is suggested that only homogeneous clusters of pixels be sampled (Figure 6.6). Samples that occur on the boundaries between map classes are difficult to assess, because positional issues here could easily impact the map and reference data labels. Sampling a homogeneous cluster of pixels as a single sample unit solves this problem. However, such restrictions may result in a biased sample that avoids heterogeneous areas that are a function of a mix of pixels (e.g., a homogeneous area such as a lake vs. a heterogeneous area of sparse vegetation including a mingling of some trees and grass), as depicted in Figure 6.7.

It is important to remember that the sample unit dictates the level of detail for the accuracy assessment. If the assessment is performed on a 3×3 cluster of pixels, then nothing can be said about an individual pixel in that cluster, nor can anything be said about the polygons (i.e., management areas, forest stands, agricultural fields, etc.) that contain the cluster. Additionally, it is critical to recognize that *each sample unit must be considered a single sample*. If, for example, a 3×3 cluster of pixels is used as the sample unit, then it must be counted as one sample and not as nine samples. There are numerous examples in the literature where authors have mistakenly counted each pixel in a cluster as a separate accuracy assessment sample. The nine

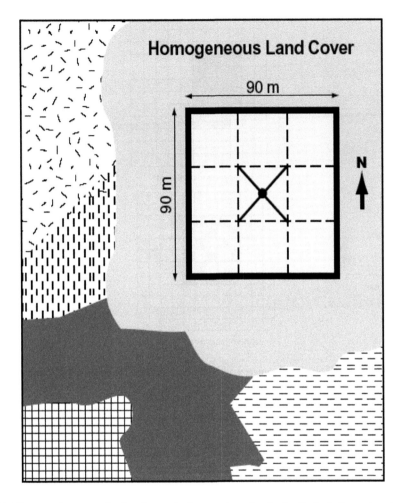

FIGURE 6.6   An example showing a single sample unit of a 3×3 pixel cluster in a homogeneous area used to minimize positional error in the reference data collection process. The various shading types represent different land cover types.

pixels together count as one sample. Also, the presence of spatial autocorrelation in most thematic maps dictates that samples should be spaced adequately apart from one another.

Extending the concept of a cluster of pixels to higher-resolution imagery requires knowledge about the positional accuracy of the imagery. As previously stated, common registration (positional) accuracies for Landsat Thematic Mapper and SPOT satellite imagery (with 10–30 meter pixels) are about half a pixel, and the GPS accuracy is 5–15 meters. Therefore, selecting a homogeneous cluster of 3×3 pixels as the sample unit ensures that the center of the sample will definitely fall within the 3×3 cluster. If the sample is homogeneous, and the collection is performed at the center of the sample, then an error that could be caused by positional issues will be eliminated. Higher–spatial resolution satellite imagery now has pixel sizes from

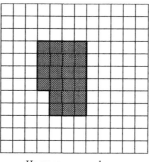

Homogeneous polygon

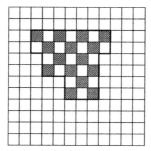

Heterogeneous polygon

FIGURE 6.7   Comparison of accuracy assessment sample units comprised of homogeneous versus heterogeneous pixels.

4 meters to below 1 meter. However, because of the off-nadir acquisition and other issues, the positional accuracy of these data are often in the range of 5–10 meters, and the GPS accuracy is still 3–10 meters. Therefore, the use of a 3 × 3 pixel cluster as the sampling unit would be too small and inappropriate in this case to compensate for the positional error. If the registration accuracy was 5 meters, the GPS accuracy was 10 meters, and the pixel size was 2 meters, then the sample unit cluster would need to be a homogeneous area of at least 8 × 8 pixels to account for this positional error. It is imperative that the positional accuracy of the sensor be considered in the selection of the sample unit cluster size, or else the thematic assessment will be flawed, as the error indicated in the error matrix will be a combination of both thematic and positional error.

POLYGON AS SINGLE SAMPLE UNIT

Most large-scale thematic maps delineate the landscape into polygons of homogeneous map classes based on an effective classification scheme, as discussed previously in this chapter. Polygons are delineated on the edges of classes where more between- than within-class variation exists. While the pixels inside the polygons may vary dramatically (as in a sparse stand of trees or a suburban area), the class label across the pixels is constant. Usually, the polygon map is created either through manual interpretation or, more commonly today, through the use of object-based image analysis (OBIA) and the creation of objects/segments. These objects themselves are polygons but are typically smaller in size than the final polygon map produced. It is the final polygon map that should be assessed rather than the objects/segments, which typically represent only an intermediate result.

If the final map to be assessed is a polygon map, then the accuracy assessment sample units could also be polygons. The resulting accuracy values inform the map's user and producer about the level of detail in which they are interested—the polygons. More and more mapping projects using remotely sensed data are generating polygon- rather than pixel-based products as a result of developments in image segmentation and OBIA, especially for high–spatial resolution imagery. As a result, the polygon is replacing the cluster of pixels as the sample unit of choice in many projects. Chapter 11 presents a discussion of the issues and considerations using polygon- or object-based accuracy assessment.

However, using polygons as sample units can cause confusion if the accuracy assessment polygons are collected during the initial training data/calibration fieldwork, which occurs before the map is created. The result can often be manually delineated accuracy assessment polygons with dramatically different delineations than the final map polygons, as illustrated in Figure 6.8. When this situation occurs, some way of creating the map label for the accuracy assessment polygon must be developed. The simplest approach is to use the majority class of the polygon to create the map label. However, this may not work well in heterogeneous conditions, where the label is more a function of the mix of the ground cover (e.g. patchy seagrass or mixed hardwood–conifer forests) rather than the majority of the ground cover.

Another approach is to run the segmentation algorithm and finalize the delineation of the segments prior to the initial field trip. The resulting segments will most certainly be contained within the final map polygons and therefore, can be used as accuracy sampling units, but they will probably be smaller than the final polygons.

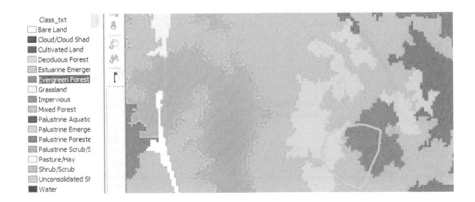

FIGURE 6.8 Mixed forest accuracy assessment reference polygon (in turquoise) over the map polygons of evergreen, mixed, and deciduous forest. Determining the map label of the reference polygon is problematic.

Because polygon- or object-based accuracy assessment has become so important as a result of the extensive use of OBIA, an entire chapter (Chapter 11) on this topic is included in this book.

CLUSTERS OF SAMPLE UNITS

It is often efficient and effective, especially when collecting reference data in the field, to collect clusters of sample units. Figure 6.9 presents such an example. The analyst collecting the reference data can stop their vehicle and collect a number of sample units around the same location. Collecting a cluster of sample units can reduce accuracy assessment costs dramatically, because travel time and/or set-up time is decreased. However, care must be taken to provide some separation between

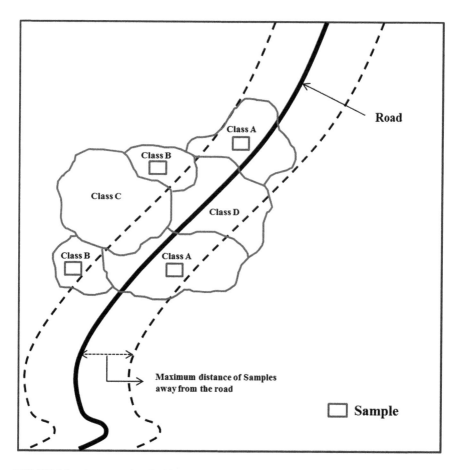

FIGURE 6.9 An example of collecting a cluster or grouping of sample units around a single location to increase the efficiency of the reference data collection. Only a single sample unit is selected in each contiguous area (i.e., forest stand, agricultural field, etc.). Also, the sample units are spaced at least a predetermined distance apart to reduce spatial autocorrelation.

sample units and to limit the number taken at each location so as to avoid spatial autocorrelation between the samples (Figure 6.9). Guidelines can be set to make sure the samples are taken some distance from the road and that only a single sample is collected within the same continuous map class (i.e., forest stand, agricultural field, etc.) (see Figure 6.9). Also, a distance value can be set between sample units of differing map classes to minimize spatial autocorrelation. Considering all of these issues maintains the balance between what is statistically valid and what can be practically achieved.

HOW MANY SAMPLES SHOULD BE TAKEN?

Accuracy assessment requires that an adequate number of samples per map class be gathered so that the assessment is a statistically valid representation of the accuracy of the map. However, the collection of reference data at each sample unit is very expensive, requiring that the sample size be kept to a minimum to be affordable. Therefore, there is a delicate balance here to obtain the necessary samples within a reasonable amount of effort and budget.

Of all the considerations discussed in this chapter, the most has probably been written about sample size. Many researchers, notably Hord and Brooner (1976), van Genderen and Lock (1977), Hay (1979), Ginevan (1979), Rosenfield et al. (1982), and Congalton (1988b), have published equations and guidelines for choosing the appropriate sample size.

The majority of work performed by early researchers used an equation based on the binomial distribution or the normal approximation to the binomial distribution to compute the required sample size. These techniques are statistically sound for calculating the sample size needed to compute the overall accuracy of a classification or even the overall accuracy of a single class. The equations are based on the proportion of correctly classified sample units and on some allowable error. However, these techniques were not designed to choose a sample size for generating an error matrix.

In the case of creating an error matrix, it is not simply a question of correct or incorrect (the binomial case). Instead, it is a matter of which error or which classes are being confused. Given an error matrix with n land cover classes, for a given class there is one correct answer and $(n - 1)$ incorrect answers (i.e., errors). Sufficient samples must be acquired to be able to adequately represent this confusion (i.e., build a statistically valid error matrix). Therefore, the use of the binomial distribution for determining the sample size for an error matrix is not appropriate. Instead, the use of the multinomial distribution is recommended (Tortora, 1978).

The appropriate sample size can and should be computed for each project using the multinomial distribution. However, in our experience, a general guideline or good "rule of thumb" suggests planning to collect a minimum of 50 samples for each map class for maps of less than 1 million acres in size and fewer than 12 classes (Congalton, 1988b). Larger-area maps or more complex maps may require more samples. In some of these cases, it may be necessary to conduct a sampling simulation (e.g., Monte Carlo simulation) to determine when the appropriate sample size has been reached. This type of simulation analysis was performed previously to determine the general guideline of 50 samples per map class (Congalton, 1988b).

In situations where the map covers a small area, it may be impossible to find 50 independent samples for some of the rare map classes (those covering small areas). In this case, all the samples that could be collected should be used (not including the samples used for training the classifier), and the documentation should state that all possible reference data samples for that class were used in the assessment. These guidelines were empirically derived over many projects, and the use of the multinomial equation has confirmed that they are a good balance between statistical validity and practicality.

Practical considerations are often a key component in the sample size determination process. For example, the number of samples for each class may be adjusted based on the relative importance of that class within the objectives of the mapping project or by the inherent variability within each of the classes. Sometimes, because of budget constraints or other factors, it is better to concentrate the sampling on selected map classes of interest and increase their number of samples while reducing the number of samples taken in the less important classes. Having a hierarchical classification scheme may facilitate generalizing some categories to reduce the number that must be assessed. Also, a decision may be made to take fewer samples in categories that show little variability, such as water or forest plantations, and increase the sampling in the categories that are more variable, such as uneven-aged forests or riparian areas. However, in most instances, some minimum number of samples (e.g., 50 samples as per the guidelines or the result of the multinomial equation calculation) should be taken in each land cover class contained in the matrix. Finally, *it may be tempting to design a sampling effort which selects many samples in categories that are most accurate and few in the confused categories, or perhaps there are other ways to artificially inflate the map accuracy*. These strategies would guarantee a high accuracy value but would not be representative of the true map accuracy. It is critical to remember that the goal when planning an assessment is to balance the statistical recommendations so as to get an adequate sample to generate an appropriate error matrix within the time, cost, and practical limitations associated with the given mapping project. However this balance is achieved, the analyst must document the exact process, so that future users of the map can know how the assessment was conducted and can therefore effectively use the map.

BINOMIAL DISTRIBUTION

As mentioned earlier, the binomial distribution or the normal approximation to the binomial distribution is appropriate for computing the sample size for determining overall accuracy or the accuracy of an individual class. Later in this book, the binomial distribution will be used to assess a change/no change map (see Chapter 14). It is appropriate for the two-case situation where only right and wrong are important. Choosing the appropriate sample size from the binomial or normal approximation is dependent on (1) the level of acceptable error one is willing to tolerate and (2) the desired level of confidence that the actual accuracy is within some minimum range. Numerous publications present look-up tables of the required sample size for a given acceptable error and desired level of confidence (e.g., Cochran, 1977; Ginevan, 1979).

For example, suppose it is decided that a map is unacceptable if the overall accuracy is 90% or less. Also, let's say that we are willing to accept a 1 in 20 chance that we will make a mistake based on our sample and accept a map that actually has an accuracy of less than 90%. Finally, let us decide that we will accept the same risk, a 1 in 20 chance, of rejecting a map that is actually correct. The appropriate look-up table would then indicate that we must take 298 samples, of which only 21 can be misclassified. If more than 21 samples were misclassified, we would conclude that the map is not acceptable.

MULTINOMIAL DISTRIBUTION

As discussed earlier in this chapter, the multinomial distribution provides the appropriate equations for determining the sample size required to generate an error matrix. The procedure for generating the appropriate sample size from the multinomial distribution is summarized here and was originally presented by Tortora (1978).

Consider a population of units divided into k mutually exclusive and exhaustive categories. Let Π_i, $i = 1, ..., k$ be the proportion of the population in the ith class, and let n_i, $i = 1, ..., k$ be the frequency observed in the ith class in a simple random sample of size n from the population.

For a specified value of α, we wish to obtain a set of intervals, $i = 1, ..., k$, such that:

$$\Pr\left\{\bigcap_{i=1}^{k}(\Pi_i \in S_i)\right\} \geq 1 - \alpha$$

That is, we require the probability that every interval S_i contains Π_i to be at least $1 - \alpha$. Goodman (1965) determined the approximate large-sample confidence interval bounds (when $n \rightarrow \infty$) as

$$\Pi_i^- \leq \Pi_i \geq \Pi_i^+$$

where:

$$\Pi_i^- = \Pi_i - \left[B\Pi_i(1-\Pi_i)/n\right]^{1/2} \tag{6.1}$$

$$\Pi_i^+ = \Pi_i + \left[B\Pi_i(1-\Pi_i)/n\right]^{1/2} \tag{6.2}$$

and B is the upper $(a/k) \times 100$th percentile of the χ^2 distribution with 1 degree of freedom. These equations are based on Goodman's (1965) procedure for simultaneous confidence interval estimation.

Examining Equations 6.1 and 6.2, we see that $\left[\Pi_i(1-\Pi_i)/n\right]^{1/2}$ is the standard deviation for the ith cell of the multinomial population. Also, it is important to realize that each marginal probability mass function is binomially distributed. If N is the total population size, then using the finite population correction factor (fpc) and the variance for each Π_i (from Cochran, 1977), the approximate confidence bounds are:

$$\Pi_i^- = \Pi_i - \left[B(N-n)\Pi_i(1-\Pi_i)/(N-1)n \right]^{1/2} \tag{6.3}$$

$$\Pi_i^+ = \Pi_i + \left[B(N-n)\Pi_i(1-\Pi_i)/(N-1)n \right]^{1/2} \tag{6.4}$$

Note that as $N \to \infty$, Equations 6.3 and 6.4 converge to Equations 6.1 and 6.2, respectively.

Next, to determine the required sample size, the precision for each parameter in the multinomial population must be specified. If the absolute precision for each cell is set to b_i, then Equations 6.1 and 6.2 become:

$$\Pi_i - b_i = \Pi_i - \left[B\Pi_i(1-\Pi_i)/n \right]^{1/2} \tag{6.5}$$

$$\Pi_i + b_i = \Pi_i + \left[B\Pi_i(1-\Pi_i)/n \right]^{1/2} \tag{6.6}$$

respectively. Similar results are obtained when the fpc is included. Equations 6.5 and 6.6 can be rearranged to solve for b_i (the absolute precision of the sample):

$$b_i = \left[B\Pi_i(1-\Pi_i)/n \right]^{1/2} \tag{6.7}$$

Then, by squaring Equation 6.7 and solving for n, the result is:

$$n = B\Pi_i(1-\Pi_i)/b_i^2 \tag{6.8}$$

or, the fpc,

$$n = BN\Pi_i(1-\Pi_i)/\left[b_i^2(N-1) + B\Pi_i(1-\Pi_i) \right] \tag{6.9}$$

Therefore, one should make k calculations to determine the sample size, one for each pair (b_i, Π_i), $i = 1, \ldots, k$, and select the largest n as the desired sample size. As functions of Π_i and b_i, Equations 6.8 and 6.9 show that n increases as $\Pi_i \to 1/2$ or $b_i \to 0$.

In rare cases, a relative precision b_i' could be specified for each cell in the error matrix and not just each class. Here, $b_i = b_i'\Pi_i$. Substituting this into Equation 6.8 gives:

$$n = B(1-\Pi_i)/\Pi_i b_i'^2 \tag{6.10}$$

A similar sample size calculation including the fpc can be computed in the same way.

Here again, one should make k calculations, one for each pair (b_i', Π_i), $i = 1, \ldots, k$. The largest n computed is selected as the desired sample size. As $\Pi_i \to 1/2$ or or $b_i' \to 0$, the sample size increases according to Equation 6.10. If $b_i' = b'$ for all i, then the largest sample size is $n = B(1 - \Pi)/\Pi b'^2$, where $\Pi = \min(\Pi_1, \ldots, \Pi_k)$.

In the majority of cases, for assessing the accuracy of remotely sensed data, an absolute precision is set for the entire classification and not each class or each cell. Therefore, $b_i = b$, and the only sample size calculation required is for the Π_i closest to ½. If there is no prior knowledge about the values of the Π_is, a "worst"-case calculation of sample size can be made assuming some $\Pi_i = \frac{1}{2}$ and $b_i = b$ for $i = 1$, ..., k. In this worst-case scenario, the sample size required to generate a valid error matrix can be obtained from this simple equation as follows:

$$n = B / 4b^2$$

This approach can be made much clearer with a numerical example. First, let's look at an example using the full equation (Equation 6.8) and then at the corresponding sample size using the worst-case or conservative sample size equation. Assume that there are eight categories in our classification scheme ($k = 8$), that the desired confidence level is 95%, that the desired precision is 5%, and that this particular class makes up 30% of the map area (Π_i=30%). The value for B must be determined from a χ^2 table with 1 degree of freedom and $1 - \alpha/k$. In this case, the appropriate value for B is $\chi^2_{(1,0.99375)} = 7.568$. Therefore, the calculation of the sample size is as follows:

$$n = B\Pi_i \left(1 - \Pi_i\right) / b_i^2$$

$$n = 7.568(0.30)(1 - 0.30) / (0.05)^2$$

$$n = 1.58928 / 0.0025$$

$$n = 636$$

A total of 636 samples should be taken to adequately fill an error matrix, or approximately 80 samples per class, given that there were eight classes in this map.

If the simplified, worst-case scenario equation is used, then the class proportion is assumed to be 50%, and the calculation is as follows:

$$n = B / 4b^2$$

$$n = 7.568 / 4(0.05)^2$$

$$n = 7.568 / 0.01 = 757$$

In this worst-case scenario, approximately 95 samples per class or 757 total samples would be required.

If the confidence interval is relaxed from 95% to 85%, the required sample sizes decrease. In the previous example, the new appropriate value for B would be

$\chi^2_{(1,0.98125)} = 5.695$, and the total samples required would be 478 and 570 for the complete equation and the worst-case scenario, respectively.

HOW SHOULD THE SAMPLES BE CHOSEN?

In addition to the considerations already discussed, the choice and distribution of samples, or sampling scheme, is an important part of any accuracy assessment. Selection of the proper scheme is critical to generating an error matrix that is representative of the entire map. First, to reach valid conclusions about a map's accuracy, the sample must be selected without bias. Failure to meet this important criterion affects the validity of any further analysis performed, because the resulting error matrix may over- or underestimate the true accuracy. Second, further data analysis will depend on which sampling scheme is selected. Different sampling schemes assume different sampling models and consequently, different variance equations to compute the required accuracy methods. Finally, the sampling scheme will determine the distribution of samples across the landscape, which will significantly affect accuracy assessment costs, especially the collection of the required reference data.

SAMPLING SCHEMES

Many researchers have expressed opinions about the proper sampling scheme to use (e.g., Hord and Brooner, 1976; Rhode, 1978; Ginevan, 1979; Fitzpatrick-Lins, 1981; Stehman, 1992). These opinions vary greatly among researchers and include everything from simple random sampling to a scheme called *stratified, systematic, unaligned sampling*.

There are five common sampling schemes that have been applied for collecting reference data: (1) simple random sampling, (2) systematic sampling, (3) stratified random sampling, (4) cluster sampling, and (5) stratified, systematic, unaligned sampling. In a simple random sample, each sample unit in the study area has an equal chance of being selected. In most cases, a random number generator is used to pick random x,y coordinates to identify samples to be collected. The main advantage of simple random sampling is the good statistical properties that result from the random selection of samples (i.e., it results in the unbiased selection of samples).

However, it is possible to use a statistically unbiased sampling scheme to assess a map and yet produce an assessment that is not of practical use. Imagine a situation in which the vast majority of the area on the map was desert or bare ground, while a small portion was developed areas, including residential, commercial, and industrial space. If 95% of the map was desert and 5% was developed, and 100 sample units were randomly selected (simple random sampling method) across the entire map, then, on average, there would be 95 sample units of desert and 5 of developed. If a map was created that showed that the entire area was desert, using the sampling approach selected here would determine a map accuracy of 95% (95 out of 100 reference sample units would be desert and agree with the map). However, if the map user was mostly interested in the developed areas (i.e., residential, commercial, and industrial areas) on this map, then this very accurate map (95%) would be completely useless to them. Therefore, great care should be taken to make sure that the sampling effort is carefully planned and implemented.

Systematic sampling is a method in which the sample units are selected at some specified and regular interval over the study area. In most cases, the first sample is randomly selected, and each successive sample is taken at some specified interval thereafter. The major advantage of systematic sampling is the ease in sampling somewhat uniformly over the entire study area.

Stratified random sampling is similar to simple random sampling; however, some prior knowledge about the study area is used to divide the area into groups or strata, and then each stratum is randomly sampled. In the case of accuracy assessment, the map has been stratified into map classes. The major advantage of stratified random sampling is that all strata (i.e., map classes), no matter how small, will be included in the sample. This factor is especially important in making sure that sufficient samples are taken in rare but important map classes.

In addition to the sampling schemes already discussed, cluster sampling has also been frequently used in assessing the accuracy of maps from remotely sensed data, especially to collect information on many samples quickly. There are clear advantages to collecting a number of sample units in close proximity to each other. However, cluster sampling must be used intelligently and with great care. Simply taking a large number of sample units (whether they be a cluster of pixels or polygons) together within the same contiguous map class is not a valid method of collecting data, because each sample unit is not independent of the others and adds very little additional information. As shown in Figure 6.9, it is effective to collect multiple, nearby sample units, especially when collecting the reference data in the field. However, spatial autocorrelation must be carefully considered, ensuring that no two sample units are in the same contiguous map class and that there is sufficient distance between the sample units. Finally, some reasonable number of sample units (between four and eight) should be collected at the location, and then the collection should be moved to another starting point sufficiently far away that the entire study area is covered with samples and they are not concentrated in only one part of the study area.

Finally, more complex sampling schemes have been suggested. For example, stratified, systematic, unaligned sampling is a method that attempts to combine the advantages of randomness and stratification with the ease of a systematic sample without falling into the pitfalls of periodicity common to systematic sampling. This method is a combined approach that introduces more randomness than just a random start within each stratum. Instead, the first sample is selected at random, then the x coordinate of the first sample is used with a new random y coordinate in the second sample, and then with another new random y coordinate in the third sample, and so forth. In the same manner, the y coordinate of the first sample is used with a new random x coordinate to locate another sample, and this continues throughout the strata. In this way, a combination of random and systematic sampling is implemented.

SAMPLING SCHEME CONSIDERATIONS

Congalton (1988b) performed sampling simulations on three spatially diverse areas (see Figure 6.5) using all five of these sampling schemes and concluded that in all cases, simple random and stratified random sampling provided satisfactory results.

It should be noted that in this analysis, the maps included only two categories (error and correct).

Simple random sampling allows reference data to be collected simultaneously for both training and assessment. Remember that these two data sets must be separated from each other so that the reference data used in the assessment remain independent. However, it is not always appropriate to use simple random sampling, because it tends to undersample rarely occurring, but possibly very important, map categories unless the sample size is significantly increased. For this reason, stratified random sampling, whereby a minimum number of samples are selected from each stratum (i.e., map class), is often recommended. However, stratified random sampling can be impractical, because stratified random samples are more easily selected after the map has been completed (i.e., when the locations of the strata are known). This can limit the accuracy assessment data to being collected late in the project instead of in conjunction with the training data collection, thereby increasing the costs of the project. In addition, in some projects, the time between the project beginning and the accuracy assessment may be so long as to cause temporal problems in collecting ground reference data. In other words, the ground may change (e.g., the crop harvested) between the time the project is started and when the accuracy assessment is begun.

The concept of randomness is a central issue when performing almost any statistical analysis, because a random sample is one in which each member of the population has an equal and independent chance of being selected. Therefore, a random sample ensures that the samples will be chosen without bias. If in-office manual interpretation of Google Earth or other high–spatial resolution imagery is used to label reference samples, then random sampling is feasible, because access to the samples will not be a problem. However, a subset of the sample should be visited on the ground to access the accuracy of the interpretation.

Despite the valuable statistical properties of random sampling, access in the field to random sample units is often problematic, because many of the samples will be difficult to reach/access. Locked gates, fences, travel distances, and rugged terrain all combine to make random field sampling extremely costly and difficult. In forested and other wildland environments, randomly selected samples may be totally inaccessible except by helicopter. The cost of getting to each of the randomly located samples can be more than the cost of the rest of the entire mapping effort. One way to deal with access constraints prior to the implementation of a sampling scheme is to create cost surfaces based on access and exclude inaccessible areas from the accuracy assessment. For example, a cost surface could be created to exclude areas that are more than a maximum distance from a road, over a prescribed percent slope, or within ownership types where access has been denied.

Obviously, one cannot spend the majority of a project's resources collecting accuracy assessment reference data. Instead, some balance must be struck. Often, some combination of random and systematic sampling provides the best balance between statistical validity and practical application. Such a system may employ systematic or simple random sampling to collect some assessment data early in a project and stratified random sampling within strata (i.e., the map classes) after the classification is completed to ensure that enough samples were collected for each class and to minimize any periodicity in the data. However, the results of Congalton (1988a) showed that periodicity in the errors as measured by the spatial autocorrelation analysis could make the use of systematic sampling risky for accuracy assessment.

An example of a combined approach could include a systematic sample tied to existing aerial photography or other high-resolution imagery with sample selection based on the center of every *n*th photo/image. Sample choices based on flight lines should not be highly correlated with any factor determining land cover unless the flight lines were aligned with a landscape feature. The choice of the number of samples per photo/image and the sampling interval between photos/images would depend on the size of the area to map and the number of samples to collect. This systematic sample would ensure that the entire mapped area gets covered.

However, if this strategy is selected, rarely occurring map classes will probably be undersampled. It may be necessary to combine this approach with stratified random sampling when the map is completed to augment the underrepresented map classes. In addition, it may be practical to bound/limit the stratified random field sample selection within some realistic distance of the roads (see Figure 6.9) to control the costs of the reference data collection. However, care must be taken, because roads tend to occur on flatter areas, in valleys along streams, and on ridge tops, which may bias the sample selection to land cover likely to exist in these places. Therefore, steps must be taken to mitigate these factors so that the most representative sample can be achieved. This type of combined approach minimizes the resources used and obtains the maximum information possible. Still, the statistical complexities of such a combination cannot be neglected. Again, a balance is desirable.

Finally, some analytic techniques assume that certain sampling schemes were used to obtain the data. For example, use of the Kappa analysis for comparing error matrices (see Chapter 8 for details on this analysis technique) assumes a multinomial sampling model. Only simple random sampling completely satisfies this assumption. If another sampling scheme or combination of sampling schemes is used, then it may be necessary to compute the appropriate variance equations for performing the Kappa analysis or other similar technique. The effects (i.e., bias) of using another of the sampling schemes discussed here and not computing the appropriate variances have not been widely considered.

An interesting project would be to test the effect on the Kappa analysis of using a sampling scheme other than simple random sampling. If the effect is found to be small, then the scheme may be appropriate to use within the conditions discussed earlier. If the effect is found to be large, then that sampling scheme should not be used to perform Kappa analysis. If that scheme is to be used, then the appropriate correction to the variance equation must be applied. Stehman (1992) performed such an analysis for two sampling schemes (simple random sampling and systematic sampling). His analysis showed that the effect on the Kappa analysis of using systematic sampling is negligible. This result adds further credence to the idea of using a combined systematic initial sample followed by a random sample to fill in the gaps.

A search of the accuracy assessment literature over the last 30 years reveals many interesting academic papers describing complex sampling strategies and other methodologies for performing an assessment. While many of these are statistically sound, most go beyond what is practically achievable for the typical remote sensing analyst looking to assess the accuracy of their map. There is always a need to balance statistical validity with what can practically be achieved. The methods and considerations described in this book endeavor to achieve such a balance.

Equal versus Minimum versus Proportional Distribution

As presented throughout this chapter, sampling for the collection of reference data to create a valid error matrix requires many considerations and careful planning. One final issue about the distribution of the samples needs to be discussed. That is, should the samples be distributed equally, or using some minimum number, or proportionally when conducting the assessment? A guideline stating that 50 samples should be collected per map class has already been discussed. Using this strategy, an approximately equal number of samples (50) are selected for each map class. Some rare categories might not quite be able to reach 50, but that is still the goal. This strategy works very well when the area of each map class is approximately the same and provides a valid error matrix.

However, what if there are a few large-area map categories, some medium-area map categories, and a few small-area map categories? Is it a good strategy to simply collect 50 samples for each map class? For example, as the organization responsible for much of the high-resolution vegetation mapping throughout the United States, the National Park Service has grappled with the trade-offs between statistical rigor and cost effectiveness for decades. Their most recent guidelines assign the number of accuracy samples per class based on both the area and the abundance of a class across the landscape, as shown in Table 6.1 (Lea and Curtis, 2010). While this recommendation begins with fewer sample units per map class than is recommended in this book, it should be noted that the National Park Service maps to very complex vegetation levels often having a classification scheme with over 60 map classes. Therefore, it is understandable why the number of sample units is so low; again, balancing statistical validity with what can practically be achieved.

What happens if the sampling is performed proportionally instead of equally (50 samples)? One would expect that the number of samples obtained for each map class would be proportional to the area of that map class, and therefore, a few categories would have many more than 50 samples, most would have about 50 samples, and a few would have many fewer samples. If the analyst is only considering overall accuracy, this result may be satisfactory. However, what if it turns out that the map

---

TABLE 6.1

An Example of Distributing the Number of Reference Sample Units by Area from the U.S. National Park Service Guidelines (Lea and Curtis, 2010)

Map Class Total Area	Number of Samples per Map Class
> 50 hectares	30
8.33 to 50 hectares	0.6 per hectare
< 8.33 hectares	5

categories of the large areas are very easy to map (e.g., water or sand), while the map categories of the small areas are much more difficult? Finally, what if it is the small-area map categories that the user of the map is most interested in knowing about? In this case, each map class is important, and user's and producer's accuracies must be computed for each one. Are there enough samples in the small-area categories to compute valid accuracies for those categories? Perhaps, in this situation, it would be relevant to set some minimum number of samples in each map class and then add additional samples to each map class proportionally to the class area. In other words, the smallest map class could be allocated a minimum of 30 samples (enough that the user's and producer's accuracy could be computed for this map class), while the rest of the map categories received more samples allocated proportionally to their area. The minimum number of samples could also be set to 50 to maintain the guideline set in this book. Ideally, the sampling could occur in such a way that each map class received enough samples to be exactly proportional to its area, such that the final sample was proportional to area, starting with the smallest area having 30 or 50 samples. Unfortunately, this ideal approach is not practical, as it could result in the large-area map categories needing hundreds of samples or more to obtain that proportionality. A more realistic approach would be to impose a minimum number of samples on each map class (30 or 50) and then collect an additional number of samples proportionally to the area of the other map categories. For example, if there were six map categories, the original goal (from the guideline earlier or the multi-nomial equation) may be to collect 300 reference samples (six map categories × 50 samples per class = 300). If each map class received 30 samples (30 samples × 6 map categories = 180 samples), then there would be 120 (300–180) samples left to be proportionately allocated to each map class except the smallest one. This approach considers that the areas of each map class are not the same while still maintaining a realistic number of total samples. It is a balance/trade-off between statistical validity and what can be practically achieved.

Finally, in some situations, a truly proportional sample may be appropriate. For example, a recent project was funded by the National Aeronautics and Space Administration (NASA), called Global Food Security Support Analysis Data (GFSAD), to map the eight major agricultural crops around the world using Landsat imagery (Teluguntla et al., 2015). Part of this project mapped cropland extent (crop/no crop) for the entire world by dividing the planet into Agricultural Ecological Zones (AEZs). Given this two–map class (crop/no crop) situation, proportional sampling was employed to generate the error matrices for assessing the cropland extent map. Figure 6.10 shows a map of cropland extent (crop/no crop) for one AEZ in Africa. Note that for this AEZ, only a very small percentage of the zone is crop (3.29%). However, this zone is atypical, in that only seven of the 73 AEZs mapped for the world had such small portions of crop (Yadav and Congalton, 2018).

A quick look at the error matrix presented in Table 6.2 demonstrates the issue that we have been discussing regarding proportional sampling. As expected, there were only a small number of crop reference data samples, since crop is such a small proportion of the map. In this case, there are nine crop reference data samples out of the 250 random samples selected to generate the matrix (3.29% of 250 samples is approximately eight expected samples). The other 67 AEZs did not experience this

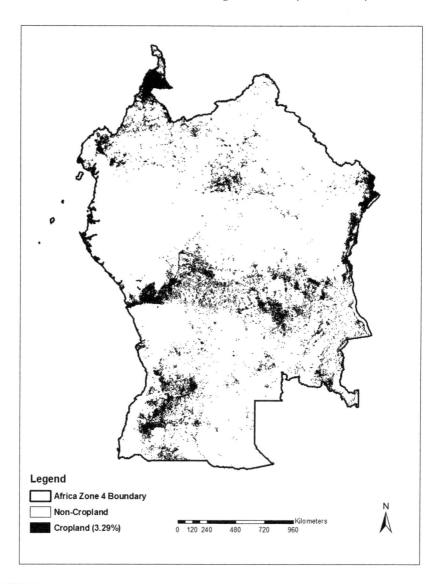

FIGURE 6.10   A map of crop extent (crop/no crop) for an Agricultural Ecological Zone (AEZ) in Africa.

same issue, as sufficient samples were allocated for both the crop and the non-crop categories as a result of the proportional sample.

CONCLUSIONS

This chapter is quite long and filled with lots of information and many considerations. Table 6.3 presents a summary of the pros and cons of the accuracy assessment sampling schemes discussed in this chapter. In addition, this chapter has attempted

TABLE 6.2

An Error Matrix for a Map of Crop Extent (Crop/No Crop) for an AEZ in Africa

		Reference Data			
		Crop	No-Crop	Total	User's Accuracy
Map Data	Crop	8	7	15	53.3%
	No-Crop	1	234	235	99.6%
Total		9	241	250	
Producer's Accuracy		88.9%	97.1%		**96.8%**

TABLE 6.3

A Summary of the Pros and Cons of Various Accuracy Assessment Sampling Schemes

Sampling Scheme	Pros	Cons
Random	Unbiased sample selection. Excellent statistical properties.	Expensive, especially for fieldwork. Does not ensure that enough samples will be taken in each class. Does not ensure good distribution of samples across the landscape.
Stratified random	Unbiased sample selection. Ensures adequate sample in each class, because a minimum number of samples are selected from each stratum (class).	Requires prior knowledge about the distribution of map classes so that strata can be developed. Expensive, especially for fieldwork. Often difficult to find enough samples in rare map classes. Does not ensure good distribution of samples across the landscape.
Systematic	Easy to implement. Less expensive than random sampling. Ensures good distribution of samples across the landscape.	Can be biased if sampling pattern is correlated with a landscape pattern (periodicity). Weaker statistically, as each sample unit does not have equal probability of selection.
Cluster	Least expensive as samples are close to one another, reducing travel time in the field and/or set-up time in the office.	Can be impacted by spatial autocorrelation. which results in the samples not being independent. If the samples are not independent from one another, then they are not distinct samples, and more independent samples must be taken.

to demonstrate the many questions that the analyst must ask when conducting a valid assessment. Again, it must be emphasized that there is no one correct way (i.e., recipe) to conduct an accuracy assessment. Many considerations discussed in this chapter are critically important so that the assessment can be as effective and efficient as possible.

Perhaps most importantly, the entire accuracy assessment process should be documented so that others can know exactly what procedures were followed. If the analyst makes certain decisions about the assessment based on specific considerations and does not document what they did and why, then it is impossible for anyone else to understand the accuracy assessment or to really know the accuracy of the map. Too often, published papers and project reports contain error matrices representing the map accuracy without sufficient explanation of the exact considerations made and processes followed. Sometimes, the error matrices themselves are not even published but rather, just some summary graphs or tables with only overall accuracy values. If there was a single recipe for conducting an assessment, this situation might be satisfactory. However, given the complexities described in this book, it is critical that complete and appropriate documentation be included along with the corresponding error matrices so that everyone can fully understand how the assessment was conducted. Failure to provide full documentation with complete error matrices results in a flawed assessment.

7 Reference Data Collection

Collection of the reference data for use in an accuracy assessment is a key component of any assessment. Failure to collect appropriate reference data dooms the assessment to produce erroneous results. The collection of accuracy assessment data requires completing the following three steps while considering both the reference data being collected and the map being assessed:

1. First, the accuracy assessment sample sites must be accurately co-located both on the reference source and on the map. This task can be a relatively simple one in an urban area or a far more difficult one in a wildland area where few recognizable landmarks exist. While the use of the global positioning system (GPS) has greatly increased our ability to efficiently and effectively locate accuracy sites, it is still possible to misidentify the location of a site. If the sample is in the wrong place on either the reference source or the map, a thematic error will result that is actually a positional error.

2. Next, the sample unit must be delineated. Sample units should represent the same area on both the reference data and the map. Usually, they are delineated once, either on the reference source data or on the map, and then transferred to the other. However, care should be taken so that the source of the reference data is accurately co-registered to the map being assessed. Significant misregistration can result in thematic errors that are actually caused by positional error.

3. Finally, the reference and map labels must be assigned to each sample unit based on the map classification scheme. In other words, the same classification scheme that was used to create the map from the remotely sensed data must be used to label the reference data sample units. The reference sample units may be collected from a variety of sources and may be labeled through either observation or measurement.

Serious oversights and problems can arise at each step of the reference data collection. To adequately assess the accuracy of the remotely sensed classification, each step must be implemented correctly on each and every sample. If the reference labels are inaccurate, then the entire assessment becomes meaningless. Four basic considerations drive all reference data collection:

1. What should the source be for the reference data samples? Can existing maps or existing field data be used as the reference data? Can the information be collected from higher–spatial resolution remotely sensed data, or is it necessary to make field visits to the sample units?

2. What type of information should be collected for each sample? Are measurements required to be taken, or are observations adequate to label the sample units?
3. When should the reference data be collected? Should they be collected during initial field investigations when the map is being made, or should they only be collected after the map is completed? What are the implications of using older data than the date of the remotely sensed imagery for accuracy assessment?
4. How can we ensure that the reference data are collected correctly, objectively, and consistently?

There are many methods for collecting reference data, some of which depend on making observations (qualitative assessments) and some of which require detailed, quantitative measurements. Given the varied reliability, difficulty, and expense of collecting reference data, it is critical to know which of these data collection techniques are valid and which are not for any given project. In most situations, the driving force behind the answer to this question is the complexity of the classification scheme used to make the map.

We all understand that maps are rarely 100% correct. Each remote sensing project requires trade-offs between the selection of the remotely sensed data used to create the map (especially the cost of the imagery) and the scale (i.e., spatial resolution) and level of accuracy required by the project. Today, much of the moderate spatial resolution imagery we use is freely available to all (e.g., Landsat and Sentinel imagery), as is dated high-resolution imagery in Google Earth or ArcGIS online. Even with the cost of up-to-date high–spatial resolution imagery, we accept some level of map error as a trade-off for the cost savings inherent in using remotely sensed data (instead of all fieldwork) to create the map. However, accuracy assessment reference labels must be correct if they are to constitute a fair assessment of the map. Thus, reference labels must be derived using source data or methods that are assumed to be more reliable than those used to make the map.

WHAT SHOULD BE THE SOURCE OF THE REFERENCE DATA?

The first decision in data collection requires determining what source will be used for the determination of reference labels. The type of source data required will depend on the complexity of the map classification scheme and the project budget. It is best to keep in mind this general rule: the simpler the classification scheme, the simpler and less expensive the reference data collection.

Sometimes, previously existing maps or ground data can be used as the reference data. More often, the reference source data are newly collected information that are at least one level more accurate than the remotely sensed data and methods used to make the map. Thus, high–spatial resolution aerial or satellite imagery is often used to assess the accuracy of maps made from moderate-resolution satellite imagery (e.g., SPOT, Landsat TM, or Sentinel 2), ground visits are often used to assess the accuracy of maps created from high-resolution imagery, and manual image interpretation has often been used to assess the accuracy of automated classification methods.

Using Existing versus Newly Collected Data

When a new map is produced, usually the first reaction is to compare the map with some existing source of information about the mapped area. Using previously collected ground information or existing maps for accuracy assessment is tempting because of the cost savings resulting from avoiding new reference data collection. While this can be a valuable qualitative tool, existing data are rarely acceptable for accuracy assessment, because:

1. Existing field inventory data usually were collected for a goal other than accuracy assessment. Often, the size of the inventory plot is too small (e.g., a 1 meter ecology site cannot be used to assess a map with a 4 meter minimum mapping unit), or the measurements made on the plots cannot be transformed into measurements that are useful for the accuracy assessment. It is especially tempting to use these ground data, since they are most likely of high quality, because they were collected from ground visits. It is imperative to be cautious in using existing field samples. Each sample must undergo quality control to ensure that the data adequately represent the classes being mapped and to evaluate whether the data collected cover a sufficiently large area:

   For example, in a project (conducted by one of the authors) to produce a fine scale vegetation map for Sonoma County, California, over 1245 classification plots were collected for building the classification scheme. A random number generator was used to select 241 plots to be held for accuracy assessment and not used in the mapping project. However after quality control was performed on the plots, only 166 were usable because some represented sub-mmu areas of vegetation, others had horizontal accuracy issues, and a few lacked complete data.

2. The classification schemes employed to create existing maps usually differ from the one being used to create the new map. Comparisons between the two maps can result in the error matrix expressing merely differences between the reference data and the map data classification schemes, rather than map error. Developing a crosswalk that specifically translates the map classification scheme to the new map classification scheme can sometimes solve this problem. However, this method rarely produces a perfect crosswalk, and therefore, some error is unavoidable.

3. Existing data are older than those being used to create the new map. Changes on the landscape (e.g., fire, urban development, wetland loss, etc.) will not be reflected in the existing data. Therefore, differences in the error matrix caused by these changes will incorrectly be assumed to be caused by map error.

4. Errors in existing maps are frequently unknown, because an accuracy assessment has often not been performed on older maps. Often, the differences caused by existing map errors are then blamed on the new map, thereby wrongly lowering the new map's accuracy.

5. Finally, as in all reference data collection, positional accuracy must be included in the considerations to ensure that the results are truly assessing the thematic accuracy of the map and not simply issues with position.

If existing information is the only available source of reference data, then consideration should be given to not performing a quantitative accuracy assessment. Instead, a qualitative comparison of the new map and existing map or field data should be performed, and the differences between the two should be identified and scrutinized. If a quantitative assessment is performed with existing data, it is vital to document the issues with the reference data so as to allow the potential user of the map to understand the limitations of such an assessment.

REMOTELY SENSED DATA VERSUS FIELD VISITS

If new data are to be collected to be used for reference samples, then a choice must be made between using field (i.e., ground) visits and airborne or satellite high-resolution imagery, video, unmanned aerial systems (UASs), or air reconnaissance as the source of the reference data. The accuracy assessment professional must assess the reliability of each data type for obtaining an accurate reference site label.

Simple classification schemes with a few (e.g., two to eight) general map classes can often be reliably assessed from air reconnaissance or interpretation of high-resolution imagery or video. As the level of detail in the map classification scheme increases, so does the complexity of the reference data collection. Eventually, even the very largest-scale imagery cannot provide valid reference data. Instead, the data must be collected on the ground.

In some situations, the use of image interpretation for generating reference data may not be appropriate. For example, aerial image interpretation is often used as reference data for assessing a land cover map generated from satellite imagery such as Landsat. The interpretation is assumed correct because the images have greater spatial resolution than the satellite imagery and because image interpretation has become a time-honored skill that is accepted as accurate. Unfortunately, errors do occur in image interpretation and air reconnaissance, depending on the skill of the interpreter and the level of detail required by the classification system. Inappropriately using interpretation as reference data could severely impact the conclusions about the accuracy of the satellite-based land cover map. In other words, one may conclude that the satellite-based map is of poor accuracy when actually, it is the interpretation that is in error.

In such situations, actual ground visitation may be the only reliable method of data collection. At the very least, a subset of data should be collected on the ground and compared with the high-resolution imagery to verify the reliability of the reference labels interpreted from the airborne imagery. Even if the majority of reference labels will come from image interpretation, it is critical that a subsample of these areas is visited in the field to verify the reliability of the interpretation. Much work is yet to be done to determine the proper level of effort and collection techniques necessary to provide this vital information. When the labels developed from image interpretation versus the ground begin to disagree regularly, it may be time to switch to ground-based reference data collection. However, the collection of ground reference data is extremely expensive, and therefore, the collection effort must be sufficient to meet the needs of the accuracy assessment while being efficient enough to meet the needs of the budget.

Field samples can also have their own problems because of the difference of perspective of someone on the ground versus that of the map that is created from imagery which is captured above the objects on the ground. People on the ground see vegetation obliquely. Most imagery used to make maps is collected from directly above the ground at a nadir view angle of 0%–3%. Because the ground view is oblique, it is common for field estimates of ground cover to be higher than estimates from the nadir view imagery. This occurs because the ground view often includes sub-canopy vegetation, which cannot be seen by the nadir imagery, and because the eyes are sometimes tricked into overestimating the vegetation stems and foliage at a higher density than is viewable from above. Alternatively, Spurr (1960) asserts that forest crown closure is overestimated from aerial imagery and underestimated from the ground. This difference in perspective can cause differences between ground reference labels and map labels that are not because of error but because of the differing view perspectives of the reference data collection personnel.

For example, a classic study by Biging et al. (1991) compared photo interpretation with ground measurements for characterizing forest structure. These characteristics included forest species, tree size class, and crown closure, which were photo-interpreted by a number of expert photo interpreters. The ground data used for comparison were a series of measurements made in a sufficient number of ground plots to characterize each forest polygon (i.e., stand). The results showed that the overall accuracy of photo interpretation of species ranged in accuracy between 75%–85%. The accuracy of size class was around 75%, and the accuracy of crown closure was less than 40%. This study reinforces the need to be careful in assuming that the results of the image interpretation are sufficient or appropriate for use as reference data in an accuracy assessment. In this study, species and size class may be interpreted for use as possible reference data. However, crown closure from above (as seen on the imagery) and crown closure as determined from the ground differ simply due to the vantage point of observation. Therefore, crown closure, in this case, may not be appropriate as reference data.

Alternatively, sometimes, objects that can be seen from remotely sensed imagery cannot be seen from the ground, with the result that the ground data can be less reliable than image interpretation. For example, in Hawaii Volcanoes National Park, there are areas where enormous, but sparse, *Metrosideros polymorpha* (Ohi'a) trees exist in the upper canopy, with the lower canopy composed of dense, impenetrable thickets of *Cibotium glaucum* (tree fern). The tree fern is so dense that the canopy of the Ohi'a trees cannot be seen from the ground. As a result, field crews would not note the existence of the Ohi'a trees, even though the crown of each individual tree is huge and clearly visible when viewed from above.

HOW SHOULD THE REFERENCE DATA BE COLLECTED?

The next decision in reference data collection involves deciding how information should be collected from the source data to obtain a reliable label for each reference sample. Reference data MUST BE labeled using the same classification scheme that was used to make the map. While using the same classification scheme for both the map and the reference data seems like a simple decision, it is surprising to see that such is not always the case. More and more recent projects have one entity producing the map and another one conducting the accuracy assessment. In this situation, it is more likely that the same classification scheme is not used by both entities, unless special care is taken to fully train the accuracy assessment crew in the use of the map. In many instances, simple observations/interpretations are sufficient for labeling a reference sample. In other cases, observation is not adequate, and actual measurements in the field are required.

The purpose of collecting reference data for a sample site is to derive the "correct" reference label for the sample for comparison with the map label. Often, the reference label can be obtained by merely observing the sample from an airplane, a car, UAS, or high-resolution imagery. For example, in most cases, a golf course can be accurately identified through observation from quite a distance away.

The decision on whether or not accuracy assessment reference data should be obtained from observations or measurements will be determined by the complexity of the landscape, the detail of the classification scheme, the required precision of the accuracy assessment, and the project budget. Reference data for simple classification schemes that distinguish homogeneous land cover types from one another (e.g., water versus agriculture) usually can be obtained from observations and/or estimations either on the ground or from larger-scale remotely sensed data. For example, distinguishing conifer forest from an agricultural field from a golf course can be determined from observation alone. Collecting reference data may be as simple as looking at high-resolution imagery or observing sites on the ground.

However, complex classification schemes may require some measurements to determine the precise (i.e., non-varying) reference sample labels. For example, a more complex forest classification scheme may involve collecting reference data for tree size class (related to the diameter of the tree trunk). Tree size class is important both as a determinant of endangered species habitat and as a measurement of wood product merchantability. Size class can be ocularly estimated on high-resolution imagery and on the ground. However, different individuals may produce different estimations based on their training and experience and therefore, introduce variability into the observation. This variation will exist not only between individuals but also within one individual. The same observer may see things differently depending on whether it is Monday or Friday; or whether it is sunny or raining; or especially, depending on how much coffee he or she has consumed. To avoid variability in human estimation, size class can be measured in the field, but a great many trees will need to be measured to precisely estimate the size class for each sample unit. In such instances, the accuracy assessment professional must decide whether to require measurements (which can be time consuming and expensive) or to accept the variation inherent in human estimation.

Whether or not measurements are required depends on the level of precision required by the map users and on the project budget. For example, information on spotted owl habitat requirements indicates that the owls prefer older, multi-storied stands that include large trees. "Large" in this context is relative, and precise measurements of trees will probably not be needed as long as the map accurately distinguishes between stands of single-storied small trees and multi-storied large trees. In contrast, many wood products mills can only accept trees within a specific size class.

Trees that are 1 inch smaller or larger than the prescribed range cannot be accepted by the machinery in the mill. In this case, measurements will probably be required.

Observer variability is especially evident in estimates of vegetation cover that cannot be precisely measured from high-resolution imagery. In addition, ground verification of aerial estimates of vegetative cover is problematic, because as discussed earlier, estimates of cover from the ground (i.e., below tree canopies) are fundamentally different from estimates made from above the canopy. Therefore, using ground estimates as reference data for above-cover estimates can be like comparing apples and oranges.

The trade-offs inherent between observation and measurement are exemplified in a pilot study conducted to determine the level of effort needed to collect appropriate ground reference data for use in forest inventory. The objective of this study was to determine whether visual calls made by trained experts in the field to observe forest polygons (i.e., stands) were sufficient to accurately label each polygon, or whether actual ground measurements needed to be made. There are obviously many factors influencing the accuracy of ground data collection, including the complexity of the vegetation itself. A variety of vegetation complexities were represented in this study. The results are enlightening to those remote sensing specialists who routinely collect forest ground data only by visual observation. The pilot study was part of a larger project aimed at developing the use of digital remotely sensed data for commercial forest inventory (Biging and Congalton, 1989).

Commercial forest inventory involves much more than creating a land cover map derived from digital remotely sensed data. Often, the map is used only to stratify the landscape, which is followed by a field inventory conducted on the ground to determine tree volume statistics for each type of stand of trees. A complete inventory requires that the forest type, the size class, and the crown closure of a forested area be known to determine the volume of the timber in that area. If a single species dominates, the forest type is commonly named by that species (Eyre, 1980). However, if a combination of species is present, then a mixed label is used (e.g., the mixed conifer type). The size of the tree is measured by the diameter of the tree at 4.5 feet above the ground (i.e., diameter at breast height [DBH]) and then is divided into size classes such as poles, small saw timber, and large saw timber. This measure is obviously important, because large-diameter trees contain a greater volume of high-quality wood (i.e., valuable timber) than small-diameter trees. Crown closure, as measured by the amount of ground area the tree crowns occupy (canopy closure), is also an important indicator of tree size and numbers. Therefore, in this pilot study, it was necessary to collect ground reference data not only on tree species/type, but on crown closure and size class as well.

Ground reference data were collected using two approaches. In the first approach, a field crew of four entered a forest stand (i.e., polygon), observed the vegetation, and came to a consensus for a visual call of dominant species/type, size class of the dominant species, crown closure of the dominant size class, and crown closure of all tree species combined. Dominance was defined as the species or type comprising the majority of forest volume. In the second approach, measurements were conducted on a fixed-radius plot to record the species, DBH, and height of each tree falling within the plot. A minimum of two plots (1/10th or 1/20th acre) were measured for each forested polygon. Because of the difficulty of making all the required measurements (precise location and crown width for each tree in the plot) to estimate crown closure on the plot, an approach using transects was developed to determine crown closure. A minimum of four 100 foot long transects randomly located within the polygon were used to collect crown closure information. The percentage of crown closure was determined by the presence or absence of tree crown along transects at 1 foot intervals. All the measurements were input into a computer program that summarized the results into the dominant species/type, the size class of the dominant species/type, the crown closure of the dominant size class, and the crown closure of all tree species for each forested area. The results of the two approaches were compared using an error matrix.

Table 7.1 shows the results of the field measurements versus visual calls as expressed in an error matrix for the dominant species. This table indicates that species can be fairly well determined from a visual call, because there is strong agreement between the field measurements and the visual call. Of course, this conclusion requires one to assume that the field measurements are a better measure of ground reference data; a reasonable assumption in this case. Therefore, ground reference data collection of species information can be effectively maximized using visual calls, and field measurements appear to be unnecessary.

Table 7.2 presents the results of comparing the two ground reference data collection approaches for the dominant size class. As in species, the overall agreement is relatively high, with most of the confusion occurring between the larger classes. The greatest inaccuracies result from visually classifying the dominant size class (i.e., the one with the most volume) as size class three (12–24 inch DBH), when in fact, size class four (>24 inch DBH) trees contained the most volume. This visual classification error is easy to understand. Tree volume is directly related to the square of DBH. There are numerous cases when a small number of large trees contribute the majority of the volume in the stand, while there may be many more medium-size trees present.

The dichotomy between the prevalence of medium-size trees and dominance in volume by a small number of large trees can be difficult to assess visually. It is likely that researchers and practitioners would confuse these classes in cases where the size class with the majority of volume was not readily evident. In cases like this, simply improving one's ability to visually estimate diameter would not improve one's ability to classify size class. The ability to weigh numbers and sizes to estimate volume requires considerable experience and would certainly require making plot and tree measurements to gain and retain this ability.

Tables 7.3 and 7.4 show the results of comparing the two collection approaches for crown closure. Table 7.3 presents the crown closure of the dominant size class results, while Table 7.4 shows the results of overall crown closure. In both matrices, there is very low agreement (46%–49%) between the observed estimate and the field measurements. Therefore, it appears that field measurements may be necessary to obtain precise measures of crown closure, and visual calls, although less expensive and quicker, may vary at an unacceptable level.

In conclusion, it must be emphasized that this is only a small pilot study. Further work needs to be conducted in this area to evaluate ground reference data collection methods and to include the validation of aerial methods (i.e., image interpretation and videography). The results demonstrate that making visual calls of species is relatively

TABLE 7.1

Error Matrix Showing the Field Measurements versus Visual Calls for Dominant Species

		TF	MC	LP	DF	PP	PD	OAK	Row Total	Species
		Field Measurement							**Row Total**	
	TF	14	0	0	0	0	0	0	14	
	MC	0	10	0	0	0	2	0	12	TF = true fir
										MC = mixed conifer
Visual Call	LP	0	0	1	0	0	0	0	1	LP = lodgepole pine
										DF = Douglas fir
	DF	0	1	0	8	0	0	0	9	PP = Ponderosa pine
										PD = PP and DF
	PP	1	1	0	0	0	0	0	2	OAK = oaks
	PD	0	0	0	1	0	0	0	1	
	OAK	0	0	0	0	0	0	0	0	
	Column Total	15	12	1	9	0	2	0	39	OVERALL ACCURACY = 33/39 = 85%

PRODUCER'S ACCURACY	USER'S ACCURACY
TF = 14/15 = 93%	TF = 14/14 = 100%
MC = 10/12 = 83%	MC = 10/12 = 83%
LP = 1/1 = 100%	LP = 1/1 = 100%
DF = 8/9 = 89%	DF = 8/9 = 89%
PP = 0/0 = —	PP = 0/2 = 0%
PD = 0/2 = 0%	PD = 0/1 = 0%
OAK = 0/0 = —	OAK = 0/0 = —

easy and accurate except where many species occur simultaneously. Size class is more difficult to assess than species because of the implicit need to estimate the size class with the majority of volume. Crown closure is by far the toughest to determine. It is most dependent on where one is standing when the call is made. Field measurements, such as the transects used in this study, provide an alternative means of determining crown closure. This study has shown that at least some ground data must be collected using measurements, and it has suggested that a multi-level effort may result in the most efficient and practical method for collection of ground reference data.

Table 7.5 presents the pros and cons of selecting the different sources of reference data.

WHEN SHOULD THE REFERENCE DATA BE COLLECTED?

The world's landscape is constantly changing. If change occurs between the date of capture of the remotely sensed data used to create a map and the date of the reference data collection, accuracy assessment reference sample labels may be affected. When

TABLE 7.2
Error Matrix Showing the Field Measurements versus Visual Calls for Dominant Size Class

Visual Call	Field Measurement 1	2	3	4	Row Total
1	1	0	0	0	1
2	1	3	1	0	5
3	0	0	17	5	22
4	0	0	1	11	12
Column Total	2	3	19	16	40

Size Classes

1 = 0–5″ dbh
2 = 5–12″ dbh
3 = 12–24″ dbh
4 = >24″ dbh

OVERALL ACCURACY
= 32/40 = 80%

PRODUCER'S ACCURACY

1 = 1/2 = 50%
2 = 3/3 = 100%
3 = 17/19 = 89%
4 = 11/16 = 69%

USER'S ACCURACY

1 = 1/1 = 100%
2 = 3/5 = 60%
3 = 17/22 = 77%
4 = 11/12 = 92%

TABLE 7.3
Error Matrix Showing the Field Measurements versus Visual Calls for Density (Crown Closure) of the Dominant Species

Visual Call	Field Measurement O	L	M	D	Row Total
O	10	8	3	0	21
L	2	8	1	1	12
M	0	3	1	1	5
D	0	1	0	0	1
Column Total	12	20	5	2	39

Density Classes

O = Open (0–10% crown closure)
L = Low (11–25% crown closure)
M = Medium (26–75% crown closure)
D = Dense (> 75% crown closure)

OVERALL ACCURACY
= 19/39 = 49%

PRODUCER'S ACCURACY

O = 10/12 = 83%
L = 8/20 = 40%
M = 1/5 = 20%
D = 0/2 = 0%

USER'S ACCURACY

O = 10/21 = 48%
L = 8/12 = 67%
M = 1/5 = 20%
D = 0/1 = 0%

TABLE 7.4

Error Matrix Showing the Field Measurements versus Visual Calls for Overall Density (Crown Closure)

		Field Measurement				Row Total
		O	L	M	D	
Visual Call	O	0	1	1	0	2
	L	1	3	7	0	11
	M	0	0	8	10	18
	D	0	0	0	6	6
Column Total		1	4	16	16	37

Density Classes

O = Open
L = Low
M = Medium
D = Dense

OVERALL ACCURACY
= 17/37 = 46%

PRODUCER'S ACCURACY	USER'S ACCURACY
O = 0/1 = 0%	O = 0/2 = 0%
L = 3/4 = 75%	L = 3/11 = 27%
M = 8/16 = 50%	M = 8/18 = 44%
D = 6/16 = 38%	D = 6/6 = 100%

TABLE 7.5

Comparison of Sources of Reference Data

Source of Reference Data	Pros	Cons
Existing maps/data	Least expensive and quickest.	Can be out of date if change has occurred on the landscape. Must ensure that the minimum mapping unit and classification scheme used to label the existing data are identical to the scheme.
New office-interpreted data from remote sensing	Less expensive and time consuming than field collected data. Provides the same perspective as the remotely sensed data used to make the map (i.e., view from above).	Less accurate for vegetation species identification than field collected data. Can be out of date if change has occurred on the landscape since the capture of the remotely sensed reference.
New field-collected data	More accurate for vegetation species identification.	Most expensive. Does not offer the same perspective as captured by the remotely sensed data (i.e., view from below vs. view from above). Often difficult to establish because of terrain or access.

a crop is harvested, a wetland drained, or a field developed into a shopping mall; the error matrix may show a difference between the map and the reference label that is not caused by map error but rather, by landscape change.

As previously noted, high-resolution aerial or satellite imagery is often used as reference source data for accuracy assessment of maps created from moderate resolution (e.g., Landsat, Sentinel, or SPOT) satellite data. While new high-resolution imagery can be very expensive to obtain, dated global high-resolution imagery is easily accessible on Google Earth and ArcGIS online as well as other websites. However, if an area has changed since the data of the imagery because of fire, disease, harvesting, or growth, the resulting reference labels in the changed areas will be incorrect. Harvests and fire are clearly visible on most moderate satellite imagery, making it possible to detect the changes by looking at the imagery used to make the map. However, stand growth and partial defoliation from disease or pests may not be as readily observable on the imagery used to make the map, making the use of older high-resolution imagery as the reference data source problematic.

In general, accuracy assessment reference data should be collected as close as possible to the date of the collection of the remotely sensed data used to make the map. However, trade-offs may need to be made between the timeliness of the data collection and the need to use the resulting map to stratify the accuracy assessment sample. In most, if not all, remote sensing mapping projects, it is necessary to go to the field to get familiar with what causes variation in the classes to be mapped, to calibrate the eye of the image analyst, and to collect information for training the classifier (i.e., supervised classification approaches) or to aid in labeling the clusters (i.e., unsupervised classification approaches). If the reference data for accuracy assessment can be collected independently, but simultaneously, during this trip, then a second trip to the field may be eliminated, saving costs and ensuring that reference data collection is occurring close to the time the remotely sensed data are captured.

However, if accuracy assessment reference data are collected at the beginning of the project before the map is generated, then it is not possible to stratify the samples by map class, since the map has yet to be created. It is also not possible to have a proportional-to-area allocation of the samples, since the total area of each map class is still unknown. Therefore, there is much to consider here, and choices need to be made for the assessment to be as effective and efficient as possible.

An example helps illustrate these points. In the late 1990s, The USDI Bureau of Reclamation mapped the crops of the Lower Colorado River Region four times a year using Landsat data. Farmland in this region is so productive and valuable that growers plant three to four crops per year and will plow under a crop to plant a new one in response to the futures market. With so much crop change, ground data collection and accuracy assessment must occur at very nearly the same time as the imagery is collected. Therefore, the Bureau would send a ground data collection crew to the field for 2 weeks surrounding the date of image acquisition. A random number generator was used to determine the fields to be visited, and the same fields were visited during each field effort regardless of the crops being grown. Therefore, the accuracy assessment sample was random but not stratified by crop type. As Table 7.6 illustrates, some crops were oversampled each time, and others were undersampled. In this particular assessment, the Bureau believed it was more important to ensure

TABLE 7.6
Error Matrix Showing Number of Samples in Each Crop Type

MAP DATA	REFERENCE DATA															
		A	C	SG	CN	L	M	BG	CS	T	SU	O	CR	F	D	S	Total
	A	157		8				3						3			171
	C		1			1	1										3
	SG	3		163		6						12	2	1			187
	CN																0
	L			4		3					1		1				9
	M						5							1			6
	BG	1						10									11
	CS								69								69
	T																0
	SU																0
	O			1		3						7					11
	CR												2				2
	F													224			224
	D														11		11
	S																0
	Total	161	1	176	0	13	6	13	69	0	1	19	5	229	11	0	704

LEGEND			Producer's Accuracy	User's Accuracy
A	= Alfalfa	A	98%	92%
C	= Cotton	C	100%	33%
SG	= Small Grains	SG	93%	87%
CN	= Corn	CN	—	—
L	= Lettuce	L	23%	33%
M	= Melons	M	83%	83%
BG	= Bermuda Grass	BG	77%	91%
CS	= Citrus	CS	100%	100%
T	= Tomatoes	T	—	—
SU	= Sudan Grass	SU	0%	—
O	= Other Veg.	O	37%	64%
CR	= Crucifers	CR	40%	100%
F	= Fallow	F	98%	100%
D	= Dates	D	100%	100%
S	= Safflowers	S	—	—

correct crop identification than it was to ensure that enough samples were collected in rarely occurring crop types. This decision is reasonable, and as long as it is documented as part of the process, then the map user will understand what they see in the resulting error matrix.

Table 7.7 compares and contrasts the trade-offs required when deciding when to collect reference data.

ENSURING OBJECTIVITY AND CONSISTENCY

For accuracy assessment to be useful, map users must have faith that the assessment is a realistic representation of the map's accuracy. They must believe that the assessment is objective and the results are repeatable. Maintaining the following three conditions will ensure objectivity and consistency:

1. Accuracy reference data must always be kept independent of any training data.
2. Data must be collected consistently from sample site to sample site.
3. Quality control procedures must be developed and implemented for all steps of data collection.

DATA INDEPENDENCE

It was not uncommon for early accuracy assessments to use the same information to assess the accuracy of a map as was used to create the map. This unacceptable procedure obviously violates all assumptions of independence and biases the assessment in favor of the map. The independence of the reference data can be ensured in

---

TABLE 7.7
The Pros and Cons of the Timing of Reference Data Collection

When Should the Reference Data Be Collected?	Pros	Cons
When the remotely sensed data are collected	Eliminates any chance of landscape change between the date of the mapping and the date of the reference data. Cost effective as information needed to make and assess the map is collected at the same time.	Because the map has not been made, there is no way to ensure that enough samples will be taken in each map class. Can also result in spatial dependence between samples if they are collected too close together.
After the map has been made	Because the map has been made, it is possible to ensure that enough samples for each map class are collected.	Can be more expensive. Introduces the possibility of landscape change occurring between the date of the map and the date of the reference data collection.

one of two ways. First, the reference and training data collection can be performed at completely different times and/or by different people. However, collecting information at different times is expensive and can introduce landscape change problems, as discussed earlier. Using different people can also be expensive, as more personnel need to be thoroughly trained in the details of the project, and individual biases must be controlled.

The second method for ensuring independence involves collecting reference and training data simultaneously and then using a random number generator to select and remove the accuracy assessment sites from the training data set. The accuracy assessment sites are not reviewed again (i.e., kept separate from the mapping) until it is time to perform the assessment. In addition, other factors such as sample size per map class as well as spatial autocorrelation must be considered when dividing the reference and training data. Regardless of which method is used to ensure independence, accuracy assessment reference data must be kept absolutely separate from any training/labeling data, and they must not be accessible during manual map editing.

DATA COLLECTION CONSISTENCY

Data collection consistency can be ensured through personnel training and the development of objective data collection procedures. Training should occur simultaneously for all personnel at the initiation of data collection. Intensive training for 1–3 days is often necessary and must include reference collection on numerous example sites that represent the broad array of variation both between and within map classes. Trainers must ensure that reference data collection personnel are (1) applying the classification scheme correctly and (2) accurately identifying characteristics of the landscape that are inherent in the classification scheme. For example, if a classification scheme depends on the identification of plant species, then all reference data personnel must be able to accurately identify species on the reference source data. The classification scheme used in accuracy assessment must also use the same minimum mapping unit as was applied to create the map.

In addition to personnel training, objective data collection procedures are critical for consistent data collection. The more measurement (as opposed to estimation) involved in reference data collection, the more consistent and objective the collection. However, measurement increases the cost of accuracy assessment, so most assessments rely heavily on ocular estimation. If ocular estimates are to be used, then the variance inherent in estimation must be accepted as an unavoidable part of the assessment, and some method of assessing the variance must be included in the assessment. Several of these methods are discussed in Chapter 10, Fuzzy Accuracy Assessment.

An important mechanism for imposing objectivity is the use of a reference data collection form to force all data collection personnel through the same collection process. The complexity of the reference data collection form will depend on the level of the complexity of the classification scheme. The form should lead the collector through a rule-based process to a definitive reference label from the classification scheme. Forms also provide a means of performing a quality assessment/quality control check on the collection process. This control check is performed by recording intermediate results that lead to the determination of a specific map class. For example, the form might have a place to record the canopy closure of an area. If the data collector records a canopy closure of 20% trees and a final map class of conifer forest, and yet the definition of forest in the classification scheme states that forests must have greater than or equal to 25% canopy closure, then a quality control analysis would reveal that the sample site cannot be conifer forest. Figure 7.1 is an example data collection form for a relatively simple classification scheme. An important portion of this form is the dichotomous key that leads data collection personnel to the land cover class label based solely on the classification scheme rules.

Site Name: ____________________ Site ID #: ________

Quad name: ____________________ Satellite Scene ID #: ________

Latitude: ________ Longitude: ________ Elevation range: ________

Method for determining position: ____________________

Comments on Position: ____________________

Comments on Weather: ____________________

Crew names: ____________________ Date: ________

CATEGORIES		
Background	**Forest**	**Palustrine forest**
Developed	**Scrub/Shrub**	**Estuarine emergent**
Crop/Grass	**Exposed Land**	**Palustrine emergent**
	Water	

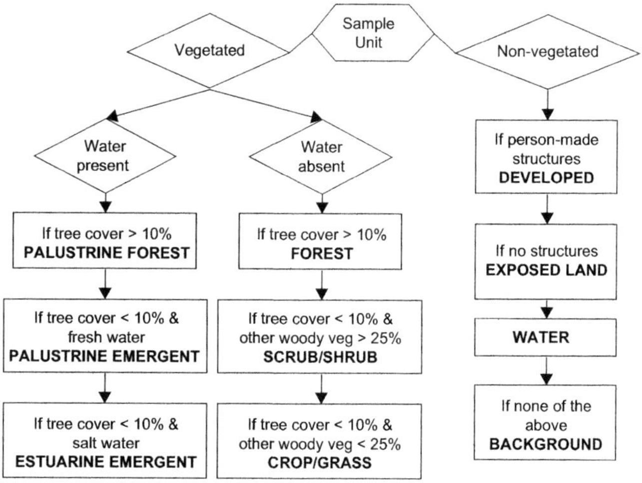

Actual Category as determined from flow chart: ____________________

Comments on anomalies, variability, or interesting findings: ____________________

FIGURE 7.1 Example of a reference data collection form for a simple classification scheme.

Reference data collection forms, regardless of their complexity, have some common components. These include (1) the name of the collector and the date of the collection; (2) locational information about the site; (3) some type of table or logical progression that represents what the collector is seeing; (4) a place to fill in the actual reference label from the classification scheme; and (5) a place to describe any anomalies, any variability, or interesting findings at the site.

These days, it is more common to have the form on a laptop computer, data logger, or tablet then as a piece of paper. Such automated forms provide more detail and information to the data collector. Images of sample areas that are representative of various map classes can be viewed at the collector's discretion. A data library can be created, allowing the collector to only input values within given ranges, or pull-down menus can list the possible choices. Regardless of how the form is represented, it is vital to make use of some form to ensure objectivity.

QUALITY CONTROL

Quality control is necessary at every step of data collection. Each error in data collection can translate into an incorrect indication of map accuracy. Data collection errors result in both over- and underestimations of map accuracy.

The following text discusses some of the most common quality control problems in each step of accuracy assessment data collection. Because accuracy assessment requires collecting information from both the reference source data and the map, each step involves two possible occasions of error: during collection from the map and during collection from the reference source data:

1. *Location of the accuracy assessment sample site.* It is not uncommon for accuracy assessment personnel to collect information at the wrong location, because inadequate procedures were used to locate the site on either the map or the reference data. Even in this age of GPS, locational errors persist. The authors have seen numerous recent accuracy assessment data sets where all the samples were collected on land using GPS equipment, but somehow, several samples landed on water in the database. As discussed in Chapter 4, any errors in the position of the location of the accuracy assessment sample site either on the reference data or on the map will result in a thematic error. Positional accuracy cannot be ignored when conducting a thematic accuracy assessment.

   A common method for locating accuracy assessment sites on reference high-resolution imagery is to view the site on the map and then "eyeball" the location onto the imagery based on similar patterns of land cover and terrain in both the map and the reference data. In this situation, it is critical to provide the reference personnel with as many tools and information as possible to help them locate the site. GPS equipment has become critical for ensuring location during fieldwork. Helpful information includes digitized flight line maps and other ancillary data such as stream, road, or ownership coverages. Phones, field computers, or tablets linked to a GPS and loaded with GIS software, the imagery, and ancillary data can reduce field time and increase reference location accuracy immeasurably.

Field location is always problematic, but especially in wildlands (e.g. tundra, open water, wilderness areas, etc.) with few recognizable landscape characteristics. GPS is extremely helpful and should always be used to ensure the correct location of field sample sites. However, it should be understood that GPS locations are not always correct and that there are issues that directly impact the GPS signal and therefore, the location. These issues include dense canopies, satellite position, and multi-path. Field personnel must be familiar with these issues when collecting reference data in the field and make every effort to compensate for them.

2. *Sample unit delineation.* Both the reference site and the map accuracy assessment site must represent exactly the same location. Thus, not only must the sites be properly located, but they must also be delineated precisely and correctly transferred to a planimetric base. For example, if an existing map is used as the reference source, and the map was not registered correctly, then all accuracy assessment reference sites will not register to the new map being assessed, and a misalignment will occur when the reference site and the map site are compared. This is not an uncommon situation when aerial photography was used to create an existing historical map, and the transfer from the photo to the map was performed ocularly without the use of photogrammetric equipment. Today, most imagery is adequately registered, which should minimize this issue. However, for historical analysis (e.g., change detection), the use of analog aerial photographs as a source of reference data is still prevalent.

Also, given the need and ability to automate this entire process today, it is possible to use map coordinates to represent both the reference data samples and map locations and to automate the process of creating the error matrix, such that the analyst is completely removed from the process. This happens completely inside the computer, and only the final error matrix is revealed. While it is very appealing for this process to be automated, it is also very dangerous if left completely unchecked, because no visual comparison of the sample units is performed, and none of the potential problems are discovered. Historically, many of these issues would have been discovered during the manual comparison of reference sample sites with map sites during the error matrix generation. Therefore, even if using a fully automated method, it is recommended that some sample of these sample units be visually examined to check for errors.

3. *Data collection and data entry* are the most common sources of quality control problems in accuracy assessment. Data collection errors occur when measurements are done incorrectly, variables of the classification scheme are misidentified (e.g., species), or the classification scheme is misapplied. In addition, weak classification schemes (i.e., those that have been poorly defined) will also create ambiguity in data collection. Unfortunately, the first indication of a weak classification system often occurs during accuracy

assessment, when the map is already completed, and refinement of the classification scheme is not possible unless the entire project is to be redone.

Data collection errors are usually monitored and detected by selecting a subsample of the accuracy assessment sites and collecting reference data at these sites simultaneously by two different personnel. Often, the most experienced personnel are assigned to the subsample. When differences are detected, the source of the differences needs to be immediately identified so that they can be corrected. If data collection errors are the source of the differences, then personnel are either retrained or removed from reference data collection. Even the most experienced personnel can make data entry errors.

When higher–spatial resolution imagery is used as the reference source data, it is extremely useful and informative if a ground assessment of a subsample of the imagery interpretation is conducted. In other words, a test to see how effective the imagery is as a reference data source is important. Some subsample of the study area can be ground visited and the samples compared with the image interpretation of the same samples to create an error matrix. High accuracy between the ground and image samples confirms the use of the high–spatial resolution imagery as the source of the reference data. Any errors (i.e., off-diagonal values) in the matrix reveal issues where confusion between map classes occurs and could possibly be corrected to further improve the image interpretation. Similarly, reference samples chosen from an existing map must also be assessed for accuracy in the same way as just described by comparing a subset of these samples with some ground/field collection.

Quality control of data entry errors can be easily reduced by using digital data entry forms that restrict each field of the form to an allowable set of characters. Data can also be entered twice and the two data sets compared to identify differences and errors. Data entry errors also can occur when the site is digitized. Quality control must include a same-scale comparison of the digitized site with the source map.

Finally, although no reference data set may be completely accurate, it is important that the reference data have high accuracy; otherwise, the assessment is not a fair characterization of map accuracy. Therefore, it is critical that reference data collection is carefully considered in any accuracy assessment. Determining the proper level of effort and the appropriate collection techniques necessary to provide this vital information is an ongoing effort. This effort will continue as long as remotely sensed data and geospatial analysis continue to improve in quality and quantity.

8 Basic Analysis Techniques

As presented in the previous two chapters, the goal of thematic map accuracy assessment is to generate a statistically valid, but practically achievable, error matrix that is representative of the accuracy of the map of interest. Generating such a matrix involves many considerations, including the various sources of error, the classification scheme to be used, and the many statistical factors required to collect the appropriate reference data. Once the matrix has been successfully created, descriptive statistics such as overall, producer's, and user's accuracies can be computed directly from the matrix. A large number of additional measures have also been suggested by the remote sensing/geospatial community (e.g., Foody, 1992; Gopal and Woodcock, 1994; Liu et al., 2007; Foody, 2009; Pontius and Millones, 2011). Liu et al. (2007) performed a comparative analysis of 34 different measures that can be computed from the error matrix. Each of these measures provides information about some aspect of the map accuracy. If the error matrix is readily available to the map user, then any of these measures can be computed at any time. However, if only some summary measures are available instead of the complete matrix, then the map accuracy can only be evaluated based on these measures. The value of having the full error matrix cannot be overstated. Only with the full matrix will the map user be able to compute whatever additional descriptive and analytical statistics they choose.

This chapter presents two widely accepted, basic analysis techniques (Kappa and Margfit) that can be performed once an error matrix has been properly generated. While there has been some discussion in the literature about the shortcomings of these two techniques, both have been widely applied to many accuracy assessments, and both are standard components of many remote sensing software packages. Therefore, these techniques are presented here, including a discussion of the issues with using them. Like the many other less-used measures proposed by the remote sensing community, these techniques can be applied only if the full error matrix is available for analysis.

KAPPA

The Kappa analysis is a discrete multivariate technique used in accuracy assessment for statistically determining whether one error matrix is significantly different from another (Bishop et al., 1975). It is extremely powerful to be able to test whether one error matrix is significantly different from another, and Kappa is an effective way to perform this test. For example, if one wanted to show that a new classification algorithm was better than a commonly used algorithm, a map of the same area could be generated using each classification algorithm and then an error matrix generated for each map. The Kappa analysis could then be used to statistically determine whether the new algorithm is indeed the better approach. Similar analyses could be performed between interpreters, between different sources of imagery, between multi-temporal and single-date analysis, and almost any other comparison imaginable. In each case, a Kappa analysis could be used to determine which error matrix (and therefore, map) is significantly better than the other. This is the power of computing a Kappa statistic.

The result of performing a Kappa analysis is a KHAT statistic (actually $\hat{K}$, an estimate of Kappa), which can also be used as another measure of agreement or accuracy (Cohen, 1960). This measure of agreement is based on the difference between the actual agreement in the error matrix (i.e., the agreement between the remotely sensed classification and the reference data as indicated by the major diagonal) and the chance agreement, which is indicated by the row and column totals (i.e., marginals). In this way the KHAT statistic is similar to the more familiar χ^2 analysis.

Although this analysis technique has been in the sociology and psychology literature for many years, the method was not introduced to the remote sensing community until 1981 (Congalton, 1981) and not published in a remote sensing journal before Congalton et al. (1983). Since then, many papers have been published recommending this technique. Consequently, the Kappa analysis has become a standard component of almost every accuracy assessment (Congalton et al., 1983; Rosenfield and Fitzpatrick-Lins, 1986; Hudson and Ramm, 1987; Congalton, 1991) and is a standard component of most image analysis software packages that include accuracy assessment procedures. In fact, hundreds if not thousands of papers have been published reporting the Kappa statistic as part of the thematic map accuracy assessment (Jensen, 2016). However, it must be remembered that the power of Kappa is in its ability to test whether one error matrix is statistically significantly different from another and not in simply reporting this value as another measure of accuracy.

The following equations are used for computing the KHAT statistic and its variance.

Let:

$$p_o = \sum_{i=1}^{k} p_{ii}$$

be the actual agreement, and

$$p_c = \sum_{i=1}^{k} p_{i+}p_{+j}$$

where p_{i+} and p_{+j} are as previously defined at the end of Chapter 5, the "chance agreement."

Assuming a *multinomial sampling model*, the maximum likelihood estimate of Kappa is given by:

$$\hat{K} = \frac{p_o - p_c}{1 - p_c}.$$

For computational purposes,

$$\hat{K} = \frac{n\sum\limits_{i=1}^{k} n_{ii} - \sum\limits_{i=1}^{k} n_{i+}n_{+i}}{n^2 - \sum\limits_{i=1}^{k} n_{i+}n_{+i}}.$$

where n_{ii}, n_{i+}, and n_{+i} are as previously defined at the end of Chapter 5.

The approximate large sample variance of Kappa is computed using the Delta method as follows:

$$\text{vâr}(\hat{K}) = \frac{1}{n}\left\{\frac{\theta_1(1-\theta_1)}{(1-\theta_2)^2} + \frac{2(1-\theta_1)(2\theta_1\theta_2-\theta_3)}{(1-\theta_2)^3} + \frac{(1-\theta_1)^2(\theta_4-4\theta_2^2)}{(1-\theta_2)^4}\right\}$$

where:

$$\theta_1 = \frac{1}{n}\sum_{i=1}^{k} n_{ii},$$

$$\theta_2 = \frac{1}{n^2}\sum_{i=1}^{k} n_{i+}n_{+i},$$

$$\theta_3 = \frac{1}{n^2}\sum_{i=1}^{k} n_{ii}\left(n_{i+}+n_{+i}\right),$$

$$\theta_4 = \frac{1}{n^3}\sum_{i=1}^{k}\sum_{j=1}^{k} n_{ij}\left(n_{j+}+n_{+i}\right)^2.$$

A KHAT value is computed for each error matrix and is another measure of how well the remotely sensed classification agrees with the reference data. Confidence intervals around the KHAT value can be computed using the approximate large sample variance and the fact that the KHAT statistic is asymptotically normally distributed. This fact also provides a means for testing the significance of the KHAT statistic for a single error matrix to determine whether the agreement between the remotely sensed classification and the reference data is significantly greater than 0 (i.e., better than a random classification).

It is always satisfying to perform this test on a single matrix and confirm that your classification is meaningful and significantly better than a random classification. If it is not, you know that something has gone terribly wrong during the classification process.

Finally and most importantly, there is a test to determine whether two independent KHAT values and therefore, two error matrices are significantly different. With this test, it is possible to statistically compare two analysts, the same analyst over time, two algorithms, two types of imagery, or even two dates of imagery and see which produces the higher accuracy. Both the single error matrix and paired error matrix tests of significance rely on the standard normal deviate as follows:

Let $\hat{K}_1$ and $\hat{K}_2$ denote the estimates of the Kappa statistic for error matrix #1 and #2, respectively. Let also $v\hat{a}r(\hat{K}_1)$ and $v\hat{a}r(\hat{K}_2)$ be the corresponding estimates of the variance as computed from the appropriate equations. The test statistic for testing the significance of a single error matrix is expressed by:

$$Z = \frac{\hat{K}_1}{\sqrt{v\hat{a}r(\hat{K}_1)}}.$$

Z is standardized and normally distributed (i.e., standard normal deviate). Given the null hypothesis $H_0 : K_1 = 0$ and the alternative $H_1 : K_1 \neq 0$, H_0 is rejected if $Z \geq Z_{\alpha/2}$, where $\alpha/2$ is the confidence level of the two-tailed Z test, and the degrees of freedom are assumed to be ∞ (infinity).

The test statistic for testing whether two independent error matrices are significantly different is expressed by:

$$Z = \frac{|\hat{K}_1 - \hat{K}_2|}{\sqrt{v\hat{a}r(\hat{K}_1) + v\hat{a}r(\hat{K}_2)}}.$$

Z is standardized and normally distributed. Given the null hypothesis $H_0 : (K_1 - K_2) = 0$ and the alternative $H_1 : (K_1 - K_2) \neq 0$, H_0 is rejected if $Z \geq Z_{\alpha/2}$.

It is prudent at this point to provide an actual example so that the equations and theory can come alive to the reader. The error matrix presented as an example in Table 8.1 was generated from Landsat Thematic Mapper (TM) data using an unsupervised classification approach by Analyst #1. A second error matrix was generated using the exact same imagery and classification approach; however, the unsupervised clusters were labeled by Analyst #2 (Table 8.2). It is important to note that Analyst #2 was not as ambitious as Analyst #1 and did not collect as much reference data for comparison with the map for conducting the accuracy assessment (i.e., generating an error matrix).

Table 8.3 presents the results of the Kappa analysis on the individual error matrices. As previously presented, the KHAT values can be used in addition to overall accuracy as another measure of agreement or accuracy. This measure is different from overall accuracy, because it is indicative of different information contained within the error matrix. Overall accuracy is simply the sum of the major diagonal of the error matrix divided by the total number of samples, while the KHAT value indirectly incorporates the off-diagonal elements through using the row and column totals (i.e., marginal). The values can range from +1 to −1. However, since there should be a positive correlation between the remotely sensed classification and the reference data, positive KHAT values are expected. Landis and Koch (1977) characterized the possible ranges for KHAT into three groupings: a value greater

TABLE 8.1
Error Matrix Produced Using Landsat Thematic Mapper Imagery and an Unsupervised Classification Approach by Analyst #1

		Reference Data				Row Total	
		D	C	AG	SB		Land Cover Categories
	D	65	4	22	24	115	D = deciduous
	C	6	81	5	8	100	C = conifer
Classified Data	AG	0	11	85	19	115	AG = agriculture SB = shrub
	SB	4	7	3	90	104	
	Column Total	75	103	115	141	434	

OVERALL ACCURACY =
(65+81+85+90)/434 =
321/434 = 74%

PRODUCER'S ACCURACY	USER'S ACCURACY
D = 65/75 = 87%	D = 65/115 = 57%
C = 81/103 = 79%	C = 81/100 = 81%
AG = 85/115 = 74%	AG = 85/115 = 74%
SB = 90/141 = 64%	SB = 90/104 = 87%

TABLE 8.2
An Error Matrix Using the Same Imagery and Classification Algorithm as in Table 8.1 Except That the Work Was Performed by a Different Analyst

		Reference Data				Row Total	
		D	C	AG	SB		Land Cover Categories
	D	45	4	12	24	85	D = deciduous
	C	6	91	5	8	110	C = conifer
Classified Data	AG	0	8	55	9	72	AG = agriculture SB = shrub
	SB	4	7	3	55	69	
	Column Total	55	110	75	96	336	

OVERALL ACCURACY =
(45+91+55+55)/336 =
246/336 = 73%

PRODUCER'S ACCURACY	USER'S ACCURACY
D = 45/55 = 82%	D = 45/85 = 53%
C = 91/110 = 83%	C = 91/110 = 83%
AG = 55/75 = 73%	AG = 55/72 = 76%
SB = 55/96 = 57%	SB = 55/69 = 80%

TABLE 8.3
Individual Error Matrix Kappa Analysis Results

Error Matrix	KHAT	Variance	Z Statistic
Analyst #1	0.65	0.0007778	23.4
Analyst #2	0.64	0.0010233	20.0

than 0.80 (i.e., 80%) represents strong agreement; a value between 0.40 and 0.80 (i.e., 40%–80%) represents moderate agreement; and a value below 0.40 (i.e., 40%) represents poor agreement.

Table 8.3 also presents the variance of the KHAT statistic and the Z statistic used for determining whether the classification is significantly better than a random result. At the 95% confidence level, the critical value would be 1.96. Therefore, if the absolute value of the test Z statistic is greater than 1.96, the result is significant, and you would conclude that the classification is better than random. The Z statistic values for the two error matrices in Table 8.3 are both 20 or more, and so, both classifications are significantly better than random. In other words, the analysts were using their expertise effectively when making their maps.

Table 8.4 presents the results of the Kappa analysis that compares the error matrices, two at a time, to determine whether they are significantly different. This test is based on the standard normal deviate and the fact that although remotely sensed data are discrete, the KHAT statistic is asymptotically normally distributed. The results of this pairwise test for significance between two error matrices reveal that these two matrices are not significantly different. This is not surprising, since the overall accuracies were 74% and 73% and the KHAT values were 0.65 and 0.64, respectively. Therefore, it could be concluded that these two analysts may work together, because they produce approximately equal classifications. If two different techniques or algorithms were being tested, and if they were shown to be not significantly different, then it would be best to use the cheaper, quicker, or more efficient approach.

TABLE 8.4
Kappa Analysis Results for the Pairwise Comparison of the Error Matrices

Pairwise Comparison	Z Statistic
Analyst #1 vs. Analyst #2	0.3087

MARGFIT

In addition to the Kappa analysis, a second technique called Margfit can be applied to "normalize" or standardize the error matrices for comparison purposes. Margfit uses an iterative proportional fitting procedure, which forces each row and column (i.e., marginal) in the matrix to sum to a predetermined value; hence the name Margfit—marginal fitting. If the predetermined value is 1, then each cell value is a proportion of 1 and can easily be multiplied by 100 to represent percentages or accuracies. The predetermined value could also be set to 100 to obtain percentages directly or to any other value the analyst chooses.

In this normalization process, differences in sample sizes used to generate the matrices are eliminated, and therefore, individual cell values within the matrix are directly comparable. In addition, because as part of the iterative process, the rows and columns are totaled (i.e., marginals), the resulting normalized matrix is more indicative of the off-diagonal cell values (i.e., the errors of omission and commission). In other words, all the values in the matrix are iteratively balanced by row and column, thereby incorporating information from that row and column into each individual cell value. This process then changes the cell values along the major diagonal of the matrix (correct classifications), and therefore, a normalized overall accuracy can be computed for each matrix by summing the major diagonal and dividing by the total of the entire matrix.

TABLE 8.5
Normalized Error Matrix from Analyst #1

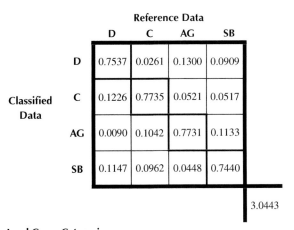

		Reference Data			
		D	C	AG	SB
	D	0.7537	0.0261	0.1300	0.0909
Classified Data	C	0.1226	0.7735	0.0521	0.0517
	AG	0.0090	0.1042	0.7731	0.1133
	SB	0.1147	0.0962	0.0448	0.7440

3.0443

Land Cover Categories

D = deciduous
C = conifer
AG = agriculture
SB = shrub

NORMALIZED ACCURACY =
0.7537+0.7735+0.7731+0.7440 =
3.0443/4.0 = 76%

Consequently, one could argue that the normalized accuracy is a better representation of accuracy than is the overall accuracy computed from the original matrix, because it contains information about the off-diagonal cell values. Table 8.5 presents the normalized matrix generated from the original error matrix presented in Table 8.1 (an unsupervised classification of Landsat TM data by Analyst #1) using the Margfit procedure. Table 8.6 presents the normalized matrix generated from the original error matrix presented in Table 8.3, which used the same imagery and classifier, but was performed by Analyst #2.

In addition to computing a normalized accuracy, the normalized matrix can also be used to directly compare cell values between matrices. For example, we may be interested in comparing the accuracy each analyst obtained for the conifer category. From the original matrices, we can see that Analyst #1 classified 81 sample units correctly, while Analyst #2 classified 91 correctly.

Neither of these numbers means much, because they are not directly comparable due to the differences in the number of samples used to generate the error matrix by each analyst. Instead, these numbers would need to be converted into percentages or user's and producer's accuracies so that a comparison could be made.

Here, another problem arises; do we divide the total correct by the row total (user's accuracy) or by the column total (producer's accuracy)? We could calculate both and compare the results, or we could use the cell value in the normalized matrix. Because of the iterative proportional fitting routine, each cell value in the matrix

---

TABLE 8.6
Normalized Error Matrix from Analyst #2

| | | Reference Data | | | |
		D	C	AG	SB
	D	0.7181	0.0312	0.1025	0.1488
Classified	C	0.1230	0.7607	0.0541	0.0619
Data	AG	0.0136	0.1017	0.7848	0.0995
	SB	0.1453	0.1064	0.0587	0.6898

2.9534

Land Cover Categories

D = deciduous
C = conifer
AG = agriculture
SB = shrub

NORMALIZED ACCURACY =
0.7181+0.7607+0.7848+0.6898 =
2.9534/4.0 = 74%

TABLE 8.7
Comparison of the Accuracy Values for an Individual Category

Error Matrix	Original Cell Value	Producer's Accuracy	User's Accuracy	Normalized Value
Analyst #1	81	79%	81%	77%
Analyst #2	91	83%	83%	76%

has been balanced by the other values in its corresponding row and column. This balancing has the effect of incorporating producer's and user's accuracies together. Also, since each row and column add to 1, an individual cell value can quickly be converted to a percentage by multiplying by 100. Therefore, the normalization process provides a convenient way of comparing individual cell values between error matrices regardless of the number of samples used to derive the matrix (Table 8.7).

Table 8.8 provides a comparison of the overall accuracy, the normalized accuracy, and the KHAT statistic for the two analysts. In this particular example, all three measures of accuracy agree about the relative ranking of the results. However, it is possible for these rankings to disagree, simply because each measure incorporates various levels of information from the error matrix into its computations. Overall accuracy only incorporates the major diagonal and excludes the omission and commission errors. As already described, normalized accuracy directly includes the off-diagonal elements (omission and commission errors) because of the iterative proportional fitting procedure. As shown in the KHAT equation, KHAT accuracy indirectly incorporates the off-diagonal elements as a product of the row and column marginals. Therefore, depending on the amount of error included in the matrix, these three measures may not agree.

It is not possible to give clear-cut rules as to when each measure should be used. Each accuracy measure incorporates different information about the error matrix, and therefore, they must be examined as different computations attempting to explain the error. Our experience has shown that if the error matrix tends to have a great many off-diagonal cell values with zeros in them, then the normalized results tend to disagree with the overall and Kappa results.

Many zeros occur in a matrix when an insufficient sample has been taken or when the classification is exceptionally good. Because of the iterative proportional fitting routine, these zeros tend to take on positive values in the normalization process, showing that some error could be expected. The normalization process then tends to reduce the accuracy because of these positive values in the off-diagonal

TABLE 8.8
Summary of the Three Accuracy Measures for Analysts #1 and #2

Error Matrix	Overall Accuracy	KHAT	Normalized Accuracy
Analyst #1	74%	65%	76%
Analyst #2	73%	64%	74%

cells. If a large number of off-diagonal cells do not contain zeros, then the results of the three measures tend to agree. There are also times when the Kappa measure will disagree with the other two measures. Because of the ease of computing all three measures, and because each measure reflects different information contained within the error matrix, we recommend an analysis such as the one performed here to glean as much information from the error matrix as possible. Of course, other measures can also be computed from the matrix that may further inform the analyst or the map user about the map accuracy. As long as the error matrix is provided, it is fully possible to compute whatever measures the analyst or map user finds appropriate.

CONDITIONAL KAPPA

In addition to computing the Kappa coefficient for an entire error matrix, it may be useful to look at the agreement for an individual category within the matrix. Individual category agreement can be tested using the conditional Kappa coefficient. The maximum likelihood estimate of the Kappa coefficient for conditional agreement for the ith category is given by:

$$\hat{K}_i = \frac{nn_{ii} - n_{i+}n_{+i}}{nn_{i+} - n_{i+}n_{+i}},$$

where n_{i+} and n_{+i} are as previously defined at the end of Chapter 5 and the approximate large sample variance for the ith category is estimated by:

$$\text{vâr}(\hat{K}_i) = \frac{n(n_{i+} - n_{ii})}{\left[n_{i+}(n - n_{+i})\right]^3}\left[(n_{i+} - n_{ii})(n_{i+}n_{+i} - nn_{ii}) + nn_{ii}(n - n_{i+} - n_{+i} + n_{ii})\right].$$

The same comparison tests available for the Kappa coefficient apply to this conditional Kappa for an individual category.

WEIGHTED KAPPA

The Kappa analysis is appropriate when all the error in the matrix can be considered to be of equal importance. However, it is easy to imagine a classification scheme where errors may vary in their importance. In fact, this latter situation is really the more realistic approach. For example, it may be far worse to classify a forested area as water than to classify it as shrub. In this case, the ability to weight the Kappa analysis would be very powerful (Cohen, 1968). The following section describes the procedure to conduct a weighted Kappa analysis.

Let w_{ij} be the weight assigned to the i, jth cell in the matrix. This means that the proportion p_{ij} in the i, jth cell is to be weighted by w_{ij}. The weights should be restricted to the interval $0 \leq w_{ij} \leq 1$ for $i \neq j$, and the weights representing the maximum agreement are equal to 1; that is, $w_{ii} = 1$ (Fleiss et al., 1969).

Therefore, let:

$$p_o^* = \sum_{i=1}^{k} \sum_{j=1}^{k} w_{ij} p_{ij}$$

be the weighted agreement, and

$$p_c^* = \sum_{i=1}^{k} \sum_{j=1}^{k} w_{ij} p_{i+} p_{+j}$$

(where p_{ij}, p_{i+}, and p_{+j} are as previously defined at the end of Chapter 5) the weighted "chance agreement."

Then, the weighted Kappa is defined by

$$\hat{K}_w = \frac{p_o^* - p_c^*}{1 - p_c^*}.$$

To compute the large sample variance of the weighted Kappa, define the weighted average of the weights in the ith category of the remotely sensed classification by:

$$\overline{w}_{i+} = \sum_{j=1}^{k} w_{ij} p_{+j},$$

where p_{+j} is as previously defined at the end of Chapter 5 and the weighted average of the weights in the jth category of the reference data set by:

$$\overline{w}_{+j} = \sum_{i=1}^{k} w_{ij} p_{i+}.$$

where p_{i+} is as previously defined at the end of Chapter 5.

The variance may be estimated by:

$$\hat{var}(\hat{K}_w) = \frac{1}{n\left(1 - p_c^*\right)^4} \left\{ \sum_{i=1}^{k} \sum_{j=1}^{k} p_{ij} \left[w_{ij}\left(1 - p_c^*\right) - \left(\overline{w}_{i+} + \overline{w}_{+j}\right)\left(1 - p_o^*\right) \right]^2 \right.$$

$$\left. - \left(p_o^* p_c^* - 2p_c^* + p_o^*\right)^2 \right\}$$

The same tests of significant difference described previously for the Kappa analysis apply to the weighted Kappa. An individual weighted Kappa value can be evaluated to see whether the classification is significantly better than random. Two independent weighted Kappas can also be tested to see whether they are significantly different.

Although the weighted Kappa has been in the literature since the 1960s and was even suggested to the remote sensing community by Rosenfield and Fitzpatrick-Lins

(1986), it has not received widespread attention. The reason for this lack of use is undoubtedly the need to select appropriate weights. Manipulating the weighting scheme can significantly change the results. Therefore, comparisons between different projects using different weighting schemes would be very difficult. The subjectivity of choosing the weights is always hard to justify. Using the unweighted Kappa analysis avoids these problems.

COMPENSATION FOR CHANCE AGREEMENT

Some researchers and scientists have objected to the use of the Kappa coefficient as a measure for assessing the accuracy of remotely sensed classifications, because the degree of chance agreement may be overestimated (e.g., Foody, 1992; Pontius and Millones, 2011). Remember from the equation for computing the kappa coefficient

$$\hat{K} = \frac{p_o - p_c}{1 - p_c}$$

that p_o is the observed proportion of agreement (i.e., the actual agreement), and p_c is the proportion of agreement that is expected to occur by chance (i.e., the chance agreement). However, in addition to the chance agreement, p_c also includes some actual agreement (Brennan and Prediger, 1981) or agreement for cause (Aickin, 1990). Therefore, since the chance agreement term does not consist solely of chance agreement, the Kappa coefficient may underestimate the classification agreement.

This problem is known to occur when the marginals are free (not fixed *a priori*), which is most often the case with remotely sensed classifications. Foody (1992) presents a number of possible solutions to this problem, including two Kappa-like coefficients that compensate for chance agreement in different ways. Others have suggested additional measures, as previously discussed in this chapter. However, the power of the Kappa analysis is not in using it as a measure of agreement or accuracy but rather, in statistically testing for significant differences between error matrices. Therefore, despite this potential limitation as a measure of accuracy, it must still be considered a vital accuracy assessment measure for performing this statistical comparison. Jensen (2016) calls this *the Kappa debate* and suggests that we must wait to see what will happen with the use of Kappa. However, despite the claim that Kappa is dead in 2011 (Pontius and Millones, 2011), the use and value of Kappa seem very much alive in the remotely sensed literature today.

CONFIDENCE LIMITS

Confidence intervals are extremely common and are an expected component of any statistical estimate. However, computing confidence intervals for values in an error matrix is significantly more complex than simply computing a confidence interval for a traditional statistical analysis. The following example illustrates the calculations derived from the error matrix (Card, 1982). This example is designed assuming simple random sampling. If another sampling scheme is used, the variance equations change slightly.

The same error matrix as in Table 8.1 will be used to compute the confidence intervals. However, the map marginal proportions, π_j, computed as the proportion of the map falling into each map category, are also required (Table 8.9). The map marginal proportions are not derived from the error matrix but are simply the proportion of the total map area falling into each category. These proportions can quickly be obtained by dividing the area of each category by the total map area.

Given this matrix, the first step is to compute the individual cell probabilities using the following equation:

$$\hat{p}_{ij} = \pi_j n_{ij}/n_{.j}$$

The individual cell probabilities are simply the map marginal proportion multiplied by the individual cell value, all divided by the row marginal. The results of these computations are shown in Table 8.10.

The true marginal proportions, $\hat{p}_i$, can then be computed using the equation:

$$\hat{p}_i = \sum_{j=1}^{r} \pi_j n_{ij}/n_{.j}$$

The true marginal proportions can also be computed simply by summing the individual cell probabilities in each column. For example, $\hat{p}_1 = 0.170 + 0.024 + 0.000 + 0.008 = 0.202$, $\hat{p}_2 = 0.357$, $\hat{p}_3 = 0.157$, and $\hat{p}_4 = 0.285$.

The third step is to compute the probability correct given the true class i; in other words, the producer's accuracy. It should be noted that the values here differ

---

TABLE 8.9
Error Matrix Showing Map Marginal Proportions

		True (i) Reference Data				Row	Map Marginal
		D	**C**	**AG**	**SB**	**Total**	**Proportions, π_j**
	D	65	4	22	24	115	0.3
Map (j) Classified Data	**C**	6	81	5	8	100	0.4
	AG	0	11	85	19	115	0.1
	SB	4	7	3	90	104	0.2
Column Total		75	103	115	141	434	OVERALL ACCURACY = (65+81+85+90)/434 = 321/434 = 74%

TABLE 8.10

Error Matrix of Individual Cell Probabilities, $\hat{P}_{ij}$

		True (i) Reference Data			
		D	C	AG	SB
	D	0.170	0.101	0.057	0.063
Map (j) Classified Data	C	0.024	0.324	0.020	0.032
	AG	0.000	0.010	0.074	0.017
	SB	0.008	0.013	0.006	0.0173

somewhat from those computed in the error matrix discussion, because these values have been corrected for bias by incorporating the true marginal proportions, as shown in the following equation:

$$\hat{\theta}_{ii} = (\pi_i/\hat{p}_i)(n_{ii}/n_{.i}) \quad \text{or} \quad \hat{p}_{ii}/\hat{p}_i$$

As expected, the producer's accuracy is computed by taking the diagonal cell value from the cell probability matrix (Table 8.10) and dividing by the true marginal proportion. For example, $\theta_{11}=0.170/0.202=0.841$ or 84%, $\theta_{22}=0.908$, $\theta_{33}=0.471$, and $\theta_{44}=0.607$.

The next step is to compute the probability correct given map class j; in other words, the user's accuracy. This computation is made exactly as described in the error matrix discussion by taking the diagonal cell value and dividing by the row (j) marginal. The equation for this calculation is as follows:

$$\hat{l}_{jj} = \frac{n_{jj}}{n_{.j}}$$

Therefore, $\hat{l}_{11}=65/115=0.565$ or 57%, $\hat{l}_{22}=0.810$, $\hat{l}_{33}=0.739$, and $\hat{l}_{44}=0.865$.

Step five is to compute the overall correct by summing the major diagonal of the cell probabilities or using the equation:

$$\hat{P}_c = \sum_{j=1}^{r} \pi_j n_{jj}/n_{.j}$$

Therefore, in this example, $\hat{P}_c=0.170+0.324+0.074+0.173=0.741$ or 74%.

We have now made essentially the same calculations as described in the error matrix discussion, except that we have corrected for bias by using the true marginal proportions. The next step is to compute the variances for those terms (overall, producer's, and user's accuracies) for which we wish to calculate confidence intervals.

Variance for overall accuracy, $\hat{P}_c$

$$V(\hat{P}_c) = \sum_{i=1}^{r} p_{ii}(\pi_i - p_{ii})/(\pi_i n)$$

Therefore, in this example,

$$\hat{P}_c = [0.170(0.3 - 0.170)/(0.3)(434)$$
$$+\ 0.324(0.4 - 0.324)/(0.4)(434)$$
$$+\ 0.074(0.1 - 0.074)/(0.1)(434)$$
$$+\ 0.173(0.2 - 0.173)/(0.2)(434)]$$
$$=\ .00040$$

Confidence interval for overall accuracy, $\hat{P}_c$

$$\hat{P}_c = 2\left[V(\hat{P}_c)\right]^{\frac{1}{2}}$$

Therefore, in this example, the confidence interval for

$$\hat{P}_c = 0.741 \pm 2(0.0004)^{1/2}$$
$$=\ 0.741 \pm 2(0.02)$$
$$=\ 0.741 \pm 0.04$$
$$=\ (0.701, 0.781) \text{ or } 70\% \text{ to } 78\%$$

Variance for producer's accuracy, $\hat{\theta}_{ii}$

$$V(\hat{\theta}_{ii}) = p_{ii}p_i^{-4}\left[p_{ii}\sum_{j\neq1}^{r} p_{ij}(\pi_j - p_{ij})/\pi_j n + (\pi_i - p_{ii})(p_i - p_{ii})^2/\pi_i n\right]$$

Therefore, in this example,

$$V(\hat{\theta}_{11}) = 0.170(0.202)^{-4}\{0.170[0.024(0.4 - 0.024/(0.4)(434)$$
$$+\ 0.008(0.2 - 0.008)/(0.2)(434)]$$
$$+\ (0.3 - 0.170)(0.202 - 0.170)^2/(0.3)(434)\}$$
$$=\ 0.00132$$

Confidence interval for producer's accuracy, $\hat{\theta}_{ii}$

$$\hat{\theta}_{ii} \pm 2 \left[V(\hat{\theta}_{ii}) \right]^{\frac{1}{2}}$$

Therefore, in this example, the confidence interval for

$$\hat{\theta}_{11} = 0.841 \pm 2(0.00132)^{1/2}$$

$$= 0.841 \pm 2(0.036)$$

$$= 0.841 \pm 0.072$$

$$= (0.768, 0.914) \text{ or } 77\% \text{ to } 91\%$$

Variance for user's accuracy, $\hat{l}_{ii}$

$$\hat{\theta}_{ii} V(\hat{l}_{ii}) = p_{ii} (\pi_i - p_{ii}) / \pi_i^2 n$$

Therefore, in this example,

$$V(\hat{l}_{11}) = 0.170(0.3 - 0.170) / (0.3)^2 (434)$$

$$= 0.00057$$

Confidence interval for

$$\hat{l}_{ii} \pm 2 \left[V(\hat{l}_{ii}) \right]^{\frac{1}{2}}$$

Therefore, in this example, the confidence interval for

$$\hat{l}_{ii} = 0.565 \pm 2(0.00057)^{1/2}$$

$$= 0.565 \pm 2(0.024)$$

$$= 0.741 \pm 0.048$$

$$= (0.517, \ 0.613) \text{ or } 52\% \text{ to } 61\%$$

It must be remembered that these confidence intervals are computed from asymptotic variances. If the normality assumption is valid, then these are 95% confidence intervals. If not, then by Chebyshev's inequality, they are at least 75% confidence intervals.

AREA ESTIMATION/CORRECTION

In addition to all the uses of an error matrix already presented, it can also be used to update the areal estimates of the map categories. The map derived from the remotely sensed data is a complete enumeration of the ground. However, the error matrix is an indicator of where misclassification occurred between what the map said and what is actually on the ground. Therefore, it is possible to use the information from the error matrix to revise the estimates of total area for each map category. It is not possible to update the map itself or to revise a specific location on the map, but it is possible to revise total area estimates. Updating in this way may be especially important for small, rare categories whose estimates of total area could vary greatly depending on even small misclassification errors.

Czaplewski and Catts (1990) and Czaplewski (1992) have reviewed the use of the error matrix to update the areal estimates of map categories. They propose an informal method, both numerically and graphically, to determine the magnitude of bias introduced in the areal estimates by the misclassification. They also review two methods of statistically calibrating the misclassification bias. The first method is called the *classical estimator* and was proposed to the statistical community by Grassia and Sundberg (1982) and used in a remotely sensed application by Prisley and Smith (1987) and Hay (1988). The classical estimator uses the probabilities from the omission errors for calibration.

The second method is the inverse estimator, and it uses the probabilities from the commission errors to calibrate the areal estimates. Tenenbein (1972) introduced this technique in the statistical literature, and Chrisman (1982) and Card (1982) have used it for remote sensing applications. The confidence calculations derived in the previous section are from Card's (1982) work using the inverse estimator for calibration. More recently, Woodcock (1996) has proposed a modification of the Card approach incorporating fuzzy set theory into the calibration process.

Despite all this work, not many users have picked up on these calibration techniques or the need to perform the calibration. From a practical standpoint, overall total areas are not that important. We have already discussed this in terms of non-site-specific accuracy assessment. However, as more and more work is done looking at change, and especially changes of small, rare categories, the use of these calibration techniques may grow in importance.

9 Analysis of Differences in the Error Matrix

The previous chapter presented some of the measures and analytical techniques that can be applied to the error matrix. This chapter delves more deeply into the matrix itself and demonstrates what can be learned from a more thorough investigation of all components of the error matrix, including exploring why some of the map labels do not match the reference labels. While much attention is commonly placed on overall accuracy, or perhaps producer's and user's accuracies, by far the more interesting analysis concerns discovering why some of the accuracy assessment samples did not fall on the diagonal of the error matrix (i.e., why were there specific omission and commission errors?). To both effectively use the map and make better maps in the future, we need to know what causes these off-diagonal samples or differences in the matrix to occur.

Not all the values in the off-diagonals of the error matrix actually represent true error. Instead, all off-diagonal samples or differences in the error matrix will be the result of one of four possible sources:

1. errors in the reference data
2. sensitivity of the classification scheme to observer variability
3. inappropriateness of the remote sensing data employed for mapping a specific land cover class
4. mapping error

This chapter reviews each one of these sources and discusses the impacts of each one on the accuracy assessment results.

ERRORS IN THE REFERENCE DATA

A major and required assumption of the error matrix is that the label from the reference data represents the "correct" label of the sample site, and that all differences between the map and the reference label are due to classification and/or delineation error (i.e., mapping error). While this assumption is necessary, the reference data will never be perfect. As previously discussed, the term *ground truth* should be avoided for exactly this reason. Throughout this book, the authors prefer the use of *reference data* or *reference label* to refer to the sample data set being compared with the map that is being assessed. Unfortunately, error matrices can be inadequate indicators of map error, because they are often confused by errors in the reference data (Congalton and Green, 1993). Errors in the reference data can be a function of:

- *Registration differences.* Registration differences between the reference data and the remotely sensed map classification are caused by delineation and/or digitizing errors. For example, if the global positioning system

(GPS) is not used in the field during accuracy assessment, it is possible for field personnel to collect data in the wrong area. Then again, even the GPS can be off by a considerable distance, especially if great care is not used when collecting the GPS information. Other registration errors can occur when an accuracy assessment site is incorrectly delineated or digitized, or when an existing map used for reference data is not precisely registered to the map being assessed. Even when all these factors are controlled, positional error must still be carefully considered, as presented in Chapter 4 and applied in Chapter 6.

- *Data entry errors.* Data entry errors are common in any database project and can be controlled only through rigorous quality control. Developing digital data entry forms for use with a data logger, laptop, tablet, or cell phone that will only allow a certain set of characters for specific fields can catch errors during data entry. One of the best (yet expensive) methods for catching data entry errors is to enter all data twice and then compare the two data sets. Differences usually indicate an error.

- *Classification scheme errors.* Every accuracy assessment map and reference site must have a label derived from the classification scheme used to create the map. Classification scheme errors occur when personnel misapply the classification scheme to the map or reference data; a common occurrence with complex classification schemes. If the reference data are in a database, then such errors can be avoided, or at least highlighted, by programming the classification scheme rules and using the program to determine the actual label of the accuracy assessment sites. Classification scheme errors also occur when the classification scheme used to label the reference site is different from the one used to create the map—a common occurrence when existing data or maps are used as reference data.

- *Change.* Changes in land cover may occur between the date of the remotely sensed imagery collection and the date of the reference data. As previously discussed in Chapter 7, land cover change can have a profound effect on accuracy assessment results. Tidal differences, crop or tree harvesting, urban development, natural disasters, fire, and pests all can cause the landscape to change in the time period between capturing the remotely sensed data and the accuracy assessment reference data collection.

- *Mistakes in labeling reference data.* Labeling mistakes usually occur because inexperienced and/or poorly trained personnel are used to collect reference data. Even with experienced personnel, the more detailed the classification scheme, the more likely it is that an error in labeling the reference data will occur. For example, some conifer and hardwood species are difficult to distinguish on the ground, much less from aerial photography. Young crops of broccoli, Brussels sprouts, and cauliflower are easily confused. Thus, an accuracy assessment must also be performed on the reference data. If manual image interpretation is used to assess a map created through semi-automated methods, then a sample of the interpreted sites should be visited on the ground. If field data are used, then some sample of the sites must be visited by two different personnel and their answers compared. If the answers mostly agree, then the collection is satisfactory. If the answers mostly disagree, then a problem exists in the reference data collection method.

Table 9.1 summarizes the reference data errors discovered during the quality control process for an actual accuracy assessment. Only six out of 125 of the differences between the map and reference labels were caused by actual errors in the map. Over two-thirds of the differences (85 sites) were caused by mistakes in the reference data. The most significant error occurred from using different classification schemes (50 sites). In this project, National Wetlands Inventory (NWI) maps were used exclusively to map wetlands (i.e., wetlands were defined in the classification scheme to be those areas identified by NWI data as wetlands). However, when the accuracy assessment was performed, the image interpreters collecting the reference data used a different definition of wetlands and disagreed with all the NWI wetlands labels. The remaining differences were caused by observer variation, as discussed in the next section of this chapter.

SENSITIVITY OF THE CLASSIFICATION SCHEME TO OBSERVER VARIABILITY

The natural world that we attempt to map is much more continuous than it is broken into distinct boundaries (e.g., vegetation cover, soil type, or land use). However, the classification scheme rules we use for mapping often impose discrete boundaries on these continuous conditions. In situations where breaks in the classification scheme represent artificial distinctions along a continuum, observer variability is often difficult to control and, while unavoidable, can have profound effects on accuracy assessment results (Congalton, 1991; Congalton and Green, 1993). Analysis of the error matrix must include exploring how many of the matrix differences are the result of observers being unable to precisely distinguish between these map classes when the accuracy assessment site is on the margin between two or more map classes in the classification scheme.

TABLE 9.1
Analysis of Map and Reference Label Differences

Map vs. Reference Difference	Number of Sites Different	Map Error	Reference Label Error	Date Change	Classification Scheme Difference	Variation in Estimation
Barren vs. water	19	0	6	8	0	5
Hardwood vs. water	6	0	0	0	0	6
Herb vs. forested	50	6	17	4	0	23
Wetland vs. all other types	50	0	0	0	50	0
Total	125	6	23	12	50	34

Plato's parable of the shadows in the cave is useful for thinking about observer variability. In the parable, Plato describes prisoners who cannot move:

> Above and behind them a fire is blazing in the distance, and between the fire and the prisoners there is a ... screen which marionette players have in front of them over which they show puppets ... [the prisoners] see only their own shadows, or the shadows of one another which the fire throws on the opposite wall of the cave ... To them ... the truth would be literally nothing but the shadows of the images.
>
> **(Plato, *The Republic*, Book VII, 515-B, from Benjamin Jowett's translation as published in Vintage Classics, Random House, New York)**

Like Plato's prisoners in the cave, we all perceive the world within the context of our experience. The difference between reality and perceptions of reality is often as fuzzy as Plato's shadows. Our observations and perceptions vary from day to day and depend on our training, experience, or mood.

The analysis in Table 9.1 shows the impact that variation in interpretation can have on accuracy assessment. In the project, two image interpreters were asked to label the same accuracy assessment reference sites. Almost 30% (34 of 125) of the differences between the map and the reference label were caused by variation in interpretation/estimation.

Consider, for example, the assessment of a map of tree crown closure with classification scheme rules defining classes as:

Un-vegetated: 0%–10% crown closure
Sparse: 11%–30%
Light: 31%–50%
Medium: 51%–70%
Heavy: 71%–100%

An accuracy assessment reference site from image interpretation (in this case, aerial photography) estimated the tree crown cover to be 45%, and therefore, according to the classification scheme, the site would be labeled as Light. However, since it is recognized that crown closure can only be interpreted on aerial photos to ±10% (Spurr, 1960), it is also feasible that the proper label could be Medium. Either the label of Light or the label of Medium is within the variability of the reference data collection. The map user would be much more concerned with a difference caused by a map label of Un-vegetated versus a reference label of Heavy tree crown cover. Differences on class margins are both inevitable and far less significant to the map user than other types of differences.

Classification schemes that employ estimates of percentage of vegetative cover are particularly susceptible to this type of confusion in the error matrix. Appendix 9.1 presents a set of very complex classification scheme rules for mapping the forest and woodlands of Haleakalā National Park in Hawaii. Looking at these classification rules, it is apparent that very small differences in vegetative cover estimates by species result in multiple options for the map label. This sensitivity calls for making measurements to obtain more precise estimates or for incorporating techniques that better account for this variability.

Several researchers have noted the impact of the variation in human interpretation on map results and accuracy assessment (Gong and Chen, 1992; Lowell, 1992; Congalton and Biging, 1992; Congalton and Green, 1993). Woodcock and Gopal (1992) state: "The problem that makes accuracy assessment difficult is that there is ambiguity regarding the appropriate map label for some locations. The situation of one category being exactly right and all other categories being equally and exactly wrong often does not exist." Lowell (1992) calls for "a new model of space which shows transition zones for boundaries, and polygon attributes as indefinite." As Congalton and Biging (1992) conclude in their study of the validation of photo-interpreted stand-type maps, "The differences in how interpreters delineated stand boundaries were most surprising. We were expecting some shifts in position, but nothing to the extent that we witnessed. This result again demonstrates just how variable forests are and the subjectiveness of photo interpretation."

While it is difficult to control observer variation, it is possible to measure the variation and to use the measurements to compensate for differences between reference and map data that are caused not by map error but by variation in interpretation. One option is to measure each reference site precisely to reduce observer variance in reference site labels. This method can be prohibitively expensive, usually requiring extensive field sampling. The second option incorporates fuzzy logic into the reference data to compensate for non-error differences between reference and map data and is discussed in Chapter 10.

INAPPROPRIATENESS OF THE REMOTE SENSING DATA EMPLOYED TO MAKE THE MAP

Early satellite remote sensing projects were primarily concerned with testing the viability of various remote sensing data for mapping certain types of land cover. Researchers tested the hypotheses of whether or not the imagery could be used to detect land use, or crop types, or forest types. Many accuracy assessment techniques were developed primarily to test these hypotheses.

Recently, accuracy assessment has been more focused on learning about the reliability of a map for land management or policy analysis. However, some of the differences in the error matrix will be because the map producer was attempting to use remotely sensed data or methods that were incapable of distinguishing certain land cover or vegetation class types. Even with today's high–spatial resolution imagery and sophisticated classification methods, there is some information that is still unattainable remotely. For example, map classes that are differentiated only by the vegetation species in the understory of a closed canopy tree stand are not distinguishable using optical remote sensing data, which cannot penetrate the canopy. Understanding what differences are caused by the technology is useful to the map producer when the next map is being made.

In a project to map Wrangell-St. Elias National Park, Landsat TM data were employed as the primary remotely sensed data, with 1:60,000 aerial photography as ancillary data. The classification scheme included distinguishing between pure and mixed stands of black and white spruce. Accuracy assessment analysis consistently showed success at differentiating *pure* stands of black versus white spruce. However, consistently differentiating these species in mixed or occasional hybrid stands was found to be unreliable. This phenomenon is not surprising, considering the difficulty often associated with differentiating these species in mixed and hybrid stands on the ground. In summary, optical remotely sensed data, at the scales (i.e., combination of spatial and spectral resolutions) employed, cannot be used to reliably and consistently differentiate between mixed classes of these two tree species.

To make the map more reliable, the map user can collapse the classification scheme across specific map classes into a single, more general map class. In this example, the non-pure spruce classes of Closed, Open, and Woodland were collapsed into an "Unspecified Interior Spruce" class. In the resulting collapsed error matrix, "Unspecified Interior Spruce" map labels were considered to be mapped correctly if they corresponded to a pure or mixed white spruce or black spruce reference site demonstrating the same density class of "Closed", "Open," or "Woodland". For example, a map label of "Open Unspecified Interior Spruce" was considered to be correctly mapped if its corresponding reference label for the site was "Open black spruce," "Open white spruce," or "Open black/white spruce" mix. While less information is displayed on the map, the remaining information is more reliable.

MAPPING ERROR

The final cause of differences in the error matrix is mapping error (i.e., actual real errors). Often, these are difficult to distinguish from an inappropriate use of remote sensing data. Usually, they are errors that are systematic, particularly obvious, and unacceptable. For example, it is not uncommon for an inexperienced remote sensing professional to produce a map of land cover from satellite data that misclassifies northeast-facing forests on steep slopes as water. Because water and shadowed wooded slopes both absorb most energy, this type of error is explainable, but it is unacceptable and must be avoided. Many map users will be appalled at this type of error and are not particularly interested in having the electromagnetic spectrum explained to them as an excuse. However, careful editing, comparison with other imagery, modeling that all water exists in areas without slope, and comparison with existing maps of waterways and lakes will reduce the possibility of this type of map error.

Understanding the causes of true error can point the map producer to additional methods to improve the accuracy of the map. Perhaps other bands or band combinations will improve accuracy, including computing band ratios, indices, and transformations (e.g., Normalized Difference Vegetation Index (NDVI), principal components analysis (PCA), etc.). The incorporation of ancillary data such as slope, aspect, elevation, or lidar-derived digital surface models may be useful. In the Wrangell-St. Elias example, confusion existed between the "Dwarf Shrub" classes and the "Graminoid" class. The confusion was addressed through the use of unsupervised classifications and park-wide models using digital elevation data, field-based data, and aerial photography. First, an unsupervised classification with 20 classes was run for only those areas of the imagery classified as "Dwarf Shrub" in the map. A digital elevation coverage was used to stratify the study area for subsequent relabeling of unsupervised classes previously mapped as "Dwarf Shrub" but actually representing areas of "Graminoid" cover on the ground. From the unsupervised classification, two spectral classes were found to consistently represent "Graminoid" cover throughout the study area, while another spectral class was found to represent "Graminoid" cover in areas below 3500 feet elevation. These spectral classes were subsequently recoded to the "Graminoid" class.

SUMMARY

Analysis of the causes of differences in the error matrix can be one of the most important and interesting steps in the creation of a map from remotely sensed data. In the past, too much emphasis has been placed on the overall accuracy of the map without delving into the conditions that give rise to that accuracy. By understanding what caused the reference and map data to differ, we can use the map more reliably and produce both better maps and better accuracy assessments in the future.

APPENDIX 9.1
Forest and Woodland Map Class Key of Haleakalā National Park

1a. Forests, woodlands, and scrub (dwarf woodland) dominated by native tree species or one of several tree species thought to be early Polynesian introductions (*Aleurites moluccana, Cocos nucifera, Hibiscus tiliaceus*, or *Pandanus tectorius*). Introduced non-native trees such as species *Psidium* or *Schinus* may be present but not dominant 2

1b. Forests, woodlands, and scrub (dwarf woodland) dominated by non-native species of Eucalyptus, *Grevillea, Mangifera, Phyllostachys, Prosopis, Psidium, Schinus, Spathordea, Syzygium, Terminalia, Thevetia*, or other non-native species 27

2a. Forests and woodlands with canopies dominated or co-dominated by *Aleurites moluccana, Erythrina sandwicensis, Hibiscus tiliaceus, Myrsine lessertiana*, or *Pandanus tectorius* 3

2b. Forests, woodlands, and scrub (dwarf woodland) dominated or co-dominated by *Acacia koa, Metrosideros polymorpha*, or other native tree species 8

3a. Forests and woodlands with canopies dominated or co-dominated by *Aleurites moluccana*, often with *Syzygium cumini* or species of *Psidium* .. ***Aleurites moluccana* Lowland Wet Forest (F_ALMO)**

3b. Forests and woodlands with canopies dominated or co-dominated by *Erythrina sandwicensis, Hibiscus tiliaceus, Myrsine lessertiana, Pandanus tectorius*, or *Cocos nucifera* 4

4a. Forests and woodlands with canopies dominated or co-dominated by *Pandanus tectorius*, often with *Schinus terebinthifolius* or *Terminalia catappa* .. ***Pandanus tectorius* Coastal Mesic Forest (F_PATE_M)**

4b. Forests and woodlands with canopies dominated or co-dominated by *Erythrina sandwicensis, Hibiscus tiliaceus, Myrsine lessertiana*, or *Cocos nucifera* 5

5a. Forests and woodlands with canopies dominated by *Hibiscus tiliaceus. Aleurites moluccana* and other wet and riparian tree species may be present in canopy .... ***Hibiscus tiliaceus* Lowland Wet Forest [Provisional] (F_HITE_L)**

5b. Forests and woodlands with canopies dominated or co-dominated by *Erythrina sandwicensis, Myrsine lessertiana*, or *Cocos nucifera* 6

6a. Forests and woodlands with tree canopy dominated by *Erythrina sandwicensis* ....***Erythrina sandwicensis* Lowland Dry Woodland (W_ERSA)**

6b. Forests and woodlands with canopies dominated by *Myrsine lessertiana* or *Cocos nucifera* 7

7a. Forests and woodlands with tree canopies dominated or co-dominated by *Myrsine lessertiana*.

Metrosideros polymorpha may be present to co-dominant. Understory is typically dominated by *Coprosma foliosa* and *Dodonaea viscosa* ....**Myrsine lessertiana** – (*Metrosideros polymorpha*)/*Coprosma foliosa* – *Dodonaea viscosa* **Montane Mesic Forest (F_MYMECODO)**

7b. Forests and woodlands with tree canopies dominated by *Cocos nucifera* **Cocos nucifera Strand Woodland (W_CONU)**

8a. Forests and woodlands with canopies dominated or co-dominated by *Acacia koa*. If *Metrosideros polymorpha* is present then cover of *Acacia koa* is 1/3 or more cover of *Metrosideros polymorpha* in the tree canopy **9**

8b. Forests and woodlands with canopies dominated or co-dominated by *Metrosideros polymorpha*. If *Acacia koa* is present then cover is low (<1/3 cover) of *Metrosideros polymorpha* tree canopy..**16**

9a. Forest and woodlands with canopies dominated by *Acacia koa* and/or *Cheirodendron trigynum*. If *Metrosideros polymorpha* is present then cover of *Metrosideros polymorpha* is <1/3 the cover of *Acacia koa* in the tree canopy. Non-native trees such as *Psidium cattleianum* may be present to co-dominant **10**

9b. Forests and woodlands with canopies co-dominated by *Acacia koa* and *Metrosideros polymorpha*. Each species has at least 1/3 cover of the other species...**13**

10a. Forests and woodlands with mixed tree canopy co-dominated by *Acacia koa* and *Cheirodendron trigynum* with *Antidesma platyphyllum* and/or *Syzygium sandwicensis* trees often present to co-dominant. Non-native trees such as *Psidium cattleianum* may be present to co-dominant with native trees in disturbed stands of this diverse, wet lowland forest. Non-native shrubs *Clidemia hirta* or *Rubus rosifolius* may dominate the understory in disturbed stands...........**Acacia koa** – **Cheirodendron trigynum** – (*Antidesma platyphyllum, Syzygium sandwicensis*)/(*Clidemia hirta*) **Lowland Wet Forest (F_ACCHANSYCL)**

10b. Forests not as above...**11**

11a. Forests and woodlands with upper canopies dominated by >10% *Acacia koa* and with sub-canopy strongly to exclusively dominated by *Psidium cattleianum* ....**Acacia koa** – **Psidium cattleianum Semi-natural Forest (F_ACPS)**

11b. Forests and woodlands with canopies strongly dominated by *Acacia koa*...**12**

12a. Montane mesic forests dominated by *Acacia koa* with *Coprosma foliosa* and *Dodonaea viscosa* present to co-dominant in the understory ....
Acacia koa/*Coprosma foliosa* – *Dodonaea viscosa* **Montane Mesic Woodland (W_ACCODO)**

12b. Montane mesic forests dominated by *Acacia koa* with *Leptecophylla tameiameiae* dominant or co-dominant with *Dodonaea viscosa* in the understory. *Coprosma foliosa* is typically absent .....
Acacia koa/*Leptecophylla tameiameiae* – (*Dodonaea viscosa*) **Montane Woodland1 (F_ACLEDO)**

13a. Understory dominated by *Hedychium gardnerianum*. This is a disturbed wet forest type .**Acacia koa** – **Metrosideros polymorpha**/*Hedychium gardnerianum* **Semi-natural Woodland [Park Special] (W_ACMEHE)**

13b. Understory not dominated by *Hedychium gardnerianum* **14**

14a. Wet montane forests and woodlands with canopies co-dominated by *Metrosideros polymorpha* and *Acacia koa* with *Melicope clusiifolia* and *Cheirodendron trigynum* trees often present. Native trees dominate the stand, but nonnative trees such as *Psidium cattleianum* may be present to co-dominant. The diverse understory is composed of an open to dense shrub layer with wet indicator species such as *Alyxia oliviformis, Astelia menziesiana, Broussaisia arguta, Clermontia arborescens* ssp. *waihiae, Freycinetia arborea, Hedyotis terminalis*, or *Smilax melastomifolia*. The open to moderately dense, mixed herbaceous layer composed of diverse ferns and high cover of moss. *Dicranopteris linearis* may co-dominate the herbaceous layer with up to 30% cover, but does not strongly dominate the understory. *Diplazium sandwichianum* had high cover of at least one plot. Epiphytes are common ....***Acacia koa – Metrosideros polymorpha* Wet Montane Woodland (W_ACME_M)**

14b. Lowland forests and woodlands generally <1000 m elevation **15**

15a. Lowland wet forests co-dominated by *Metrosideros polymorpha* and *Acacia koa. Dicranopteris linearis* strongly dominates the understory forming a dense mat (generally >40% of *Dicranopteris linearis* with low diversity of other understory species). Shrubs are generally absent or have low cover ....***Acacia koa – Metrosideros polymorpha/Dicranopteris linearis* Lowland Wet Forest (F_ACMEDI)**

15b. Lowland diverse mesic forests co-dominated by *Metrosideros polymorpha* and *Acacia koa. Antidesma platyphyllum* and *Melicope clusiifolia* trees may be present to co-dominant. The sampled stand occurred at 780 m and was disturbed with moderate cover of the non-native tree *Psidium cattleianum* and understory strongly dominated by the non-native shrub *Clidemia hirta* ....***Acacia koa – Metrosideros polymorpha* Lowland Mesic Forest (F_ACME_L)**

16a. Forests and woodlands with tree canopies co-dominated by *Myrsine lessertiana. Metrosideros polymorpha* may co-dominate. The understory is typically dominated by *Coprosma foliosa* and *Dodonaea viscosa*...
   ***Myrsine lessertiana – (Metrosideros polymorpha)/Coprosma foliosa – Dodonaea viscosa* Montane Mesic Forest (F_MYMECODO)**

16b. Forests and woodlands with canopies dominated by *Metrosideros polymorpha* and/or *Melicope clusiifolia* **17**

17a. Wet montane forests with tree canopies co-dominated by *Metrosideros polymorpha* and *Melicope clusiifolia*. The understory often dominated by *Elaphoglossum wawrae*. Other ferns *Diplazium sandwichianum* and *Dryopteris wallichiana* are present, but not co-dominant............
   ***Metrosideros polymorpha – Melicope clusiifolia* Montane Wet Forest [Park Special] (F_MEME)**

17b. Forests and woodlands dominated or co-dominated by *Metrosideros polymorpha. Melicope clusiifolia* is absent or if present then <10% cover **18**

18a. Wet, upper montane forests and woodlands with tree canopies dominated by *Metrosideros polymorpha* and *Rubus hawaiensis* dominating the shrub layer. *Vaccinium calycinum* may be present to co-dominant. Ferns such as *Dryopteris wallichiana, Diplazium sandwichianum* or *Sadleria cyatheoides*, and mosses are often abundant ....***Metrosideros polymorpha/Rubus hawaiensis* Montane Wet Forest (F_MERU)**

18b. Forests and woodlands dominated or co-dominated by *Metrosideros polymorpha. Rubus hawaiensis* is absent or if present then with <10% cover **19**

19a. Wet, montane forests and woodlands with tree canopies dominated by *Metrosideros polymorpha* and diverse shrub layer dominated or co-dominated by *Vaccinium calycinum*, often with *Broussaisia arguta* and/or *Coprosma foliosa* present to co-dominant. *Athyrium microphyllum* is present in herbaceous layer. Other wet forest indicator species may include *Astelia menziesiana, Clermontia arborescens, Coprosma granadensis, Melicope clusiifolia,* and *Sadleria pallida. Dicranopteris linearis* or *Diplazium sandwichianum* are often present but do not strongly dominate. Other ferns and mosses are typically abundant. Epiphytes are usually abundant ....***Metrosideros polymorpha/Vaccinium calycinum – (Broussaisia arguta, Coprosma foliosa)/Athyrium microphyllum*...Montane Wet Forest (F_MEVABRCOAT)**

19b. Forests and woodlands dominated or co-dominated by *Metrosideros polymorpha. Vaccinium calycinum* does not dominate or co-dominate the understory **20**

20a. Wet montane forests and woodlands with tree canopies co-dominated by *Metrosideros polymorpha* and *Cheirodendron trigynum* ssp. *trigynum* with both species generally having at least 10% cover, although *Metrosideros polymorpha* often has higher cover. If *Ilex anomala* or *Myrsine lessertiana* is present, then cover of *Cheirodendron* spp. is greater in cover than either. *Cibotium* spp. may be present to dominant in tree sub-canopy and understory. Understory is often strongly dominated by *Diplazium sandwichianum* (ferns) and moss. Epiphytes are abundant… .... …***Metrosideros polymorpha – Cheirodendron trigynum/(Cibotium* spp.)** **Montane Wet Forest (F_MECHCI)**

20b. Forests and woodlands dominated or co-dominated by *Metrosideros polymorpha*. *Cheirodendron* spp. is absent or if present then with low cover <5% cover....................................**21**

21a. Forests and woodlands with tree canopies dominated by *Metrosideros polymorpha* and *Sadleria cyatheoides* strongly dominating the understory ....***Metrosideros polymorpha/Sadleria cyatheoides* Forest [Park Special] (F_MESA)**

21b. Forests and woodlands dominated or co-dominated by *Metrosideros polymorpha*. *Sadleria cyatheoides* is absent or if present then with relatively low cover **22**

22a. Forests and woodlands with open to moderately dense tree canopies dominated by *Metrosideros polymorpha* and a wet, boggy herbaceous understory dominated or co-dominated by *Carex alligata* and/or *Machaerina angustifolia*. Disturbed stands may be co-dominated by non-native species *Ageratina adenophora*, *Paspalum urvillei* or *Tibouchina herbacea* …............ ***Metrosideros polymorpha/Carex alligata – (Machaerina angustifolia*) Wet Woodland (W_MECAMA)**

22b. Forests and woodlands dominated or co-dominated by *Metrosideros polymorpha*. *Carex alligata* and *Machaerina angustifolia* absent or if present then with low cover **23**

23a. Wet forests and woodlands dominated or co-dominated by *Metrosideros polymorpha* with the ferns *Dicranopteris linearis* or *Dryopteris wallichiana* dominating the understory.........................**24**

23b. Forests and woodlands not as above...........**Unclassified Native Forest or Woodland at HALE (FW_UNNA)**

24a. Wet forests and woodlands dominated or co-dominated by *Metrosideros polymorpha* with the fern *Dicranopteris linearis* strongly dominating the understory with 25% or more cover...............**25**

24b. Forests and woodlands dominated or co-dominated by *Metrosideros polymorpha*. The fern *Dicranopteris linearis* is absent or if present does not strongly dominate the understory.................**26**

25a. Wet lowland forests and woodlands generally below 1000 m elevation. Non-native species such as *Psidium cattleianum*, *Clidemia hirta*, and *Tibouchina herbacea* are more common in lowland stands ...***Metrosideros polymorpha/Dicranopteris linearis* Lowland Mesic Woodland (W_MEDI_L)**

25b. Wet montane forests and woodlands generally above 1000 m elevation. Native wet species such as *Broussaisia arguta*, *Elaphoglossum paleaceum*, *Hedyotis terminalis*, and *Lycopodiella cernua* are common in these montane stands . ***Metrosideros polymorpha/Dicranopteris linearis* Montane Wet Woodland (W_MEDI_M)**

26a. Mesic subalpine forests and woodlands dominated by *Metrosideros polymorpha* with the fern *Dryopteris wallichiana* dominating the understory. *Lycopodium venustulum* may be present to co-dominant..........***Metrosideros polymorpha/Dryopteris wallichiana* Subalpine Mesic Forest (F_MEDR)**

26b. Montane mesic forest with tree canopy dominated by *Metrosideros polymorpha* and open to dense shrub layer dominated or co-dominated by *Leptecophylla tameiameiae* and *Dodonaea viscosa* ..***Metrosideros polymorpha/Leptecophylla tameiameiae – Dodonaea viscosa* Montane Woodland1 (W_MELEDO_M)**

27a. Forests and woodlands with canopies dominated by *Prosopis pallida* often with a tall shrub layer dominated by *Leucaena leucocephala*. Understory is typically strongly dominated by introduced grasses ....***Prosopis pallida* Coastal Dry Semi-natural Woodland (W_PRPA)**

27b. Forests and woodlands not as above..**28**

28a. Forests and woodlands with canopies dominated by *Terminalia catappa*. *Syzygium cumini* may be present to co-dominant. Common along coast lines, where stands tend to be small.................
Terminalia catappa Semi-natural Woodland [Park Special] (W_TECA)

28b. Forests and woodlands not as above...**29**

29a. Forests and woodlands with tree canopy strongly dominated by *Thevetia peruviana*. Native tree *Erythrina* sandwicensis was subdominant (12% cover) in the single sampled stand ......***Thevetia peruviana* Semi-natural Woodland [Park Special] (W_THPE)**

29b. Forests and woodlands not as above...**30**

30b. Former plantation areas dominated or co-dominated by *Eucalyptus globulus* and/or *Eucalyptus saligna* **Eucalyptus spp. – Mixed Alien Semi-natural/Plantation Forest [Provisional] (F_EUCA)**

30b. Forests and woodlands not as above...**31**

31a. Forests and woodlands with canopies dominated by *Grevillea robusta*.............***Grevillea robusta* Lowland Mesic Semi-natural Forest (F_GRRO)**

31b. Forests and woodlands not as above...**32**

32a. Forests and woodlands with canopies strongly dominated by *Phyllostachys nigra* **Phyllostachys nigra Semi-natural Thicket (T_PHNI)**

32b. Forests and woodlands not as above...**33**

33a. Forests and woodlands with canopies dominated by often emergent *Mangifera indica*. Non-native *Ficus microcarpa*, *Psidium cattleianum*, *Schinus terebinthifolius*, and *Syzygium cumini* trees may be present to co-dominant...............***Mangifera indica* Semi-natural Forest (F_MAIN)**

33b. Forests and woodlands not as above...**34**

34a. Forests and woodlands with canopies dominated by *Spathodea campanulata*. The single stand sampled had moderate cover of *Aleurites moluccana* and *Psidium guajava* and high cover of the non-native grass *Oplismenus hirtellus*.............................***Spathodea campanulata* – *Psidium* spp. Semi-natural Forest [Park Special] (F_SPPS)**

34b. Forests and woodlands not as above...**35**

35a. Canopy ranges from moderately dense tall woodland to dense scrub woodland strongly dominated by *Schinus terebinthifolius* (usually 2× cover or more of other tree species). *Psidium guajava* trees are often present and may be co-dominant **Schinus terebinthifolius Semi-natural Scrub Woodland (W_SCTE_S)**

35b. Forests and woodlands not as above...**36**

36a. Forests and woodlands with tree canopy dominated by *Syzygium cumini*. Single sampled stands had *Mangifera indica* sub-dominant ... **Syzygium cumini Lowland Mesic Semi-natural Forest (F_SYCU)**

36b. Forests and woodlands not as above...**37**

37a. Forests and woodlands with tree canopy dominated or co-dominated by *Syzygium jambos*. *Psidium guajava* trees may be present to co-dominant.......... **Syzygium jambos Lowland Mesic Semi-natural Forest (W_SYJA)**

37b. Forests and woodlands not as above...**38**

38a. Forests and woodlands with canopies strongly dominated by *Psidium cattleianum*. May have an emergent canopy of *Acacia koa*...**39**

38b. Forests and woodlands not as above...**40**

39a. Forests and woodlands dominated by *Psidium cattleianum* with an emergent canopy dominated by >10% *Acacia koa* ....***Acacia koa -Psidium cattleianum* Semi-natural Forest (F_ACPS)**

39b. Forests and woodlands with canopies strongly dominated by *Psidium cattleianum*. *Psidium guajava*, *Schinus terebinthifolius*, and *Syzygium cumini* are often present sometimes with remnant native tree species.....***Psidium cattleianum* Lowland Wet Semi-natural Forest (F_PSCA)**

40a. Forests and woodlands with canopies dominated by *Psidium guajava*. Other introduced trees are common. Understory is often disturbed and dominated by introduced species ....
   ***Psidium guajava*/Disturbed Understory Semi-natural Forest (F_PSGU)**
40b. Forests and woodlands not as above..**41**
41a. Plantation Forests and woodlands with canopies dominated by *Cryptomeria japonica*, *Melaleuca quinquenervia*, *Pinus radiata*, or *Syncarpia glomulifera*. Other introduced trees are common ....**Mixed Forest Plantation [Park Special] (F_MIXE)**
41b. Forests and woodlands not as above**Unclassified Semi-natural Forest or Woodland at HALE (FW_UNSN)**

10 Fuzzy Accuracy Assessment

As our use of remotely sensed data and maps has continued to grow in complexity (i.e., increased spatial and spectral resolution), so have the classification schemes associated with these efforts. The classification scheme then becomes an even more important factor influencing the accuracy of the entire project, especially as it becomes more difficult to maintain mutual exclusivity. A review of the accuracy assessment literature points out some of the limitations of using only the traditional error matrix approach to the accuracy assessment of a map with a complex classification scheme. Congalton and Green (1993) recommend the error matrix as a jumping-off point for identifying sources of confusion (i.e., differences between the map created from remotely sensed data and the reference data) and not simply the "error." As presented in Chapters 7 and 9, the variation in human interpretation can have a significant impact on the collection of reference data and on what is considered correct. If image interpretation is used as the source of the reference data, and that interpretation is biased, inconsistent, or even wrong, then the results of the accuracy assessment could be very misleading. The same holds true for observations made in the field. As classification schemes become more complex, more variation in the ability for humans to precisely and accurately interpret the reference data is introduced. In addition, in situations where the breaks (i.e., divisions between classes) in the classification system represent artificial distinctions along a continuum, variation in human interpretation is often very difficult to control and, while unavoidable, can have profound effects on the accuracy assessment results (Congalton, 1991; Congalton and Green, 1993). Several researchers have noted the impact of the variation in human interpretation on map results and accuracy assessment (Gong and Chen, 1992; Lowell, 1992; McGuire, 1992; Congalton and Biging, 1992).

Gopal and Woodcock (1994) proposed the use of fuzzy sets to "allow for explicit recognition of the possibility that ambiguity might exist regarding the appropriate map label for some locations on the map. The situation of one category being exactly right and all other categories being equally and exactly wrong often does not exist." In this fuzzy set approach, it is recognized that instead of a simple system of correct (agreement) and incorrect (disagreement), there can be a variety of responses, such as absolutely right, good answer, acceptable, understandable but wrong, and absolutely wrong.

Fuzzy set theory or fuzzy logic is a form of set theory. While initially introduced in the 1920s, fuzzy logic gained its name and its algebra in the 1960s and 1970s from Zadeh (1965), who developed fuzzy set theory as a way to characterize the ability of the human brain to deal with vague relationships. The key concept is that membership in a class is a matter of degree. Fuzzy logic recognizes that on the margins of classes that divide a continuum, an item may belong to both classes. As Woodcock and Gopal (1992) state, "The assumption underlying fuzzy set theory is that the transition from membership to non-membership is seldom a step function." Therefore, while a 100% hardwood stand can be labeled hardwood, and a 100% conifer stand may be labeled conifer, a 49% hardwood and 51% conifer stand may be acceptable if labeled either conifer or hardwood.

Lowell (1992) calls for "a new model of space which shows transition zones for boundaries, and polygon attributes as indefinite." As Congalton and Biging (1992) conclude in their study of the validation of photo-interpreted stand-type maps, "The differences in how interpreters delineated stand boundaries were most surprising. We were expecting some shifts in position, but nothing to the extent that we witnessed. This result again demonstrates just how variable forests are and is a strong indicator of human variation in photo interpretation."

There are a number of techniques that have been proposed to incorporate ambiguity or fuzziness into the accuracy assessment process. Three methods are presented in this chapter: (1) expanding the major diagonal of the error matrix, (2) measuring map class variability, and (3) using a fuzzy error matrix approach.

EXPANDING THE MAJOR DIAGONAL OF THE ERROR MATRIX

The simplest and most straightforward method for incorporating fuzziness into the accuracy assessment process is to expand the major diagonal of the error matrix. Remember that the major diagonal of the error matrix represents agreement between the reference data and the map and is represented by a single cell in the matrix for each map class. By acknowledging some fuzziness in the classification, the class boundaries may be expanded to accept as correct plus or minus one class of the actual class. In other words, the major diagonal is no longer just a single cell for each map class, but rather, is wider. This method works well if the classification scheme is continuous, such as elevation or tree size class or forest crown closure. If the classification scheme is discrete, such as in a vegetation or land cover mapping project, then this method is not appropriate, since expanding beyond the major diagonal in the matrix would have no reasonable meaning.

Table 10.1 presents the traditional error matrix for a classification of forest crown closure (a continuous classification scheme divided into six discrete classes). Only exact matches are considered correct and are tallied along the major diagonal. The overall accuracy of this classification is 40%. Table 10.2 presents the same error matrix, only the major diagonal has been expanded to include plus or minus one crown closure class. In other words, for crown closure class 3, both crown closure classes 2 and 4 are also accepted as correct. This revised major diagonal then results in a tremendous increase in overall accuracy to 75%.

The immediate advantage of using this method of accounting for fuzzy class boundaries is obvious; the accuracy of the classification can increase dramatically. The disadvantage is that if the reason for accepting plus or minus one class cannot be adequately justified or does not meet the map user's requirements, then it may be viewed that you are simply expanding the major diagonal to try to get higher

TABLE 10.1

Error Matrix Showing the Ground Reference Data versus the Image Classification for Forest Crown Closure

		Ground Reference						Row Total
		1	2	3	4	5	6	
Image Classification	1	2	9	1	2	1	1	16
	2	2	8	3	6	1	1	21
	3	0	3	3	4	9	1	20
	4	0	0	2	8	7	10	27
	5	0	1	2	1	6	16	26
	6	0	0	0	0	3	31	34
Column Total		4	21	11	21	27	60	144

Crown Closure Categories

Class 1 = 0% CC
Class 2 = 1–10% CC
Class 3 = 11–30% CC
Class 4 = 31–50% CC
Class 5 = 51–70% CC
Class 6 = 71–100% CC

OVERALL ACCURACY =
58/144 = 40%

PRODUCER'S ACCURACY	USER'S ACCURACY
Class 1 = 2/4 = 50%	Class 1 = 2/16 = 13%
Class 2 = 8/21 = 38%	Class 2 = 8/21 = 38%
Class 3 = 3/11 = 27%	Class 3 = 3/20 = 15%
Class 4 = 8/21 = 38%	Class 4 = 8/27 = 30%
Class 5 = 6/27 = 22%	Class 5 = 6/26 = 23%
Class 6 = 31/60 = 52%	Class 6 = 31/34 = 91%

accuracies (i.e., cheating). Therefore, although this method is very simple to apply, it should only be used when agreement exists that it is a reasonable course of action. The other techniques described next may be more difficult to apply but easier to justify.

MEASURING MAP CLASS VARIABILITY

The second method for incorporating fuzziness into the accuracy assessment process is not as simple as expanding the major diagonal of the error matrix. While it is difficult to control variation in human interpretation, it is possible to measure that variation and to use these measurements to compensate for differences between reference and map data that are caused not by map error but by variation in interpretation. There are two options available to control the variation in human interpretation to reduce the impact of this variation on map accuracy. One is to measure each reference site with great precision to minimize the variance in the reference site labels. This method can be prohibitively expensive, usually requiring extensive field sampling and detailed measurements. The second option is to measure the variance and use these measurements to compensate for non-error differences between reference and map data. Measuring the variance requires having multiple analysts assess

TABLE 10.2

Error Matrix Showing the Ground Reference Data versus the Image Classification for Forest Crown Closure within Plus or Minus One Tolerance Class

		Ground Reference						Row Total	
		1	2	3	4	5	6		
Image Classification	1	2	9	1	2	1	1	16	
	2	2	8	3	6	1	1	21	
	3	0	3	3	4	9	1	20	
	4	0	0	2	8	7	10	27	
	5	0	1	2	1	6	16	26	
	6	0	0	0	0	3	31	34	
Column Total		4	20	11	21	27	60	144	

Crown Closure Categories

Class 1 =   0%   CC
Class 2 = 1–10%   CC
Class 3 = 11–30%   CC
Class 4 = 31–50%   CC
Class 5 = 51–70%   CC
Class 6 = 71–100% CC

OVERALL ACCURACY =
108/144 = 75%

PRODUCER'S ACCURACY	USER'S ACCURACY
Class 1 = 4/4 = 100%	Class 1 = 11/16 = 69%
Class 2 = 20/21 = 95%	Class 2 = 13/21 = 62%
Class 3 = 8/11 = 73%	Class 3 = 10/20 = 50%
Class 4 = 13/21 = 62%	Class 4 = 17/27 = 63%
Class 5 = 16/27 = 59%	Class 5 = 23/26 = 88%
Class 6 = 47/60 = 78%	Class 6 = 34/34 = 100%

each reference site. This assessment could be done by field visitation or by using image interpretation and requires an objective and repeatable method to capture the impacts of human variation. The collection of reference data for accuracy assessment is an expensive component of any mapping project; multiple visits to every reference site to capture variation may be exorbitantly expensive. Therefore, while it is theoretically possible, measuring reference data variability is not a viable component of most remotely sensed mapping projects. In other words, while this approach may be taken as an interesting academic exercise to get a better understanding of the variation in a particular situation, from a practical standpoint, this approach would not be employed.

THE FUZZY ERROR MATRIX APPROACH

The previous approaches of expanding the major diagonal to incorporate fuzziness in the accuracy assessment process may be hard to justify, and the effort needed to measure the variability using repeat observations may be cost prohibitive. Therefore, another method is required to incorporate fuzziness into the map accuracy assessment process. As mentioned earlier, one challenge in using fuzzy logic is the development of the specific rules for its application that everyone can agree on. Fuzzy systems often rely on experts for the development of these rules. Hill (1993) developed an arbitrary but practical fuzzy set rule that determined "sliding class widths" for assessing the accuracy of maps produced for The California Department of Forestry and Fire Protection of the Klamath Province in Northwestern California. Woodcock and Gopal (1992) relied on experts in their application of fuzzy sets to assess the accuracy of maps generated for Region 5 of the U.S. Forest Service. While both of their methods incorporated fuzziness into the accuracy assessment process, neither uses an error matrix approach. Instead, a number of other metrics to represent map accuracy and agreement were computed.

THE FUZZY ERROR MATRIX

Given the wide acceptance of the error matrix as the standard for reporting the accuracy of thematic maps, it would be far better to employ some approach that combines both the error matrix and some measure of fuzziness. Such a technique, called the *fuzzy error matrix app*roach, was introduced by Green and Congalton (2004) and is described here. The use of the fuzzy error matrix is a very powerful tool in the accuracy assessment process, because the fuzzy error matrix allows the analyst to compensate for situations where classification scheme breaks represent artificial distinctions along a continuum of land cover and/or where observer variability is often difficult to control. In other words, fuzziness can be used when it is difficult to define or interpret the map classes to maintain the characteristic of mutual exclusivity in the classification scheme. While one of the assumptions of the traditional or deterministic error matrix used in the rest of this book is that a reference data sample site can have only one label, this is not the case with the fuzzy error matrix approach.

Let's continue with the example used so far in this chapter. Table 10.3 presents a fuzzy error matrix generated from a set of fuzzy rules applied to the same classification as was used to generate the deterministic (i.e., non-fuzzy or traditional) error matrix that was presented in Table 10.1. In this case, the classification was defined using the following fuzzy rules:

- Class 1 was defined as always 0% crown closure. If the reference data indicated a value of 0%, then only a map classification of 0% was accepted.
- Class 2 was defined as acceptable if the reference data were within 5% of those of the map classification. For example, if the reference data indicated that a sample had 15% crown closure, and the map classification put it in Class 2 (1–10% crown closure), the answer would not be absolutely correct but would be considered acceptable.
- Classes 3–6 were defined as acceptable if the reference data were within 10% of those of the map classification. For example, a sample classified as Class 4 on the image but found to be 55% crown closure on the reference data would be considered acceptable.

As a result of these fuzzy rules, off-diagonal elements in the matrix contain two separate values. The first value in the off-diagonal represents those labels that,

TABLE 10.3

Error Matrix Showing the Ground Reference Data versus the Image Classification for Forest Crown Closure Using the Fuzzy Logic Rules

		\multicolumn{6}{c}{Ground Reference}	Row Total
		1	2	3	4	5	6	
Image Classification	1	2	6,3	1	2	1	1	16
	2	0,2	8	2,1	6	1	1	21
	3	0	2,1	3	4,0	9	1	20
	4	0	0	0,2	8	5,2	10	27
	5	0	1	2	1,0	6	12,4	26
	6	0	0	0	0	2,1	31	34
Column Total		4	21	11	21	27	60	144

Crown Closure Categories

Class 1 =	0%	CC
Class 2 = 1–10%		CC
Class 3 = 11–30%		CC
Class 4 = 31–50%		CC
Class 5 = 51–70%		CC
Class 6 = 71–100%		CC

OVERALL ACCURACY =
92/144 = 64%

PRODUCER'S ACCURACY	USER'S ACCURACY
Class 1 = 2/4 = 50%	Class 1 = 8/16 = 50%
Class 2 = 16/21 = 76%	Class 2 = 10/21 = 48%
Class 3 = 5/11 = 45%	Class 3 = 9/20 = 45%
Class 4 = 13/21 = 62%	Class 4 = 13/27 = 48%
Class 5 = 13/27 = 48%	Class 5 = 19/26 = 73%
Class 6 = 43/60 = 72%	Class 6 = 33/34 = 97%

although not absolutely correct, are considered acceptable labels within the fuzzy rules. The second value indicates those labels that are still unacceptable (i.e., wrong). The major diagonal still only tallies those labels considered to be absolutely correct. Therefore, to compute the accuracies (overall, producer's, and user's), the values along the major diagonal (i.e., absolutely correct) and those deemed acceptable (i.e., those in the first value) in the off-diagonal elements are combined. In Table 10.3, this combination of absolutely correct and acceptable answers results in an overall accuracy of 64%. This overall accuracy is significantly higher than the original error matrix (Table 10.1) but not as high as Table 10.2. This approach is also easier to justify, as the fuzzy rules are defined to effectively compensate for the observer variability.

It is much easier to justify the fuzzy rules used in generating Table 10.3 than it is to simply extend the major diagonal to plus or minus one whole class, as was done in Table 10.2. For crown closure, it is recognized that mapping typically varies by plus or minus 10% (Spurr, 1948). Therefore, it is reasonable to define as acceptable a range within 10% for classes 3–6. Classes 1 and 2 take an even more conservative approach and therefore, are even easier to justify.

In addition to this fuzzy set theory working for continuous variables such as crown closure, it also applies to more categorical data. All that is required is a set of fuzzy rules to explain or capture the variation. For example, in the hardwood range area of California, many land cover types differ only by which hardwood species is dominant. In many cases, the same species are present, and the specific land cover type is determined by which species is most abundant. Also, in some of these situations, these species look very much alike on aerial imagery and even on the ground. It is clear that there is a great deal of acceptable and unavoidable variation in mapping the hardwood range.

In a worldwide mapping effort funded by the National Geospatial-Intelligence Agency (NGA) using Landsat imagery, no ground visitation was possible for collecting the reference data. Because this project was conducted prior to the advent of Google Earth, the imagery used for the reference data was of such low resolution in some areas of the world as to make interpretation of the individual classes very difficult. The use of this fuzzy error matrix approach was the only viable solution in this case (Green and Congalton, 2004). A traditional, deterministic accuracy assessment conducted with such highly variable reference data would have unfairly represented the accuracy of this mapping effort.

Therefore, in many situations, the use of these fuzzy rules, which allow the incorporation of acceptable reference labels in addition to the absolutely correct reference labels into the construction of the error matrix, makes a great deal of sense. Using fuzzy rules in an error matrix approach combines all the established descriptive and analytical power of the error matrix while incorporating variation into the assessment.

As also discussed for the approach of expanding the major diagonal presented earlier in this chapter, some analysts or map users may decide that using a fuzzy error matrix approach is just a way to increase map accuracy (i.e., cheat). There are several factors that may be considered here to convince any skeptic that this approach is valid and necessary in some situations. First, a simple exercise can help. Gather those interested in the particular mapping project. Show a particularly complex reference data site (complex means that it is difficult to maintain mutual exclusivity of the classification scheme here) and ask each to write down what they think the appropriate map label is based on the classification scheme used for the project. Make sure each person has the full classification scheme with not only the map labels but also the definitions of each map class. (Better yet, if possible, take the group to a site and conduct this same process.) Then, have each person reveal their answer and explain why they chose that answer. It will quickly become apparent that a number of possible answers have been proposed, each with strong justification. Again, this result occurs because the differences between the map classes are subtle, are subject to interpretation, and lack strong mutual exclusivity. Conducting this exercise in a few areas usually convinces everyone that the use of fuzziness is justified in this project. Second, the argument can be made that the map classes that are very similar to each other should be collapsed, given a hierarchical classification scheme, to a more general class. For example, in a project to create a fine-scale map of the vegetation of Haleakalā National Park, the associations of *Sophora chrysophylla* - (*Coprosma montana*, *Leptecophylla tameiameiae*) Subalpine Shrubland and *Sophora chrysophylla* - *Dodonaea viscosa* - (*Leptecophylla tameiameiae*) Montane Shrubland were found to be indistinguishable as separate map classes, because they were differentiated only by the presence or absence of the small shrub *Dodonaea viscosa*. As a result, just one plant of *Dodonaea viscosa* could impact the vegetation label. Because the associations were indistinguishable, they were combined into the *Sophora chrysophylla* - (*Coprosma montana* - *Leptecophylla tameiameiae* - *Dodonaea viscosa*) Subalpine Shrubland map class (Green et al., 2015).

However, collapsing the map classes into more general classes reduces the number of map classes and the detail in the map. The analyst or map user may want to retain as much detail as possible and is willing to have some fuzziness in the assessment process to retain the finest level of information. Finally, as discussed earlier, a situation may arise when it is simply not possible to obtain/collect reference data at a spatial resolution that allows precise labeling of each map class. In this case, it would be problematic to assume that the reference data are highly accurate without compensating for observer variability. Failure to use a fuzzy assessment approach would result in poor map accuracy not necessarily because the map is in error but perhaps, because the reference data are.

IMPLEMENTATION OF THE FUZZY ERROR MATRIX

The implementation of the fuzzy error matrix approach is greatly simplified with the use of a special reference data collection form such as the one in Figure 10.1.

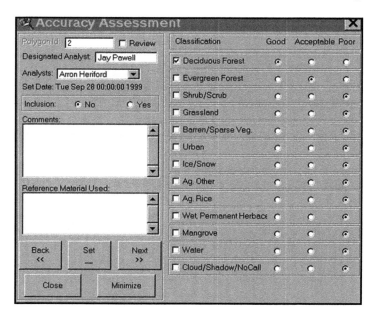

FIGURE 10.1   Form for labeling accuracy assessment reference sites.

Each reference site can be evaluated for the likelihood of being identified as each of the possible map classes given the fuzzy rules for that project. First, the analyst determines the most appropriate ("good") label for the site and enters this label in the appropriate box under the "classification" column on the form. This label determines which row of the matrix the site will be tallied in and is also the value used for calculation of the deterministic error matrix. After the most appropriate label has been assigned to the site, the remaining possible map labels are evaluated as either "acceptable" or "poor" candidates for the site's label, again as indicated by the fuzzy rules. For example, a site might fall near the classification scheme margin between deciduous forest and evergreen forest because of the exact mix of species and/or the difficulty of interpreting the exact mixture on the reference data imagery. In this instance, the analyst might rate deciduous forest as the most appropriate label (i.e., "good"), but also rate evergreen forest as "acceptable" (see Figure 10.1). No other map classes would be acceptable, and all the others would be rated as "poor."

Figure 10.2 shows another example of an accuracy assessment form created for a fine-scale mapping project in Sonoma County, California. Rather than evaluating the appropriateness of all 83 possible classes in the map for each accuracy assessment reference sample, the reference data personnel fill in only what they believe the

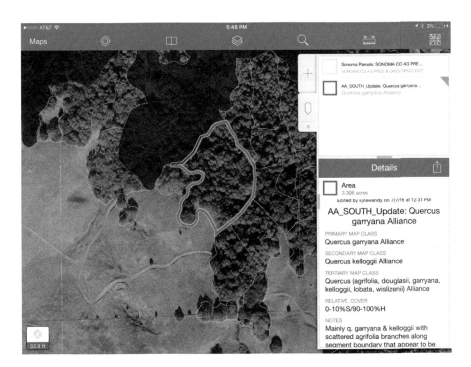

FIGURE 10.2 Accuracy assessment form developed in ArcGIS Collector by Tukman Geospatial for the Sonoma County mapping project. The form incorporates fuzzy logic by allowing the reference data collection personnel to enter up to three possible labels for the accuracy assessment sample site.

most probable (i.e., primary), second most probable (i.e., secondary), and third most probable (i.e., tertiary) estimates of what the reference label could be. If the map class obviously only has one possible label (e.g., water), then no secondary or tertiary label is entered. In situations with complex classification schemes, this approach can greatly minimize the amount of time taken to label each accuracy assessment reference sample.

The fuzzy error matrix approach allows the analyst to compensate for interpreter variability and difficulty in determining only a single label for each reference data sample site. While this method can be used for any accuracy assessment, it works best when (1) there are issues in collecting good reference data because of limitations in the reference data collection methods, including the quality/spatial resolution of the reference data imagery, (2) when interpreter variability cannot be controlled, or (3) when the ecosystem being mapped is highly heterogeneous, as represented by a highly complex classification scheme. If all these factors can be controlled (i.e., there is little variation or fuzziness in the classification scheme, or detailed measurements can be taken to minimize the variation), then there may be little need for this approach. However, in most projects creating maps from remotely sensed imagery, the use of the fuzzy error matrix approach can significantly help to incorporate the variation inherent in the project and should be judiciously employed. It also provides additional information to the map user, who can use the fuzzy error matrix to assess the ambiguity associated with each map label.

ANOTHER FUZZY ERROR MATRIX EXAMPLE

Table 10.4 shows a fuzzy error matrix for a categorical classification scheme (i.e., a land cover mapping project). Again, the power of this approach is the ability to compute the same descriptive metrics as in the traditional deterministic error matrix. The computation of the overall, producer's, and user's accuracy statistics for the fuzzy error matrix follows the same methodology as the traditional deterministic error matrix, with the following additions. Non-diagonal cells in the matrix contain two tallies, which can be used to distinguish class labels that are uncertain, or that fall on class margins, from class labels that are most probably in error. The first number represents those sites in which the map label matched an "acceptable" reference label in the fuzzy assessment (Table 10.4). Therefore, although the label was not considered the most appropriate, it was considered acceptable given the fuzziness of the classification system and/or the minimal quality of some of the reference data. These sites are considered a "match" for estimating fuzzy accuracy. The second number in the cell represents those sites where the map label was considered poor (i.e., an error). The fuzzy assessment overall accuracy is estimated as the percentage of sites where the "good" and "acceptable" reference labels matched the map label. The producer's and user's accuracies are computed in the traditional way, but again, instead of just using the value on the major diagonal ("good"), the values in the first off-diagonal position ("acceptables") are also included (Table 10.4).

TABLE 10.4
Example Fuzzy Error Matrix Showing Both Deterministic and Fuzzy Accuracy Assessment

	REFERENCE DATA								User's Accuracies			
	Decid. Forest	EG Forest	Scrub/ Shrub	Grass	Barren/ Sparse	Urban	Agric.	Water	Deterministic Totals	Percent Deterministic	Fuzzy Totals	Percent Fuzzy
Deciduous Forest	**48**	24.7	0.1	0.3	0.0	0.1	0.11	0.18	48/113	42.5%	72/113	63.7%
Evergreen Forest	4.0	**17**	0.1	0.0	0.0	0.0	0.1	0.3	17/26	65.4%	21/26	80.8%
Shrub/Scrub	2.0	0.1	**15**	8.1	0.0	0.0	2.2	0.0	15/31	48.4%	27/31	87.1%
Grassland	0.1	0.0	5.1	**14**	0.0	0.0	3.0	0.0	14/24	58.3%	22/24	91.7%
M A P Barren/Sparse Veg.	0.0	0.0	0.2	0.0	**0**	0.0	0.1	0.0	0/3	0.0%	0/3	0.0%
Urban	0.0	0.0	0.0	0.0	0.0	**20**	2.0	0.0	20/22	90.9%	22/22	100.0%
Agriculture	0.1	0.1	7,15	18.6	0.0	2.0	**29**	1,2	29/82	35.4%	57/82	69.5%
Water	0.0	0.0	0.0	0.0	0.0	0.0	0.0	**8**	8/8	100.0%	8/8	100.0%
Producer's Accuracies												
Deterministic Totals	48/56	17/50	15/47	14/50	NA	20/24	29/51	8/33				
Percent Deterministic	85.7%	34.0%	31.9%	28.0%	NA	83.3%	56.9%	24.2%				
Fuzzy Totals	54/56	41/50	27/47	40/50	NA	22/24	36/51	10/33				
Percent Fuzzy	96.4%	82.0%	57.4%	80.0%	NA	91.7%	70.6%	30.3%				

Overall Accuracies

Deterministic	Fuzzy
151/311 48.6%	230/311 74.0%

SUMMARY

While three methods are presented in this chapter for dealing with variation or fuzziness in the accuracy assessment process, the fuzzy error matrix approach is by far the most useful and should be considered a best practice. The elegance of this approach is that it combines all of the power of the traditional (called *deterministic* here) error matrix, including computing overall, producer's, and user's accuracies, with the ability to incorporate the variation inherent in many classification schemes or resulting from the reference data collection process. Given that the matrix contains the information to compute both the traditional deterministic accuracy measures and fuzzy accuracy measures, there is strong impetus to use this approach. It is highly recommended that this approach be considered whenever map class variation or variation in the reference data collection process is a significant issue in the mapping project.

11 Object-Based or Polygon Accuracy Assessment

Until the turn of the century, most thematic maps generated from satellite remotely sensed data used a pixel-based classification approach, and maps generated from aerial photos used a polygon-based classification approach. The pixel-based method suffered, because pixels are arbitrary delineations of rectangular areas on the ground. We tend to see things as groupings, specifically noting where boundaries are between things that we perceive are different from each other. Recently, and especially with the advent of higher–spatial resolution satellite and airborne imagery, classification methods have been developed that group image pixels into meaningful segments or objects and then classify the segments (Blaschke, 2010). This method is called *object-based image analysis* (OBIA). When the OBIA approach is used to generate a thematic map, pixels with similar spectral characteristics are grouped into segments/objects, and then, the segments are classified as a group instead of pixel by pixel. The size and shape of the segments are generally determined by user inputs and the variability of the spectral characteristics of the pixels in the segments. Once created, these segments can have lots of valuable characteristics, such as size, shape, and texture, and a variety of zonal statistics from ancillary data that can be used to help classify the segments beyond just the spectral responses of each pixel. Once the segments/objects are classified, segments with the same map class that are adjacent to each other are grouped together (i.e., dissolved) into larger polygons. Because traditional and widely accepted aerial photo interpretation was also polygon based, and because the polygons represent recognizable objects on the ground and imagery, maps created using OBIA tend to be more understandable, relatable, and useful for map users than maps created using a pixel-based approach.

As in any thematic accuracy assessment, evaluating an object-based or polygon map requires sample units of reference data to compare with the map so that an error matrix can be generated. The rest of this chapter deals with the many factors that must be considered when conducting an OBIA-based accuracy assessment, including what is the appropriate sample unit for these reference data and what methods should be used for the collection of the samples. Many principles that have already been presented in this book still hold true. However, there are some additional concepts and issues that must be considered when dealing with the special case of OBIA-derived maps.

WHAT SAMPLE UNIT SHOULD BE USED?

By far the biggest consideration when conducting an object-based accuracy assessment is the determination of the sample unit. There are a number of possibilities here. First, it is possible simply to conduct the accuracy assessment as previously

described; that is, using a cluster of pixels (usually square) well within the map polygon that is sufficient in size to offset any positional error in the sample unit. In
this case, the methodology has already been presented in Chapters 6 and 7. A second option would be to select some larger-area polygon as the sampling unit. This
larger-area polygon could be (1) the actual map polygon, (2) a sub-segment within
the actual map polygon but buffered by some distance (i.e., number of pixels) inside
the polygon boundary to avoid the edge and therefore, any positional issues, or (3)
some fixed-area plot, also buffered and placed well within the boundaries of the map
polygon (the size of the fixed plot would need to be larger than the specified minimum mapping unit).

There are advantages and disadvantages to selecting any of these approaches. The
traditional cluster of pixels as a single sample unit has the major advantage of being
the most common method employed, in which the considerations and procedures for
efficiently and effectively collecting reference data are well documented and understood. A second and very powerful advantage is that since the sample units are all
the same size, the error matrix can be generated using a tally approach (see more
discussion later). The biggest disadvantage is that the map polygons are not being
assessed, and therefore, it is difficult to quantitatively evaluate their effectiveness
versus a pixel-based map. The biggest advantage of using some kind of polygon
to assess the OBIA results is clearly that the sample unit coincides better with the
map product. However, if either the entire map polygon or some buffered polygon
is used, the size of the sample units will be variable, and a tally approach to generating the error matrix may not be viable (an area-based error matrix may be more
appropriate). A fixed-area plot approach offers a compromise here by using an area
that is bigger than a small cluster of pixels while maintaining a fixed area, such that
a tally-based error matrix can still be employed. The biggest disadvantage of using
a polygon approach is the complexity of labeling the entire polygon in the reference
data. More detailed discussion of these important considerations follows in the next
sections of this chapter.

TRADITIONAL TALLY-BASED VERSUS AREA-BASED ERROR MATRIX

So far in this book, the error matrices presented have been based on a tally method.
That is, each reference data sample unit was considered as a single tally, and the
appropriate cell (intersection of row and column) in the matrix was incremented by
one. The tally method for generating an error matrix is simple and based on the fact
that each tally is of equal value. This assumption is valid if the sample units used to
generate the matrix are of equal size (area). For example, when using a 3×3 pixel
sample unit for assessing the accuracy of a thematic map created from Landsat TM
data, the sample unit is the same size for every sample, and a simple tally can be used
to generate the matrix. However, if the sample units are not the same size but rather,
a variable-sized polygon, then using a tally-based error matrix would be problematic.
Each polygon is not of equal value, since the polygons are of different sizes (areas)
and are then of different weight or importance in the error matrix. Therefore, it may
be more appropriate in this situation to use an area-based error matrix. The area-
based error matrix uses the same principles as the traditional error matrix. However,

instead of tallying each reference data sample in the appropriate cell, the area of the polygon (e.g., hectares or acres) used as the reference data sample unit is entered into the appropriate cell in the error matrix. Table 11.1 presents a comparison between the traditional tally-based error matrix used when a pixel-based classification approach is used and an area-based error matrix, which is more appropriate when an OBIA classification is employed using a variable-sized polygon as the sample unit. The only real difference between the matrices is that the values in the traditional matrix

TABLE 11.1

Comparison of the Traditional Error Matrix (a) and the Area-Based Error Matrix (b)

(a)

		Reference Data				Row Total	User's Accuracy
		1	2	...	k		
Map Data	1	n_{11}	n_{12}	...	n_{1k}	n_{1+}	n_{11}/n_{1+}
	2	n_{21}	n_{22}	...	n_{2k}	n_{2+}	n_{22}/n_{2+}
	⋮	⋮	⋮	...	⋮	⋮	⋮
	k	n_{k1}	n_{k2}	...	n_{kk}	n_{k+}	n_{kk}/n_{k+}
Column Total		n_{+1}	n_{+2}	...	n_{+k}	N	Overall Accuracy:
Producer's Accuracy		n_{11}/n_{+1}	n_{22}/n_{+2}	...	n_{kk}/n_{+k}		$\dfrac{\sum_{i=1}^{k} n_{ii}}{N}$

(b)

		Reference Data				Row Total	User's Accuracy
		1	2	...	k		
Map Data	1	S_{11}	S_{12}	...	S_{1k}	S_{1+}	S_{11}/S_{1+}
	2	S_{21}	S_{22}	...	S_{2k}	S_{2+}	S_{22}/S_{2+}
	⋮	⋮	⋮	...	⋮	⋮	⋮
	k	S_{k1}	S_{k2}	...	S_{kk}	S_{k+}	S_{kk}/S_{k+}
Column Total		S_{+1}	S_{+2}	...	S_{+k}	S	Overall Accuracy:
Producer's Accuracy		S_{11}/S_{+1}	S_{22}/S_{+2}	...	S_{kk}/S_{+k}		$\dfrac{\sum_{i=1}^{k} S_{ii}}{S}$

n_{kk} is the number of reference data sample units that fall in that particular cell, N is the total number of sample units, S_{kk} is the total area of all of the reference data sample units that fall in that particular cell, and S is the total area sampled.

represent tallies, while the values in the area-based error matrix represent the total areas of all the sample units for each cell.

COMPARISON OF THE TWO ERROR MATRIX APPROACHES

It is helpful and informative to see these two approaches demonstrated with an example. Tables 11.2 and 11.3 are the results of a mapping project used to test whether using multi-date images for creating a land cover map improved the accuracy of that map versus just using a single date of imagery (MacLean and Congalton, 2013). The project was conducted for the Coastal Watershed in southeastern New Hampshire, and all classifications were performed using an OBIA approach. Table 11.2 shows the results of simply using the traditional tally-based error matrix, while Table 11.3 shows the same assessment but using the area-based matrix. The overall accuracy of the map using the traditional error matrix approach was 70% (Table 11.2), while the accuracy computed using the area-based approach was higher, at almost 75% (Table 11.3). In these two maps, the most confused classes were the cleared/other open and the mixed forest classes. Cleared/other open was primarily confused with the active agriculture class, which for this area is quite understandable. Most agriculture in this area is hay/pasture, and the cleared/other open category encompassed areas such as golf courses and other grassy areas that are spectrally quite similar to pasture lands. The mixed forest class was confused with both the deciduous and coniferous forest categories. Given the variability of the forests in southern New Hampshire and the 30 meter pixels of the Landsat 5 TM images, it is also no surprise that mixed forest was commonly confused with other types of forest.

The issue here with using the area-based error matrix is that instead of each sample unit having equal weight (tally of 1), the matrix is now weighted by the size of each sample unit. Therefore, the proportion in the matrix is no longer a function of the number of sample units per map class but rather, a function of the combination of the number and size of the sample units. As previously discussed as a consideration in Chapter 6, in some situations, the assessment may be performed as a proportion, while in others, a minimum number of sample units may be selected per map class (stratum). If the impact of the area of the sample units is not considered, then the objective of the sampling strategy may be thwarted by the uneven sizes of the sample units of each map class. One could envision a situation in which an easy-to-map land cover class (e.g., water) also has very large polygons, and therefore, the map accuracy is disproportionally weighted by the water samples, resulting in an inflated accuracy. Therefore, it is vital to consider the impact of polygon size when accessing map accuracy from uneven-sized sample units so that the result is not unduly increased or decreased.

REFERENCE DATA COLLECTION/LABELING

As presented earlier in the book (Chapter 6), when classifying an image into a thematic map, samples are needed both for use as training data and also as an independent data set for the reference data. Reference sample units are usually collected either through photo/image interpretation or by going to the field and making

TABLE 11.2

Traditional Error Matrix (Cell Values in Tallies) for the Multi-date 2010 OBIA Classification

Map Data	Reference Data								Row Total	User's Accuracy
	Active Agriculture	Cleared/Other Open	Coniferous	Deciduous	Developed	Mixed Forest	Open Water	Wetlands		
Active Agriculture	51	20	0	0	14	0	0	0	85	60%
Cleared/Other Open	19	19	0	0	13	0	0	0	51	37%
Coniferous	0	0	49	4	0	16	0	0	69	71%
Deciduous	0	0	2	63	0	12	0	0	77	82%
Developed	10	10	0	0	70	0	0	0	90	78%
Mixed Forest	0	1	14	28	0	28	0	0	71	39%
Open Water	0	0	0	0	0	0	50	0	50	100%
Wetlands	0	0	0	0	0	0	0	50	50	100%
Column Total	80	50	65	95	97	56	50	50	543	Overall Accuracy: 70.0%
Producer's Accuracy	64%	38%	75%	66%	72%	50%	100%	100%		

TABLE 11.3

Area-Based Error Matrix (Cell Values in Hectares) for the Multi-date 2010 OBIA Classification

Reference Data (ha)

Map Data (ha)	Active Agriculture	Cleared/Other Open	Coniferous	Deciduous	Developed	Mixed Forest	Open Water	Wetlands	Row Total	User's Accuracy
Active Agriculture	471.3	189.7	0.0	0.0	113.7	0.0	0.0	0.0	774.7	61%
Cleared/Other Open	161.0	165.1	0.0	0.0	114.8	0.0	0.0	0.0	440.9	37%
Coniferous	0.0	0.0	737.1	36.7	0.0	212.0	0.0	0.0	985.8	75%
Deciduous	0.0	0.0	14.4	1080.9	0.0	141.0	0.0	0.0	1236.3	87%
Developed	82.1	68.9	0.0	0.0	582.4	0.0	0.0	0.0	733.4	79%
Mixed Forest	0.0	0.0	148.4	244.7	0.0	272.7	0.0	0.0	665.8	41%
Open Water	0.0	0.0	0.0	0.0	0.0	0.0	696.0	0.0	696.0	100%
Wetlands	0.0	0.0	0.0	0.0	0.0	0.0	0.0	525.2	525.2	100%
Column Total	714.4	423.7	899.9	1362.3	810.9	625.7	696.0	525.2	6058.1	Overall Accuracy: 74.8%
Producer's Accuracy	66%	39%	82%	79%	72%	44%	100%	100%		

observations or measurements. The accuracy and interpretability of the classification are fully dependent on the accuracy of both the training data and the reference data. The accuracy of the training data will influence the success of the classification, and the reference data are assumed to be correct in an accuracy assessment, so that any discrepancies between the thematic map labels and the reference data labels are assumed to be errors on the map. Therefore, how the reference data are collected and labeled can greatly influence the success of the land cover classification.

As previously discussed, in the traditional pixel-based classification approach, a small group of pixels (e.g., a three-by-three cluster or larger, depending on the spatial resolution and positional accuracy of the imagery) within a homogeneous land cover type is recommended as a single reference sample unit. When the imagery is of medium to high spatial resolution, and therefore, the pixels are relatively small, the area covered by such a reference sample unit is also quite small and generally covers only a small amount of variability within the landscape (Figure 11.1a). Therefore, a single observation taken on the ground within that reference sample unit (usually near the center) should be sufficient to accurately label that group of pixels/the reference sample unit. However, if the sample unit is a polygon as a result of using either

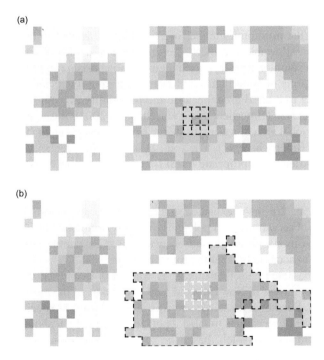

FIGURE 11.1 The figures are meant to clearly represent the landscape variability on a moderate-resolution natural color composite image (e.g., Landsat Thematic Mapper [TM] or Operational Land Imager [OLI]). (a) A three-by-three reference unit (dashed black box) as recommended for pixel-based classification. (b) In a polygon classification, if polygons are used as the reference sample unit (shown by the dashed black polygon), then a single observation (dashed white box) in the polygon may not allow accurate labeling of the entire polygon.

image interpretation or the OBIA approach, then the area is larger and/or more variable, such that a single ground observation somewhere within the polygon will often not be adequate for labeling the entire polygon (Figure 11.1b). Therefore, usually, each polygon reference sample unit in a polygon classification encompasses more variation in the landscape than in a pixel-based approach. With greater variability within the reference sample units, more than a single observation may be necessary to accurately label each sample unit (Figure 11.1b). Potentially high variability within reference sample units combined with insufficient sampling could lead to inaccurate reference labeling, especially in more complex or heterogeneous land cover classes.

In many landscapes, forested habitats provide substantial values and are the subject of intensive mapping efforts, especially for natural resource, human–environment interaction, or wildlife studies. However, forests can be quite variable in comparison to other land cover types, depending on the level of detail in classifying the forest. Therefore, more observations may be necessary to accurately label polygon reference data sample units for assessing forested land cover types in a map generated using a polygon classification approach. An analysis conducted by MacLean et al. (2013) strove to determine the number of ground-based observations that must be taken to accurately label a forest polygon in complex New England forests. This work employed a bootstrapping statistical simulation to determine the minimum number of observations needed to label the polygon within a given allowable variance. The information at each observation was collected using a common forestry-based method called *prism sampling* (i.e., horizontal point sampling or Bitterlich sampling). Prism sampling is an effective method of quantifying tree basal area, because the probability of sampling a tree is proportional to its size (Bitterlich, 1947; Husch et al., 2003). Basal area is defined as the cross-sectional area of a tree, inside the bark, at breast height (1.3 meters above the ground) (Bitterlich, 1947). The total basal area per tree species can then be used to determine the appropriate forest type label (i.e., map class) for that sample unit based on the project classification scheme. The results of this work showed that multiple observations were required to accurately label any forested polygon (MacLean et al., 2013). As expected, polygons that contained mixtures of species needed more observations than polygons that were dominated by a single type. On average, three to six ground-based observations (i.e., sample units per polygon) were needed to accurately label the forest polygons.

It should be noted here that this study was for only one section of the United States and that forests in New England tend to be some of the most complex in the United States. Forests in the south and west tend to have less species diversity and may need fewer observations to label a polygon accurately. It should also be noted that this approach was solely based on ground observations. Combining some ground observations with very high–spatial resolution imagery may reduce the need for so many ground observations. Given the availability of high-resolution imagery and most recently, what can be collected using unmanned aerial systems (UASs), there is great potential to obtain the necessary information to label the polygon with less ground effort. However, it is worth considering the effort that might be required to accurately label the reference sample unit if that unit is a polygon.

Other projects and organizations have also begun to explore the use of some type of polygon as the reference sample unit for assessing the accuracy of polygon classifications. A few examples are presented here to demonstrate some of the issues and considerations that are of key importance. It is clear that further exploration and study must be conducted before final recommendations can be made considering the accuracy assessment of polygon maps. Just like everything else in this book, there is no one method or best approach. The goals of the mapping along with statistical validity and practical application must be balanced to achieve the best assessment possible.

The U.S. National Park Service (NPS) is responsible for mapping the land cover in the National Parks throughout the United States at a very high level of specificity (see the case study in Chapter 12 for details). Over time, Park Service personnel have developed considerable experience in both mapping and assessing the accuracy of the maps generated for the parks. Lea and Curtis (2010) have published the guidelines used by the NPS for thematic map accuracy assessment. Their approach uses the minimum mapping unit (mmu) of the map to determine the size of the reference data sample unit. The sample unit is then a fixed-radius circle based on the mmu (e.g., a 0.5 ha mmu uses a fixed radius circle of 40 meters, while a 1 ha mmu uses a radius of 56 meters). Since the size of the sample unit is fixed, a tally approach for generating the error matrix can be used here.

The co-author of this book is part of a project led by Tukman GeoSpatial LLC to map the approximately 1 million acres of Sonoma County, California (www.sonomavegmap.org). As in the mapping done by the NPS, the classification scheme used in Sonoma County is quite complex, encompassing 83 map classes. The mapping was performed using an OBIA approach, and the accuracy assessment used the segments that resulted from the image segmentation process as the sampling units. Field personnel used a combination of ground observations in conjunction with high–spatial resolution imagery to label each sample unit. In this case, the sample unit size was not fixed, and both tally-based and area-based error matrices were generated as part of the accuracy assessment.

Finally, Dr. J. B. Sharma at the University of North Georgia has conducted a pilot study in which he and his team have evaluated the impact of having varying sample unit areas when using an area-based error matrix approach (Personal Communication). This work has shown that randomly selecting sample units within map classes can result in unevenly weighted (non-proportional) error matrices that may not be indicative of the actual map accuracy. His team explored using sample units of approximately the median size and achieved results that were more comparable to the tally-based method, further indicating that various sizes of sample units can impact the accuracy.

COMPUTING THE ACCURACY

In addition to the issues with collecting and labeling the reference data sample unit, assessing the accuracy of a thematic (e.g., land cover) map created using a polygon approach has some additional considerations beyond what has already been presented for the pixel-based approach. The statistics for calculating accuracy when using polygons as the reference data sample units are different from those used in a pixel-based approach, since the size of each reference sample unit varies (Radoux et al., 2010). In the traditional pixel-based approach, or when all reference units are the same size (i.e., the tally-based approach), overall accuracy is estimated using:

$$\hat{\pi} = \frac{\sum_{i=1}^{n} C_i}{n} \tag{11.1}$$

where:
- π is overall accuracy
- C_i is equal to 1 or 0 if the validation sample unit i is correctly classified (yes and no, respectively)
- n is the number of validation units collected

A quick look at this equation reveals that it is equivalent to the equation shown in Table 11.1a and in Chapter 5. Unfortunately, many researchers have employed this same equation to calculate accuracy when using variable-sized, polygon reference sample units (Radoux et al., 2010). However, this equation does not account for these variable sizes of the polygons in the accuracy assessment. The accuracy of a map created using a polygon approach should be computed using:

$$\pi = \frac{\sum_{i=1}^{N} C_i S_i}{\sum_{i=1}^{N} S_i} \tag{11.2}$$

where:
- N is the total number of segments in the image
- S_i is the area of a single sample unit i

However, accuracies for every polygon in the map are very rarely known, so Radoux et al. (2010) presented two estimates of overall accuracy. The first just replaces N with n:

$$\hat{\pi} = \frac{\sum_{i=1}^{n} C_i S_i}{\sum_{i=1}^{n} S_i} \tag{11.3}$$

which weights the accuracy assessment by the area of the validation polygons (Radoux et al., 2010). The second estimate of polygon-based accuracy includes the added information of the size of the remainder of the polygons that were not used in the validation process. While the accuracy, C_i, is not known for each polygon/segment, the size or area, S_i, is typically known for most polygon projects. Therefore, the information gained from knowing the S_i of the unsampled polygons can be used to reduce the variance of the estimate of overall accuracy using the equation

$$\hat{\pi} = \frac{1}{S_T} \left(\sum_{i=1}^{n} C_i S_i + \hat{p} \sum_{i=n+1}^{N} S_i \right)$$
(11.4)

where:
S_T is the total area of the map
$\hat{p}$ is the estimate of the probability of a polygon being classified correctly

As long as C_i is independent of S_i, $\hat{p}$ can be estimated using:

$$\hat{p} = \frac{1}{n} \sum_{i=1}^{n} C_i$$
(11.5)

Radoux et al. (2010) found that when this equation was used for $\hat{\pi}$, fewer polygons were needed as reference sample units than the number of sample units necessary for an accuracy assessment in a pixel-based approach to achieve the same accuracy and variance estimates.

Since the accuracy of maps created using a polygon approach should be weighted by the area of the reference units if polygons are used, an error matrix that incorporates area into each cell is appropriate for reporting thematic accuracy. The new area-based error matrix is set up similarly to the traditional tally-based error matrix, but instead of each reference unit having the same weight, the individual cells reflect the total area of the reference units that fall into that cell, as shown in Table 11.1b.

Using the area-based error matrix (Table 11.1b), overall accuracy can easily be computed. If overall accuracy is being computed using Equation 11.3, the same overall accuracy would be computed in the error matrix using:

$$\hat{\pi} = \frac{\sum_{i=1}^{k} S_{ii}}{S}$$
(11.6)

where the sum of S_{ii} is the sum of the major diagonal cells, similarly to how overall accuracy is computed in the traditional error matrix.

If overall accuracy is being computed using Equation 11.4, the values from the area-based error matrix can also compute the same overall accuracy using:

$$\hat{\pi} = \frac{\sum_{i=1}^{k} S_{ii} + \hat{a}(S_T - S)}{S_T}$$
(11.7)

where $\hat{a}$ is the overall accuracy of the map calculated using a traditional pixel-based error matrix.

When computing the overall accuracy using Equation 11.4, it is especially important that both the polygon-based and the traditional tally-based error matrix are reported. User's and producer's accuracies can be computed in the area-based error matrix using the same procedures as with a traditional pixel-based error matrix

(Table 11.3). Finally, Kappa statistics can also be computed using the same methods as the traditional error matrix.

CONCLUSIONS

This chapter has presented a number of approaches and the corresponding considerations to conduct an accuracy assessment when using a polygon-based classification approach for generating a thematic map. Many of the considerations previously described for a map generated using the traditional, pixel-based classification approach apply. However, care must be taken when deciding to use reference sample units of different size areas (i.e., polygons of non-equal areas). When the reference sample units are of different sizes, then a tally-based approach for generating the error matrix is no longer appropriate. Instead, an area-based error matrix can be used, in which the tallies are replaced by the actual areas of each polygon reference sample unit. The computation of the appropriate descriptive statistics (e.g., overall, producer's, and user's accuracies) and analytical statistics (e.g., Kappa statistic) is similar to the tally-based matrix. However, some efficiencies can be gained, because it is possible to know the area of all the polygons in the map, both those selected as reference sample units and those that were not. These areas can then be used in the accuracy calculations to achieve valid results with fewer samples.

Finally, labeling these larger and potentially more complex polygon reference sample units is also more difficult, as a single observation may no longer be sufficient to provide an accurate label. A pilot study in a complex forest environment showed that multiple ground observations, perhaps as many as six in a single polygon (reference sample unit), may be needed to accurately label that single sample (note that these six observations are not six samples. They are six observations used to label one polygon sample unit). Combining high-resolution imagery with ground observations may reduce these numbers. As the use of OBIA continues to develop, especially given the proliferation of high and very high–spatial resolution imagery, methods for effectively assessing the accuracy of polygon-based maps will grow in importance.

12 An Object-based Accuracy Assessment Case Study

The Grand Canyon National Park/Grand Canyon-Parashant National Monument Vegetation Classification and Mapping Project

This chapter and the next are actual case studies to demonstrate the principles and practices outlined in this book. These case studies are here to provide the reader with a thorough discussion of the thoughts, considerations, methodologies, and decisions that were made to assess the accuracy of the map in each project. Neither assessment case study is perfect, and neither should be followed as an example of exactly what to do. Rather, more details on the assessment process are provided in these two case studies than the reader would ever see in a peer-reviewed paper and even in most project final reports. The reader should use these case studies to get a good grasp of the limitations presented by each mapping project and how the assessment team dealt with these limitations in a manner that attempted to maintain statistical validity while dealing with the practical solution to the problems.

Chapter 12 presents the results of an object-based classification and accuracy assessment. This example is much more recent in the authors' experience and represents more state-of-the-art analysis of map accuracy. However, as in any assessment, there are always trade-offs to be made and lessons to be learned at the end.

Chapter 13 discusses a specific case study using both photo interpretation of one map and pixel-based classification of another map and the corresponding accuracy assessments. It is a classic example that will resonate with many readers. This case study was done early in the authors' experience with conducting such assessments and therefore, should be very appropriate for those readers who are new to or less experienced in dealing with map accuracy. Extensive details are supplied that should provide the reader with lots to consider when planning their assessment.

The combination of considerations, limitations, decisions, and discussions covered in these two case studies will provide the careful reader with insights and experience that will be of great benefit as they conduct their own accuracy assessment.

INTRODUCTION

This chapter highlights a real-world case study to demonstrate typical trade-offs and decisions that need to be made during the implementation of an object-based thematic accuracy assessment. The case study is an accuracy assessment of a fine-scale vegetation map created for the National Park Service (NPS) Grand Canyon National Park/Grand Canyon-Parashant National Monument (Kearsley et al., 2015). The first section of this chapter gives an overview of the project and briefly summarizes the mapping methods used. The next section summarizes the accuracy assessment design, data collection, and analysis tasks. The chapter concludes with a review of lessons learned during the design and implementation of the accuracy assessment.

OVERVIEW OF THE CASE STUDY

The NPS produces standardized, high-quality vegetation maps of its National Parks and NPS Units through the NPS Vegetation Inventory Program (NPS-VIP). These maps are created for the NPS Natural Resource Inventory and Monitoring (I&M) Program, whose purpose is to develop long-term baseline data in support of resource assessment, conservation initiatives, and park management. In 2007–2012, the NPS contracted with several organizations to create updated vegetation maps of Grand Canyon National Park and the NPS-administered portions of Grand Canyon-Parashant National Monument. The combined project area was approximately 1.4 million acres. The project goals were to create a hierarchical vegetation classification scheme using the National Vegetation Classification Standard (NVCS), create descriptions for the classifications at the association level, build a key to those associations, and use that information in conjunction with remote sensing methods and field visits to create a fine-scale thematic vegetation map for the project area.

The Grand Canyon National Park and Grand Canyon-Parashant National Monument are located in northern Arizona, United States (Figure 12.1). This remote region of the United States is characterized by dramatic and diverse topographical, biological, geological, and archaeological features. The Colorado River, bordered by high plateaus to the north and south, flows southwest through the canyon, dividing the eastern half and creating the southern boundary of the west half of the project area. The elevation of the Park ranges from a low of about 375 meters (1250 feet) to the highest point on the North Rim reaching over 2800 meters (9200 feet). The varied landscape and soil types support myriad habitats and a high diversity of botanical resources that derive from the convergence of flora from the Colorado Plateau, Great Basin, Mojave, and Sonoran provinces.

As in any remote sensing mapping project, four fundamental tasks were completed:

1. Understanding and characterizing the variation on the ground, which is to be mapped in a hierarchical, mutually exclusive and totally exhaustive classification scheme

FIGURE 12.1 Location of the project area in northwestern Arizona. The green lines outline Grand Canyon National Park, and the pink lines outline the NPS-administered portions of Grand Canyon-Parashant National Monument.

2. Linking variation on the ground to variation in the imagery and ancillary data used to make the map
3. Controlling variation in the imagery and ancillary data that is not correlated with the map classes
4. Capturing the variation in the imagery and ancillary data that is correlated with the map classes to create the map (Green et al., 2017)

The first step in the project involved building a hierarchical classification scheme and a dichotomous key of vegetation associations meeting the U.S. NVCS requirements (USNVCS, 2018). The scheme clearly describes the vegetation at the association level, and the key provides hierarchical rules for labeling vegetation to the association level in the field. The vegetation classification scheme and key were developed by the NPS and NatureServe (www.natureserve.org/) using two sources of vegetation classification field plot data: (1) 1508 classification plots collected by the NPS during 2007 and 2008, which were collected as 400 m^2 plots that were generally square but could vary in shape, and (2) 461 plots collected by the Grand Canyon Monitoring and Research Center (GCMRC) during the early 1980s. Vegetation classification data sampling is performed to collect data to support the classification of all the vegetation in the project area into NVCS-compliant plant associations. The selection of vegetation classification plot locations is purposely nonrandom and is usually driven by local expert knowledge of vegetation types and, in the case of this project, also the combined use of a gradsect analysis with cost surfaces (ESRI, NCGIA, and the Nature Conservancy, 1994).

The 2007/2008 classification plot data revealed 1025 taxa across the project area that were classified using quantitative techniques. The GCMRC data set was not analyzed quantitatively but was primarily used to inform and supplement the classification. During this step of the project, 217 associations were identified and described.

For the mapping task, first, the project area was divided into three phases (Figure 12.2) with each phase completed within the timeframe of a single year. Fieldwork combined with imagery and multiple ancillary data sets (e.g., slope, aspect, elevation, distance to the Colorado River, etc.) formed the base data used to create the map. First, the project imagery (National Agriculture Imaging Program [NAIP] 2007 and 2010, 1 meter, 4-band digital aerial imagery) was segmented using Trimble eCognition software. To understand how the vegetation associations varied on the ground, multiple mapping field trips were conducted to label segments to vegetation association as defined by the key. Table 12.1 summarizes the total number of mapping field samples collected per project phase.

To develop correlations between variation in vegetation associations and the variation in the imagery and ancillary data, a classification and regression tree (CART) analysis was performed between the map labels of the calibration sample segments (dependent variable) and the imagery and ancillary data (independent variables). CART analysis was done using See5 statistical software, and the resulting CART

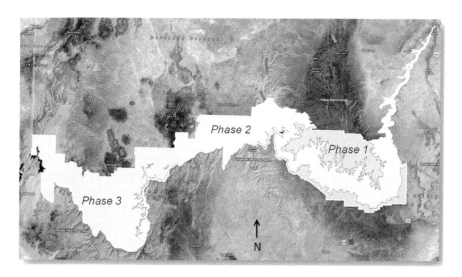

FIGURE 12.2   Project area with Grand Canyon National Park boundary outlined in yellow. The red, blue, and green shaded areas indicate the three phase areas.

TABLE 12.1
Summary of Mapping Samples Collected during the Field Visits per Phase

	Phase 1	Phase 2	Phase 3	Total
Field segments collected during mapping field trips	727	16,215	12,752	29,694

model was applied to classify all of the image segments in the project area, as illustrated in Figure 12.3.

To control variation in the imagery and ancillary data not related to the vegetation association, all the imagery and ancillary data underwent rigorous quality control, which resulted in some unreliable ancillary data sets being rejected and the discovery of large misregistration errors in portions of the 2010 imagery that were subsequently corrected by the imagery vendor. The most significant persistent issue was the deep shadows beneath the canyon rim in the 2010 imagery, as illustrated in Figure 12.4.

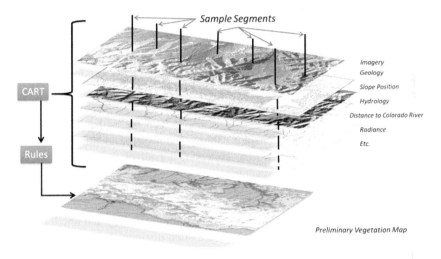

FIGURE 12.3 Illustration of using CART machine learning algorithms to develop rules to convert imagery and ancillary data into a preliminary vegetation map.

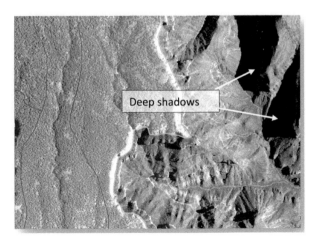

FIGURE 12.4 Example of deep shadows in the 2010 NAIP imagery beneath the Grand Canyon rim. The shadows caused variation in the imagery, which is not correlated with vegetation type.

The output of the CART analysis produced a preliminary map for each project phase. The preliminary map was manually edited, reviewed, and delivered to the NPS as a draft map. The NPS reviewed each draft map and provided input regarding areas where the maps could be improved. Final editing was performed to create the final map before the project progressed to the next phase. Figure 12.5 presents the final map for the entire project area. A link to the final map, which better shows the detail of the map, can be found at www.crcpress.com/9781498776660.

ACCURACY ASSESSMENT

The procedures used in each phase's accuracy assessment followed the 1994 NPS guidance on thematic accuracy assessment (ESRI, NCGIA, and the Nature Conservancy, 1994). This section of the chapter reviews the accuracy assessment of the project by answering the questions set forth in earlier chapters of this book.

WHAT ARE THE THEMATIC CLASSES TO BE ASSESSED?

The analysis of the classification plot data by NatureServe identified 217 vegetation associations in the project area. However, not all of the associations could be mapped. Ideally, when the vegetation association changes, the response of the remotely sensed data and the classes of the ancillary data also change, resulting in a one-to-one relationship between the associations and map classes (i.e., a strong correlation exists between vegetation association and the imagery and ancillary data). However, because vegetation associations are often distinguished from one another by sub-canopy and/or scarce indicator species, it is often impossible to map to the

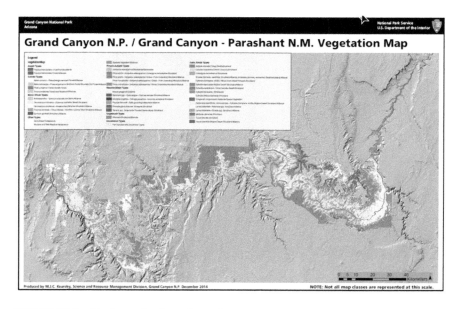

FIGURE 12.5   Final map for the entire project area.

association level, resulting in the need to collapse associations into less detailed map classes. Accordingly, the 217 vegetation associations were reduced to 87 classes that could be distinguished from the remotely sensed imagery and ancillary data. The project plant association descriptions and key can be found in the NPS final report for this project at https://irma.nps.gov/DataStore/DownloadFile/520521. The report also includes a table that relates vegetation associations to map labels. The minimum mapping unit for all map classes was 0.5 hectares.

What Is the Appropriate Sampling Unit?

The sample units for this project were field-verified segments (i.e., objects). Segments were chosen as the most appropriate sample unit, because the final map is a polygon map.

How Many Samples Should Be Taken?

The goal was to select a minimum of 30 accuracy assessment samples per map class for each phase. However, some of the map classes were extremely rare, and all of their sample segments were needed for CART modeling. All map classes that had enough samples were assessed for accuracy, resulting in 47 map classes assessed. The actual number of accuracy assessment samples per phase per map class was determined by the total number of samples collected in the field during that phase (Table 12.1). If a class shows fewer than 30 samples for a phase in Table 12.1, the class either does not exist in the area of the phase or was so rare that it was impossible to collect enough samples to provide first for the CART modeling and then, independently, for the accuracy assessment.

Also, a number of thematic classes spanned more than one project phase area, so to prevent oversampling and skewing of the accuracy assessment to more abundant classes, a subsample of the accuracy assessment phase samples was chosen for the final project accuracy assessment. For example, the *Pinus ponderosa* Forest and Woodland with Herbaceous Understory Alliance map class had a total of 158 accuracy assessment samples from all phases, so a subsample of 50 accuracy assessment sites was randomly selected and used for the final project accuracy assessment (Table 12.2).

How Should the Samples Be Chosen?

Figure 12.6 illustrates how accuracy assessment samples were selected from the NPS vegetation classification plots and mapping field segments. First, all NPS vegetation classification plots were reviewed on the imagery, and classification plot labels determined to be valid labels for the segment within which they fell were included in the set of field sample segments. The plot locations were intersected with the image segments by geographic information system (GIS) spatial analysis, and each vegetation association plot label was transferred to an image segment. Each segment was reviewed by a project analyst using the imagery to determine whether the plot label was representative of the segment. Vegetation classification plots were

TABLE 12.2

Accuracy Assessment Samples by Map Class per Phase and for Final Accuracy Assessment

Accuracy Assessment Sample Reference Label	Phase 1	Phase 2	Phase 3	Total Reference Samples Collected	Number of Reference Samples for Final Project Accuracy Assessment
Abies concolor - Pseudotsuga menziesii Dry Forest Alliance	88	0	0	88	50
Abies lasiocarpa - Picea engelmannii Southern Rocky Mountain Dry Forest Alliance	76	0	0	76	50
Acacia greggii Shrubland	0	49	74	123	50
Arctostaphylos - Quercus turbinella Shrubland Alliance	11	52	50	113	50
Artemisia bigelovii Shrubland Alliance	9	20	NA	29	29
Artemisia tridentata Shrubland Alliance	15	73	30	118	50
Baccharis spp. - Salix exigua - Pluchea sericea Shrubland Alliance	0	34	22	56	50
Brickellia longifolia - Fallugia paradoxa - Isocoma acradenia Shrubland	0	43	29	72	49
Canotia holacantha [Grand Canyon] Shrubland	0	0	5	5	5
Ceanothus fendleri/Poa fendleriana Shrub-Steppe Shrubland	4	0	0	4	4
Cercocarpus intricatus Shrubland Alliance	0	36	0	36	36
Cercocarpus montanus - Amelanchier utahensis Shrubland Alliance	0	8	0	8	8
Coleogyne ramosissima Shrublands	0	59	75	134	50
Encelia (farinosa, resinifera) Shrubland Alliance	0	43	60	103	50
*Ephedra (torreyana, viridis)/*Mixed Semi-desert Grasses Shrubland	0	45	5	50	50
Ephedra fasciculata Mojave Desert Shrubland Alliance	0	44	21	65	50
Ephedra torreyana - (Atriplex canescens, Atriplex confertifolia) Sparse Vegetation	0	46	1	47	47

(Continued)

TABLE 12.2 (CONTINUED)

Accuracy Assessment Samples by Map Class per Phase and for Final Accuracy Assessment

Accuracy Assessment Sample Reference Label	Phase 1	Phase 2	Phase 3	Total Reference Samples Collected	Number of Reference Samples for Final Project Accuracy Assessment
Ephedra torreyana - Opuntia basilaris Shrubland	0	47	0	47	47
Eriogonum corymbosum Badlands Sparse Vegetation	0	0	25	25	25
Fraxinus anomala - Rhus trilobata - Fendlera rupicola Talus Shrubland Alliance	0	45	0	45	45
Gutierrezia (sarothrae, microcephala) - Ephedra (torreyana, viridis) Mojave Desert Shrubland Alliance	0	52	55	107	50
Juniperus osteosperma Woodlands/ Savannahs	0	0	60	60	50
Larrea tridentata - Ambrosia spp. Shrubland Alliance	0	0	50	50	50
Larrea tridentata - Encelia spp. Shrubland Alliance	0	54	90	144	50
Mortonia utahensis Shrubland	0	0	50	50	50
Picea pungens/Carex siccata Forest	41	0	0	41	41
Pinus edulis - Juniperus osteosperma/Artemisia Woodland Alliance	8	54	7	69	50
Pinus edulis - Juniperus osteosperma/Cercocarpus - Quercus Woodland Alliance	9	43	12	64	50
Pinus edulis - Juniperus osteosperma/Coleogyne ramosissima Woodland	0	22	21	43	43
*Pinus edulis - Juniperus osteosperma/*Grass - Forb Understory Woodland Alliance	29	0	0	29	29
Pinus edulis - Juniperus osteosperma/Quercus turbinella Woodland	0	23	29	52	50
*Pinus edulis - Juniperus osteosperma/*Sparse Understory Woodland	0	0	30	30	30

(Continued)

TABLE 12.2 (CONTINUED)

Accuracy Assessment Samples by Map Class per Phase and for Final Accuracy Assessment

Accuracy Assessment Sample Reference Label	Phase 1	Phase 2	Phase 3	Total Reference Samples Collected	Number of Reference Samples for Final Project Accuracy Assessment
*Pinus edulis - Juniperus osteosperma/*Talus or Canyon Slope Scrub Alliance	0	8	0	8	8
*Pinus monophylla - Juniperus osteosperma/*Grass - Forb Understory Woodland Alliance	0	0	35	35	35
*Pinus monophylla - Juniperus osteosperma/*Shrub Understory Woodland Alliance	0	0	123	123	50
Pinus ponderosa Forest and Woodland with Herbaceous Understory Alliance	141	15	2	158	50
Pinus ponderosa Forest and Woodland with Shrub Understory Alliance	42	19	2	63	50
Pleuraphis rigida Herbaceous Vegetation	0	0	17	17	17
Populus fremontii - Salix gooddingii Woodland Alliance	0	20	20	40	40
Populus tremuloides - Ceanothus fendleri/Carex spp. Shrubland	37	0	0	37	37
Populus tremuloides/Carex siccata Forest	14	0	0	14	14
Populus tremuloides/Robinia neomexicana Shrubland	5	0	0	5	5
Prosopis glandulosa var. *torreyana* Shrubland	0	42	7	49	49
Pseudotsuga menziesii/ Symphoricarpos oreophilus Forest	7	0	0	7	7
Quercus gambelii Shrubland Alliance	31	52	21	104	50
Southern Rocky Mountain Montane-Subalpine Grassland Group	67	0	0	67	50
Tamarix spp. Temporarily Flooded Semi-natural Shrubland	0	32	50	82	50
Total Number of Reference Samples	**634**	**1080**	**1078**	**2792**	**1850**

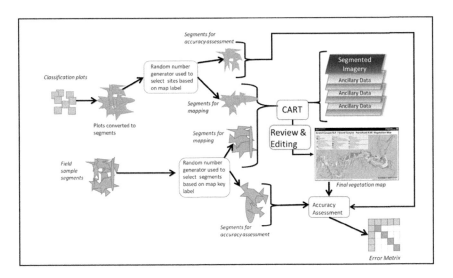

FIGURE 12.6 Illustration of process for choosing accuracy assessment samples from classification plots and field segments.

typically rejected if they represented the vegetation of a below–minimum mapping unit island within a larger polygon of a different vegetation association. On return from the field, all the classification and mapping sample segments were stratified by vegetation association, and a random number generator was used to select stratified samples from each map class, which would be used in accuracy assessment. Only map classes that had enough samples for both CART analysis and accuracy assessment were used in accuracy assessment. Accuracy assessment sample segments were held separately from the sample segments used to build the CART rules and were not available to map analysts, editors, or reviewers.

WHAT SHOULD BE THE SOURCE OF THE REFERENCE DATA?

Field-verified manual image interpretation of sample segments was the source of the reference data. Reference labels for the classification plots were determined through field interpretation and then transferred to sample segments when possible through manual interpretation of the digital aerial imagery. The reference labels for the field segments for all phases were determined through field interpretation of the segments on the imagery.

HOW SHOULD THE REFERENCE DATA BE COLLECTED?

The mapping field segment labels were collected as image segments (i.e., objects). Segments were given vegetation association labels using the following criteria:

- *Informational Homogeneity*—The segment must represent one and only one vegetation map class.

- *Spectral Homogeneity*—The segment should have less spectral variation within itself than there is between it and other segments.
- *Minimum Size*—The segment must be equal to or larger than the minimum mapping unit.
- *Project-Wide Distribution*—The segments should be collected as evenly as possible across their respective class distribution in the project area.
- *Spectral uniqueness*—The vegetation within the segment should have spectrally unique characteristics.

Field personnel also attempted to collect samples of vegetation classes that were not well represented by the classification plots and samples of areas that appeared to have spectral characteristics not yet collected.

WHEN SHOULD THE REFERENCE DATA BE COLLECTED?

Because two sources were used for accuracy assessment samples (i.e., the vegetation classification plots and the mapping field segments), the reference data were collected either prior to the mapping field trips (e.g., the NPS vegetation classification plots) or during the mapping field trips. If there had been enough classification plots, then they would have been used exclusively for accuracy assessment. However, this was not the case, and additional samples had to be collected.

Perhaps the biggest decision in the accuracy assessment of this project was to collect the mapping field segments' samples at the same time and then divide the samples into either calibration samples (for use in CART model development) or accuracy assessment samples. The decision was made because the difficult access to most of the park made two field trips per phase time consuming, enormously expensive, and beyond the budget for the project.

Collecting the samples prior to the creation of the map is cost effective and allows interim accuracy assessment as the map is being created. However, it does not ensure that an adequate number of samples per map class will be collected for all classes, so some classes may end up having fewer than the desired number of reference sample sites, as occurred in this case study (see Table 12.2).

HOW DO I ENSURE CONSISTENCY AND OBJECTIVITY IN MY DATA COLLECTION?

Consistency and objectivity were achieved by using the following procedures:

1. To ensure data collection consistency, image analysts and field crews, including NPS personnel, were simultaneously trained to collect field data. Personnel were trained in identifying vegetative cover species, recognizing ecological relationships, and using an electronic field form, which was programmed on a mobile device.
2. To ensure data collection consistency and quality control, a digital ArcMap field form linked to GPS (Figure 12.7) was used on laptops during Phase 1 and on Trimble Yuma ruggedized field computers during Phases 2 and 3.

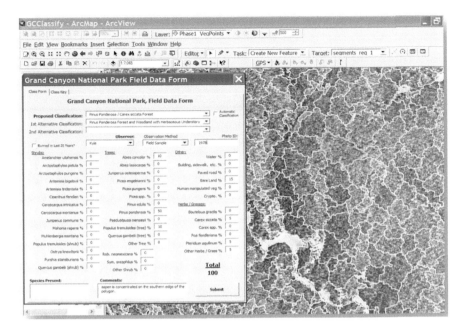

FIGURE 12.7   Digital field form with selected sample segment highlighted in cyan.

The functionality of the form included pull-down menus, automated error checking, and the classification scheme rules for easy reference.
3. To ensure quality control, all samples were reviewed at the end of the field-work to ensure that the information collected for each site was complete and correct.

To ensure data independence, accuracy assessment samples were not available to the image analysts at any time during the mapping process.

ANALYSIS

An accuracy assessment analysis was conducted for each phase area and for the final map. Because the map classes are discontinuous, the only accuracy assessment analysis technique applicable was the error matrix.

WHAT IS AN ERROR MATRIX AND HOW SHOULD IT BE USED?

The error matrices for the three project phases and the final map can be viewed at www.crcpress.com/9781498776660. Table 12.3 shows a simplified version of the final map error matrix, which was collapsed hierarchically so as to be readable in this book. The producer's and user's accuracies for all of the map classes in the final accuracy assessment are summarized in Table 12.4.

TABLE 12.3
Collapsed Final Accuracy Assessment Matrix

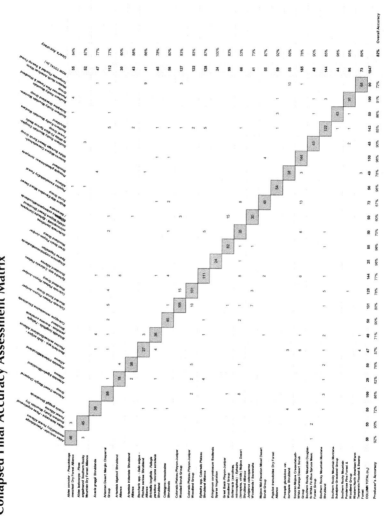

TABLE 12.4
Final Project-wide Producer's and User's Accuracies

Map Class	Producer's Accuracy (%)	User's Accuracy (%)
Canotia holacantha [Grand Canyon] Shrubland	100	100
Eriogonum corymbosum Badlands Sparse Vegetation	96	100
Pseudotsuga menziesii/Symphoricarpos oreophilus Forest	86	100
Populus tremuloides - Ceanothus fendleri/Carex spp. Shrubland	97	88
Southern Rocky Mountain Montane-Subalpine Grassland Group	86	98
Larrea tridentata - Ambrosia spp. Shrubland Alliance	86	96
Quercus gambelii Shrubland Alliance	94	85
Pinus edulis - Juniperus osteosperma/Coleogyne ramosissima Woodland	86	93
Mortonia utahensis Shrubland	88	90
Abies lasiocarpa - Picea engelmannii Southern Rocky Mountain Dry Forest Alliance	90	87
Abies concolor - Pseudotsuga menziesii Dry Forest Alliance	92	84
Ceanothus fendleri/Poa fendleriana Shrub-Steppe Shrubland	75	100
Picea pungens/Carex siccata Forest	88	86
Ephedra torreyana - Opuntia basilaris Shrubland	74	97
Coleogyne ramosissima Shrublands	90	80
Pleuraphis rigida Herbaceous Vegetation	65	100
Artemisia tridentata Shrubland Alliance	76	88
Pinus ponderosa Forest and Woodland with Shrub Understory Alliance	76	88
Populus tremuloides/Carex siccata Forest	86	75
Cercocarpus intricatus Shrubland Alliance	83	77
*Pinus monophylla - Juniperus osteosperma/*Shrub Understory Woodland Alliance	92	68
Pinus ponderosa Forest and Woodland with Herbaceous Understory Alliance	82	77
Larrea tridentata - Encelia spp. Shrubland Alliance	92	66
Fraxinus anomala - Rhus trilobata - Fendlera rupicola Talus Shrubland Alliance	76	81
Populus fremontii - Salix gooddingii Woodland Alliance	63	93
Tamarix spp. Temporarily Flooded Semi-natural Shrubland	74	80
Brickellia longifolia - Fallugia paradoxa - Isocoma acradenia Shrubland	71	78
Acacia greggii Shrublands	72	77
Ephedra torreyana - (Atriplex canescens, Atriplex confertifolia) Sparse Vegetation	62	85
Pinus edulis - Juniperus osteosperma/Artemisia Woodland Alliance	80	67
Prosopis glandulosa var. *torreyana* Shrubland	78	69
Ephedra fasciculata Mojave Desert Shrubland Alliance	64	82

(Continued)

TABLE 12.4 (CONTINUED)
Final Project-wide Producer's and User's Accuracies

Map Class	Producer's Accuracy (%)	User's Accuracy (%)
Pinus edulis - Juniperus osteosperma/Grass - Forb Understory Woodland Alliance	52	94
Pinus edulis - Juniperus osteosperma/*Quercus turbinella* Woodland	76	69
Arctostaphylos - Quercus turbinella Shrubland Alliance	80	63
Pinus edulis - Juniperus osteosperma/Sparse Understory Woodland	70	70
Populus tremuloides/*Robinia neomexicana* Shrubland	40	100
Ephedra (torreyana, viridis)/Mixed Semi-desert Grasses Shrubland	72	62
Juniperus osteosperma Woodlands/Savannahs	60	73
Encelia (farinosa, resinifera) Shrubland Alliance	76	54
Cercocarpus montanus - Amelanchier utahensis Shrubland Alliance	50	80
Pinus edulis - Juniperus osteosperma/*Cercocarpus - Quercus* Woodland Alliance	62	67
Baccharis spp. *- Salix exigua - Pluchea sericea* Shrubland Alliance	57	66
Gutierrezia (sarothrae, microcephala) - Ephedra (torreyana, viridis) Mojave Desert Shrubland Alliance	70	53
Artemisia bigelovii Shrubland Alliance	62	60
Pinus monophylla - Juniperus osteosperma/Grass - Forb Understory Woodland Alliance	46	52
Pinus edulis - Juniperus osteosperma/Talus or Canyon Slope Scrub Alliance	13	50

Note that these values are lower than those shown in Table 12.3, because Table 12.3 is a collapsed error matrix that groups several of the map classes.

WHAT ARE THE STATISTICAL PROPERTIES ASSOCIATED WITH THE ERROR MATRIX AND WHAT ANALYSIS TECHNIQUES ARE APPLICABLE?

Kappa analysis was performed on the final project-wide error matrix, and confidence intervals were calculated for the estimates of overall accuracy and Kappa using methods developed by the NPS (Lea and Curtis, 2010) (see Table 12.5).

WHAT IS FUZZY ACCURACY AND HOW CAN YOU CONDUCT A FUZZY ACCURACY ASSESSMENT?

As discussed in Chapter 10, one of the assumptions of the traditional, or deterministic, error matrix is that an accuracy assessment sample site can have only one reference label. However, classification scheme rules often impose discrete boundaries on conditions that are continuous in nature. In situations where classification scheme breaks represent artificial distinctions along a continuum of land cover, observer variability is often difficult to control and, while unavoidable, can have profound

TABLE 12.5

Overall Accuracy, Kappa, and Confidence Intervals for the Final Map

Overall Accuracy (%)	76.20
Lower Limit, 90% confidence interval (%)	73.80
Upper Limit, 90% confidence interval (%)	78.60
Kappa (%)	76.00
Lower Limit, 90% confidence interval, K (%)	74.40
Upper limit, 90% confidence interval, K (%)	77.70

Please note that the overall accuracy is lower than in the error matrix presented in Table 12.3, because Table 12.3 is a collapsed version of the full error matrix.

effects on results. While it is difficult to control observer variation, it is possible to use fuzzy logic to compensate for differences between reference and map data that are caused by variation in interpretation rather than map error (Gopal and Woodcock, 1994). In this project, both deterministic and fuzzy error matrices were compiled and analyzed for the Phase 1 area only.

Table 12.6 displays both the deterministic and fuzzy error matrices for the Phase 1 area. The overall deterministic accuracy is 85%, and overall fuzzy accuracy is 89%. The matrix can be read as follows:

- Each cell of the matrix has two rows.
- The top row is for deterministic labels, and the bottom row is for fuzzy labels.
- The diagonal (highlighted in light blue) shows the deterministic agreement.
- The red numbers in the off-diagonal cells show the number of confused sites per cell.
- The black numbers in the off-diagonal cells show the number of sites with fuzzy acceptable labels for that cell.
- Both deterministic and fuzzy user's, producer's, and overall accuracies are shown.

RESULTS

Overall accuracy for the project-wide accuracy assessment is 76%, with 1415 of the total 1847 samples in agreement and 432 confused. Approximately one-third of the confusion (128 of the 432 confused samples) is along the margins of map classes, where confusion is to be expected. For example, both the *Larrea tridentata - Encelia* spp. Shrubland Alliance and the *Encelia (farinosa, resinifera)* Shrubland Alliance are sparsely vegetated types with total shrub cover between 10% and 20%. Both types include both *Encelia* spp. and *Larrea tridentata. Fourquieria splendens* can

TABLE 12.6
Fuzzy Error Matrix for the Phase 1 Area

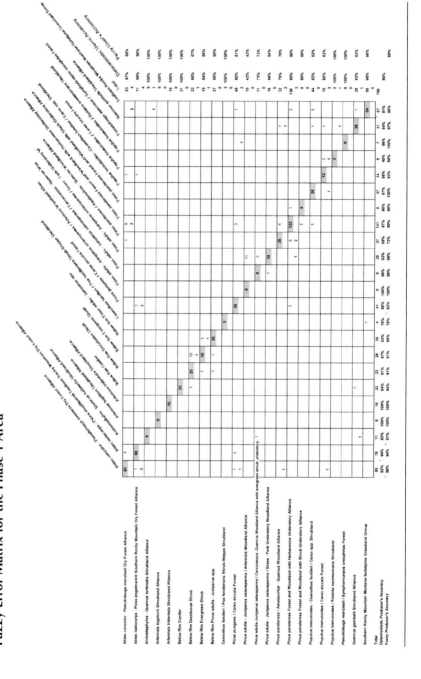

also occur in both types. The two types are distinguished primarily by the amount of *Larrea tridentata*, which occurs at up to 1% for the *Encelia* (*farinosa, resinifera*) Shrubland Alliance and averages 3%–5% in the *Larrea tridentata - Encelia* spp. Shrubland Alliance. Given that photo interpreters vary ±10% in their estimates of vegetation cover (Spurr, 1948), it is not surprising that nine of the accuracy assessment samples confuse the *Larrea tridentata - Encelia* spp. Shrubland Alliance with the *Encelia* (*farinosa, resinifera*) Shrubland Alliance.

Other similar class margin confusion occurs as follows:

- 25% of the confused samples are within the correct lifeform class with
  - 11% (48 samples) within the *Pinus edulis - Juniperus osteosperma* classes.
  - 7% (31 samples) within the *Ephedra* classes.
  - 5% (20 samples) between the two *Pinus monophylla - Juniperus osteosperma* classes.
  - 2% (eight samples) between the two *Larrea tridentata* classes.
- 3% (13 samples) of the samples confuse *Pinus monophylla - Juniperus osteosperma*/Grass - Forb Understory Woodland Alliance with *Juniperus osteosperma* Woodlands/Savannahs

Other sources of confusion are:

- Six *Baccharis* spp. - *Salix exigua - Pluchea sericea* Shrubland Alliance samples and seven *Prosopis glandulosa* var. *torreyana* Shrubland samples committed to *Tamarix* spp. Temporarily Flooded Semi-natural Shrubland. Twelve of these 13 samples are from the Phase 2 area.
- Eight *Ephedra torreyana - (Atriplex canescens, Atriplex confertifolia)* Sparse Vegetation committed to *Artemisia bigelovii* Shrubland Alliance.

LESSONS LEARNED

As in most projects, lessons are learned continually as the project progresses. Specific lessons learned during the accuracy assessment portion of the case study include the following:

1. Following the project, it was realized that nothing had been done to ensure that the segments chosen for accuracy assessment were not spatially auto-correlated with the training segments used in the CART analysis. While the accuracy assessment sample segments and the training sample segments were in different locations and kept completely separate throughout the project, there is a possibility that they were located close enough to each other to violate the independence requirement. All subsequent projects by the project team now test for spatial autocorrelation with training samples before accuracy assessment samples are chosen.

2. Fieldwork in the Phase 2 and 3 (i.e., within the Grand Canyon) areas was often physically brutal. Additionally, all sample segments collected from rafts on the Colorado River had to be entered into the mobile devices rapidly, as the raft moved swiftly down the river. As a result, the detailed field form in Figure 12.7 was often not completely filled in during Phases 2 and 3, which precluded the implementation of fuzzy accuracy assessment for those phases.

13 The California Hardwood Rangeland Project

INTRODUCTION

This case study reviews the procedures used to conduct accuracy assessments of four maps—two polygon maps created in 1981 using manual image interpretation of aerial photography and two pixel-based maps produced in 1990 from Landsat satellite imagery. While the maps considered for this case study are dated, the concepts used in their assessment are still valid today. We decided to include this case study because it contains:

- an analysis of the accuracies of maps created using manual interpretation versus semi-automated image classification
- the assessment of a pixel-based map, including the issues that arise from generating map labels for samples that are made up of groups of pixels
- a comparison of reference labels for the same area derived by different personnel, allowing the quantification of variation in human interpretation
- implementation of both deterministic and fuzzy accuracy assessment
- an in-depth analysis of the causes of differences between the reference and map labels
- numerous trade-offs between statistical rigor and practical implementation

This case study is far from being the perfect example of accuracy assessment design, implementation, and analysis. The project offered ample opportunities for learning and provided significant challenges. Yet, it is illustrative of problems typically encountered in accuracy assessment. It presents a real-world example with real-world trade-offs and considerations. The implications of each decision are analyzed and discussed. The purpose of presenting this case study is to make the reader fully aware of the obvious and subtle, yet critical, considerations in designing and implementing an accuracy assessment.

BACKGROUND

California's hardwood rangeland resource was historically characterized by low use and low value. However, over the last several decades, suburban development coupled with orchard and vineyard expansion has resulted in changes in the extent and distribution of this resource. Hardwood stocking has declined, as have the number of acres of hardwood rangeland as the resource has been converted to industrial, residential, and intensive agricultural uses. To assess and analyze the nature and implications of these changes, the California Department of Forestry and Fire Protection (CDF) instituted long-term monitoring of California's hardwood range-land in areas lower than 5000 feet in elevation.

At the end of the last century, two mapping efforts were conducted:

- a manual interpretation of 1981 aerial photography (Pillsbury et al., 1991)
- a semi-automated per-pixel image classification of 1990 Landsat satellite imagery

The accuracies of four maps created by these two efforts were assessed:

- Tree crown closure created from manual interpretation of 1981 aerial photography
- Land cover type created from manual interpretation of 1981 aerial photography
- Tree crown closure created from pixel-based classification of 1990 digital satellite imagery
- Land cover type created from pixel-based classification of 1990 digital satellite imagery

The organization of this chapter follows the organization of this book. First, the project's sample design is discussed. Next, reference data collection and methods are presented. Finally, the analysis and results are detailed.

SAMPLE DESIGN

Sample design is critical to any accuracy assessment. As with all accuracy assessments, sample design involved addressing the questions posed at the beginning of Chapter 6:

1. How is the map information distributed?
2. What is the appropriate sample unit?
3. How many samples should be taken?
4. How should the samples be chosen?

The sample design for this project was extremely complex, because it involved the assessment of four different maps under the constraint of a limited budget. As a result, trade-offs between statistical rigor and practicality are apparent throughout this case study. Because CDF could not afford to fly new photography, existing aerial photography from 1981 was used as the primary reference source data for assessment of both the 1981 and 1990 maps. The use of the 1981 photos, in turn, drove much of the sample design, including the selection of the appropriate sample unit and the methods used to select the sample units.

How Is the Map Information Distributed?

The study area is the hardwood rangeland of California, which forms a donut-shaped area around California's Central Valley and is depicted in Figure 13.1. The extent of the 1981 map coverage was defined as areas where hardwood cover types occur in California below 5000 feet in elevation. The extent of the 1990 coverage was initially defined to be that of the 1981 maps.

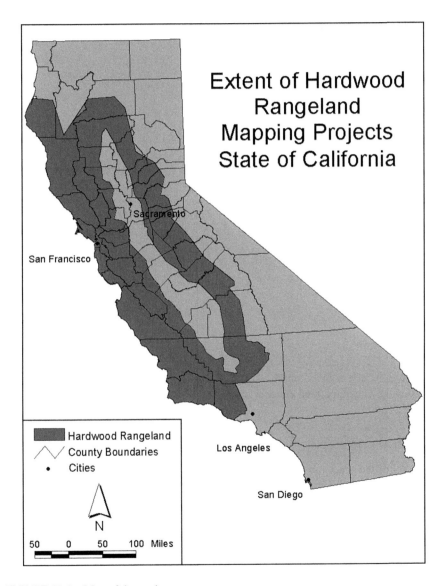

FIGURE 13.1   Map of the study area.

However, while the 1990 maps were being produced, errors of omission were discovered in the 1981 maps. Accordingly, the extent of the 1990 maps was greatly expanded to include over 30 million acres of land. To assess possible errors of omission in the 1981 maps, accuracy assessment samples were taken in locations mapped as hardwoods on the 1990 map but omitted from the 1981 map.

The classification schemes for this project characterized California's hardwood rangelands by tree crown closure and land cover type. Tree crown closure was classified into the following five classes:

1. 0% (non-hardwood)
2. 1%–9%
3. 10%–33%
4. 34%–75%
5. 76%–100%

The land cover classification system consisted of 12 classes:

1. Blue Oak Woodland
2. Blue Oak/Gray Pine Woodland
3. Valley Oak Woodland
4. Coastal Oak Woodland
5. Montane Hardwood
6. Potential Hardwood
7. Conifer
8. Shrub
9. Grass
10. Urban
11. Water
12. Other

Figure 13.2 is a dichotomous key that illustrates the rules used to distinguish between the vegetated land cover type classes.

WHAT IS THE APPROPRIATE SAMPLE UNIT?

The preliminary accuracy assessment sampling design anticipated that the 1981 vegetation type polygon coverage could be used as the sampling units for accuracy assessment of both the 1981 and the 1990 maps. However, using the 1981 polygons as the accuracy assessment sample units assumes that the polygons are homogeneous by crown closure and land cover type class, accurately delineated, and free from errors of omission. Unfortunately, during the course of the project, significant errors of omission and polygon delineation were discovered in the 1981 maps. As a result, many of the 1981 polygons had more class variation within the polygons than existed between the polygons. Many of the polygons were also several hundred acres in size, making them impractical to photo interpret or to traverse in the field.

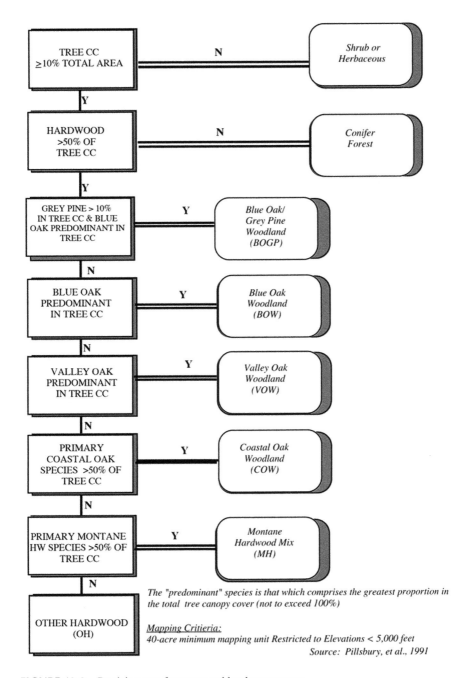

FIGURE 13.2 Decision tree for vegetated land cover types.

As a result, when the sample polygon from the 1981 map was too large, contained multiple classes, or was poorly delineated, a new sample polygon, homogeneous in crown closure and cover type, was delineated within the original polygon.

How Many Samples Should Be Taken?

A total of 640 accuracy assessment sites were sampled; 512 of these were sampled from the 1981 map, and an additional 128 samples were sampled for tree crown closure only from the 1990 map in areas that were not mapped as hardwoods in the 1981 map. Of the 512 samples from the 1981 map, 177 were labeled twice—once using manual image interpretation in the field and once using manual image interpretation in the office, which allowed a comparison of the variation between different image interpretation personnel. Tables 13.1 through 13.3 summarize the number of accuracy assessment samples by sample type, cover type, and crown closure classes. Ideally, the 640 samples would have been allocated so that at least 50 samples would have been chosen from each crown closure or cover type class. As the tables show, this goal was not met for all map classes. The reasons for these sampling deficiencies are varied and include the following practical considerations:

- The California Department of Forestry requested that the contracted sample amount be distributed equally by their management regions and then, by cover type within each region. Because not all hardwood rangeland types occur in all regions, the pre-stratification of the samples caused some types to be undersampled.
- The Valley oak (VOW) class is rare on the 1981 map, making it difficult to find enough areas to sample.
- Field access was extremely difficult, making field data collection expensive. As a result, the budget dictated that compromises be made between travel cost and sample distribution.

How Should the Samples Be Chosen and Distributed across the Landscape?

Despite known spatial autocorrelation in the distribution of hardwood rangeland cover types, clusters of sample polygons were selected for both photo-interpreted

TABLE 13.1

Numbers of Samples Selected from 1981 Map by Sample Type and Land Cover Type

Reference Cover Type Label	Sample Labeled in the Office	Sample Labeled in the Field	Total
Blue Oak Woodland	108	55	163
Blue Oak/Gray Pine	16	20	36
Valley Oak Woodland	12	11	23
Coastal Oak Woodland	71	45	116
Montane Hardwood	112	29	141
Other Hardwood	10	10	20
Non-hardwood	6	7	13
Total	335	177	512

TABLE 13.2
Numbers of Samples Selected from 1981 Map by Sample Type and Crown Closure Class

Reference Tree Crown Closure Label	Sample Labeled in the Office	Sample Labeled in the Field	Total
0%	6	8	14
1%–9%	16	10	26
10%–33%	124	54	178
34%–75%	173	78	251
76%–100%	16	27	43
Total	335	177	512

TABLE 13.3
Additional Samples Selected from 1990 Map by Sample Type and Tree Crown Closure Class

Reference Tree Crown Closure Label	Sample Labeled in the Office	Sample Labeled in the Field	Total
0%	21	11	32
1%–9%	13	2	15
10%–33%	18	9	27
34%–75%	11	10	21
76%–100%	25	8	33
Total	88	40	128

and field-visited sites because of the travel and set-up time cost savings resulting from the use of clusters of polygons. Accuracy assessment sample polygon clusters were chosen using different procedures depending on (1) whether the sample unit was chosen from the 1981 coverage or from the 1990 coverage and (2) whether the reference label was to be developed from interpretation of the aerial imagery in the office or in the field.

Samples Chosen from the 1981 Map

For samples chosen from the 1981 map to be labeled using image interpretation in the office, sample polygons were selected using a mixed random sample – cluster sampling approach. Sampling was accomplished by:

1. stratifying the map into the hardwood cover types for each of the five California management regions
2. assigning a unique number to each of the polygons
3. using a random number generator to select up to 20 sample polygons from each cover type that occurred in each region

4. transferring the chosen polygon to the 1981 aerial photography
5. selecting a cluster of two or three additional sample polygons of different cover types from the center of the photo containing the randomly selected polygon

If the sample polygons were too large or heterogeneous, a smaller homogeneous portion of the polygon was delineated as the final sample unit.

While this selection method was viable for office-interpreted samples, random sampling could not be used to select accuracy assessment polygons for labeling in the field, because accessible sites could not be determined from the aerial photography. Five test trips to the field proved that more than 50% of the randomly selected sample units lay along private ranch roads behind locked gates.

To ensure accessibility, routes were chosen that passed near or through areas that covered as much ecological variation as possible. Field-visited reference data sample polygons were selected from 1:100,000 maps of the Landsat imagery showing the 1981 map class polygons (without labels) and public roads. To reduce potential site selection bias, field personnel used dice to decide whether or not to sample an accessible map class polygon. A template was then used to delineate a sample unit on the aerial photography within the randomly selected roadside polygon. Up to two additional sample polygons on the same photo of different density or cover type class were also delineated and labeled.

Samples Chosen from the 1990 Coverage

To test the accuracy of the 1981 map's extent, accuracy assessment samples were also chosen from areas of the 1990 map that were outside the extent of the 1981 map. First, 50 hardwood pixels per management region were randomly selected as possible sample locations. Using the x,y coordinates of each randomly selected pixel, a computer program generated a sample unit by creating a 3×3 box of pixels around each randomly selected location. Fifteen of the 50 samples per management region were selected for inclusion in the assessment.

REFERENCE DATA COLLECTION

Once the complex sample design was complete, reference data collection was fairly straightforward, because the same data were collected on all reference sites. As discussed in Chapter 7, data collection required addressing four basic questions:

1. What should be the source data for the reference samples?
2. What type of information should be collected for each sample?
3. When should the reference data be collected?
4. How do we ensure that the reference data are collected correctly, objectively, and consistently?

WHAT SHOULD BE THE SOURCE DATA FOR THE REFERENCE SAMPLES?

Two sources were used for reference data—the 1981 aerial photography and 1991 field visits. Budget constraints dictated the use of the 1981 photography as the primary source data to assess the accuracy of both the 1981 and the 1990 maps. All sites were image interpreted in the office. Thus, the manually interpreted 1981 map was assessed using the same photos that were used to create the map, and the 1990 map was assessed using aerial photos that were 9 years old. Without assessment of the accuracy of the manual interpretation, the assessment of the 1981 map would have been more a comparison of two different manual interpretations rather than an accuracy assessment. Therefore, a subset of the manually interpreted sites were also field visited, and additional field samples were collected where possible.

WHAT TYPE OF INFORMATION SHOULD BE COLLECTED?

Both the 1981 and 1990 projects were concerned with mapping the extent, type, and condition of California hardwood rangeland. Each sample unit was image interpreted in the office and/or the field, and an accuracy assessment form was completed, which characterized the variation in land cover of that sample unit (see Figure 13.3). Field personnel identified primary and associate hardwood cover type species by either driving through the site or doing a partial walk-through if accessibility allowed, or viewing the site from a distance through binoculars if the site was inaccessible.

For each sample unit, the following was recorded:

1. Site information—a three-part alphanumeric accuracy assessment polygon label composed of the following codes:
   type = A (photo-interpreted in office)
        J (photo-interpreted in the field)
        P (office photo interpretation of field-verified site)
   region = CDF management region
   number = sample number
2. HWPOLY-ID—item used to identify existing site in a geographic information system (GIS)
3. Date—date of the photo interpretation
4. Observer—initials of the photo interpreter
5. Photo—photo number (identified by flightline, photo number, and United States Geological Survey [USGS] quad, respectively) of the aerial photograph on which the accuracy assessment polygon has been delineated
6. Photo source—source agency for aerial photography (e.g., CDF, National Aeronautics and Space Administration [NASA], etc.) and photo job number if available
7. Image—identifies Landsat Thematic Mapper (TM) scene(s) that the site polygon falls on using a four- or six-digit code indicating path/row(s) (e.g., 44/33, 44/32–33)

8. Observation level—used as an indicator of the potential accuracy/quality of the photo interpretation, "1" being the most accurate and "4" being the least accurate:
   1. walk through hardwood stand
   2. viewing from road adjacent to hardwood stand
   3. viewing from afar (i.e., road or ridge opposite hardwood stand)
   4. photo interpreted in office
9. Tree Crown Closure Matrix—four-letter species codes used to record percentage crown closure by primary and associated species (including Gray Pine) and "conifer;" includes numbered comments relating to species and crown closure calls in comment box
10. Other Cover Crown Closure Matrix—four-letter species codes used to record percentage crown closure occupied by the following non-tree cover types:
    Grass
    Shrub (if scrub oak, list % separate from other shrub)
    Urban
    Water
    Other (bare ground, agriculture, marsh, etc.)
11. WHR Cover Type—cover type calculated in the field or office using the Decision Tree for Mapping Hardwood Species Groups (Pillsbury et al., 1991) (see Figure 13.2) and recorded as follows:
    BOW = Blue Oak Woodland
    BOGP = Blue Oak Foothill/Gray Pine
    VOW = Valley Oak Woodland
    COW = Coastal Oak Woodland
    MH = Montane Hardwood
    OH = Other Hardwood
12. Size Class—estimated average hardwood diameter at breast height (DBH) recorded by size class:
    $S < 12''$
    $L = 12''$
13. Current Map Delineation—visual analysis of the general accuracy level of existing map polygon delineation as viewed on the computer screen by overlaying polygons on the imagery using the following descriptions:
    a. *Very poor*: existing polygon boundary does not follow hardwood stand along any of its perimeter; many unnatural contours; arbitrary polygon closure; polygon includes more than one density class, has a high level of variation in density or cover type, and has inclusions of non-hardwood cover or other hardwood cover types and densities within the 40 acre minimum mapping unit.
    b. *Poor*: existing polygon boundary shifted away from actual hardwood stand perimeter; inclusions of non-hardwood cover or other hardwood cover types and densities within the 40 acre minimum mapping unit.
    c. *Good*: existing polygon boundary generally follows hardwood stand perimeter; no inclusions of non-hardwood cover or other hardwood cover types or densities within the 40 acre minimum mapping unit.

d. *Very good*: existing polygon boundary tightly follows hardwood stand along entire perimeter; inclusions of non-hardwood cover within the 40 acres are delineated; hardwood stand has evenly distributed crown closure and homogeneous cover type throughout polygon.

14. Cover Type and Density Fuzzy Logic Matrix—each polygon evaluated for the likelihood of being identified as each of the six possible cover types and four possible existing crown closure classes. "Likelihood" is indicated using the terms "absolutely wrong," "probably wrong," "acceptable," "probably right," and "absolutely right" (Figure 13.4).

Accuracy Assessment Photo Form

Site:_____-_____-____   HWPOLY-ID:_____   Date:___/___/___   Observer:_____
     type   region   #

Photo:_____-_____-_____   Photo Source:_____   Image:_____
    flightline   photo   quad

Observation Level: 1   2   3   4

% of total area occupied by tree cover

% of total area occupied by grass, bare, shrub, etc.

C#	Species	%

C#	Species	%

NOTE: Total tree and other cover observed on photo should equal 100%.

WHR Cover Type:_____   Size:_____

Current map polygon delineation: Very Poor   Poor   Good   Very Good

Describe delineation:_

C#	Comments

FIGURE 13.3 Accuracy assessment form.

Cover Type and Density Matrix

Cover Type/ Density	Absolutely Right	Probably Right	Acceptable	Probably Wrong	Absolutely Wrong
Blue Oak					
Blue Oak/ Grey Pine					
Valley Oak Woodland					
Coastal Oak Woodland					
Montane Hardwood					
Other					
1 (<10%)					
2 (10-33%)					
3 (34-75%)					
4 (76-100%)					

FIGURE 13.4   Cover type and density fuzzy logic matrix form.

Following completion of the forms, all data were entered into a database for future analysis. In addition, on completion of the field data collection, field-verified accuracy assessment site boundaries were captured using heads-up digitizing.

WHEN SHOULD THE REFERENCE DATA BE COLLECTED?

As mentioned previously, the only reference data available for the accuracy assessment were the 1981 1:24,000, panchromatic photography used to create the 1981 map. While forest land typically does not change as quickly as agricultural or urban land, the 9 year difference between the 1981 photos and the 1990 imagery caused problems in the assessment. Fires, harvesting, and urban development changed several accuracy assessment sites between the date of the photos (1981), the date of the Landsat imagery used to create the 1990 map, and the date of the field visits (1991). Only those sites that had not changed significantly in the field were included in the field sample. However, it is impossible to know how many of the sites that were not field visited also changed between 1981 and the date of the imagery.

Quality Control

Data Independence

Because they were completed by two different organizations, the assessment of the 1981 map was completely independent of the effort to create the 1981 map. In addition, independence was also imposed in the assessment of the 1990 map. Accuracy assessment data were always kept separate from any information used to make the 1990 map. At no time did the accuracy assessment reference personnel have any knowledge of the map labels for either the 1981 or the 1990 map.

Data Consistency

Data consistency was imposed in several ways. First, an accuracy assessment manual was developed, which clearly explained all data collection procedures. Second, as illustrated in Figures 13.3 and 13.4, personnel used forms to collect all accuracy assessment data. Finally, all personnel were trained simultaneously, and the project manager frequently reviewed their work.

Data Quality

The map location of accuracy assessment samples was derived directly from the 1981 map, because the sample units were chosen from the 1981 map polygons. Location of the site on the reference data (the 1981 photos) was accomplished by viewing the polygon's boundaries over the satellite imagery and then, transferring the site location onto the photo by matching flightline location, roads, streams, and patterns of vegetation.

To minimize interpretation error, personnel most familiar with the vegetation in each region interpreted the sample units. Species identification in the office was enhanced through the use of ancillary data, including extensive field notes and ecological information concerning the distribution of hardwood types (Griffin and Critchfield, 1972).

Data entry was done once. To check the quality of the entry, a subset of the database fields were selected and compared with the original information on the forms. However, the data entry was not perfect and caused later problems in analysis of the error matrices.

ANALYSIS

DEVELOPMENT OF THE ERROR MATRICES

The first step in accuracy assessment analysis requires the development of error matrices. Error matrices, in turn, require labeling the samples. As introduced in Chapter 2, each accuracy assessment sample in an error matrix has two labels:

- The reference site label refers to the label derived from data collected from either field or office photo interpretation that makes up the reference data (the data against which the map is compared) during accuracy assessment.
- The map site label refers to the map label of the accuracy assessment site. In this project, the map label is derived either from the existing 1981 photo-interpreted map or from decision rules applied to the pixel composition of the site on the 1990 satellite map.

Reference labels included both (1) deterministic labels for analysis in a traditional error matrix and (2) labels that accounted for variation in interpretation. Fuzzy labels were created to (1) account for variation in estimates and (2) deal with the imprecision in the cover type classification system. Both expert and measured approaches to fuzzy set theory in accuracy assessment were implemented. The measured approach determines the variance from paired interpretations of the same site and removes that variance from the difference matrix. Two independent interpretations exist for each accuracy assessment reference site that was photo-interpreted both in the field and in the office. Because the site was held constant while the interpreter varied, these pairs of interpretations can be used to measure variation in interpretation. This method is fairly simple to implement with vegetation class characteristics such as crown closure, which are represented by discrete breaks in a continuum of a single variable. The algorithms for implementing this method on class characteristics represented by discrete breaks in multiple variables (e.g., cover type as a function of percent crown closure of several hardwood species types) are less defined and more difficult to implement. For this reason, the methods used by Gopal and Woodcock (1994) were implemented for the labeling of cover type reference sites.

Map accuracy sample labels for the 1981 map were taken directly from the map label for each sample. Map labels for the 1990 map were calculated for each site by applying algorithms based on the classification system's decision rules to the sample's pixel composition obtained from the crown closure and cover type raster data layers. Thus, sample polygons received labels that were the result of the mixture of pixels in the polygon. For example, an accuracy assessment sample polygon comprised of a mixture of closed canopy (76%–100%) and open (1%–10%) pixels would receive a crown cover label that was the average of the pixel values (e.g. 35%–75%).

Once the labels were created, the matrices were built. Tables 13.4 through 13.7 show the initial matrices for the four maps assessed. As with most accuracy assessments, the first matrices are often very different from the final matrices. In fact, it is probably more correct to name the initial matrices *difference matrices*, because they indicate that differences (and not necessarily map errors) exist between the reference and map labels.

Two types of analysis should be carried out on the error matrices. First, we must determine whether the results in the matrix are statistically valid (see Chapter 8). Next, we need to learn what causes samples to fall off the diagonal (see Chapter 9).

STATISTICAL ANALYSIS

Statistical analyses, including normalizing the matrices using the iterative proportional fitting procedure (i.e., Margfit) and the Kappa measure of agreement, were performed on these difference matrices. The normalization process allows individual cell values within matrices to be directly compared without regard for sample size differences. A normalized accuracy was computed for each matrix. Tables 13.8 and 13.9 present the results of the normalization for the crown closure maps from the 1981 and 1990 maps. Note that comparisons of each cell value can now be made between the matrices, and more interestingly, the values in the major diagonal are a single measure of map class accuracy instead of looking at producer's and user's accuracy as in the original matrix.

TABLE 13.4
Crown Closure Difference Matrix, 1981 Map

	Class	0%	1%–9%	10%–33%	34%–75%	76%–100%	Total
				Reference Data			
1981	0%	**0**	1	9	8	1	19
Map	1%–9%	4	**17**	78	34	3	136
	10%–33%	1	2	**59**	69	7	138
	34%–75%	2	1	21	**93**	11	128
	76%–100%	3	0	2	37	**17**	59
	Total	10	21	169	241	39	**480**

Producer's Accuracy		User's Accuracy	
Reference	**Percent**	**Map**	**Percent**
0%	0	0%	0
1%–9%	81	1%–9%	13
10%–33%	35	10%–33%	43
34%–75%	39	34%–75%	73
76%–100%	44	76%–100%	29

Overall agreement = 186/480 = 39%

TABLE 13.5
Crown Closure Difference Matrix, 1990 Map

	Class	0%	1%–9%	10%–33%	34%–75%	76%–100%	Total
				Reference Data			
1990	0%	**4**	1	1	7	5	18
Map	1%–9%	1	**11**	42	12	0	66
	10%–33%	1	7	**97**	84	4	193
	34%–75%	4	2	29	**135**	22	192
	76%–100%	0	0	0	3	**8**	11
	Total	10	21	169	241	39	**480**

Producer's Accuracy		User's Accuracy	
Reference	**Percent**	**Map**	**Percent**
0%	40	0%	22
1%–9%	52	1%–9%	17
10%–33%	57	10%–33%	50
34%–75%	56	34%–75%	70
76%–100%	21	76%–100%	73

Overall agreement = 255/480 = 53%

TABLE 13.6
Cover Type Difference Matrix, 1981 Map

	Class	NH	BOGP	BOW	COW	MH	VOW	Total
				Reference Data				
1981	NH	**0**	1	8	5	4	2	20
Map	BOGP	1	**19**	40	4	22	5	91
	BOW	2	6	**68**	4	12	7	99
	COW	2	1	22	**54**	10	5	94
	MH	3	5	8	27	**81**	1	125
	VOW	1	0	7	14	8	**2**	32
	Total	9	32	153	108	137	22	**461**

Producer's Accuracy		User's Accuracy	
Reference	**Percent**	**Map**	**Percent**
NH	0	NH	0
BOGP	59	BOGP	21
BOW	25	BOW	69
COW	5	COW	63
MH	59	MH	65
VOW	9	VOW	6

Overall agreement = 224/461 = 49%

TABLE 13.7
Cover Type Difference Matrix, 1990 Map

	Class	NH	BOGP	BOW	COW	MH	VOW	Total
				Reference Data				
1990	NH	**4**	0	1	2	9	0	16
Map	BOGP	1	**18**	27	3	18	3	70
	BOW	0	9	**98**	7	18	10	142
	COW	2	0	12	**71**	9	6	100
	MH	1	5	8	20	**75**	1	110
	VOW	1	0	7	5	8	**2**	23
	Total	9	32	153	108	137	22	**461**

Producer's Accuracy		User's Accuracy	
Reference	**Percent**	**Map**	**Percent**
NH	44	NH	25
BOGP	56	BOGP	26
BOW	64	BOW	69
COW	66	COW	71
MH	55	MH	68
VOW	9	VOW	9

Overall agreement = 268/461 = 58%

TABLE 13.8
Normalized Crown Closure Difference Matrix, 1981 Map

	Class	0%	1%–9%	10%–33%	34%–75%	76%–100%
				Reference Data		
1981	0%	**0.1672**	0.3171	0.2460	0.1383	0.1319
Map	1%–9%	0.1858	**0.4568**	0.2510	0.0693	0.0380
	10%–33%	0.1150	0.1212	**0.3533**	0.2593	0.1512
	34%–75%	0.1969	0.0747	0.1312	**0.3584**	0.2382
	76%–100%	0.3351	0.0303	0.0185	0.1747	**0.4407**

Normalized accuracy = 36%

TABLE 13.9
Normalized Crown Closure Difference Matrix, 1990 Map

	Class	0%	1%–9%	10%–33%	34%–75%	76%–100%
				Reference Data		
1990	0%	**0.5670**	0.1446	0.0276	0.0948	0.1665
Map	1%–9%	0.0838	**0.4919**	0.3470	0.0701	0.0067
	10%–33%	0.0483	0.1848	**0.4586**	0.2729	0.0348
	34%–75%	0.1514	0.0644	0.1450	**0.4573**	0.1819
	76%–100%	0.1494	0.1143	0.0218	0.1049	**0.6101**

Normalized accuracy = 52%

TABLE 13.10
Individual Error Matrix Kappa Analysis Results

Error Matrix	KHAT	Variance	Z Statistic
Table 13.4	0.17	0.0010371	5.4
Table 13.5	0.28	0.0011688	8.1
Table 13.6	0.34	0.0008001	12.1
Table 13.7	0.45	0.0008323	15.6

The results of the Kappa analysis are shown in Tables 13.10 and 13.11. A test of significance of an individual matrix was performed to see whether the classification process was significantly better than a random assignment of map class labels. Table 13.10 shows that these results were significant for all four matrices. Table 13.11 presents the results of the appropriate pairwise comparisons. This test determines whether one error matrix is statistically significantly different from another.

TABLE 13.11
Kappa Analysis Results for the Pairwise
Comparison of the Error Matrices

Pairwise Comparison	Z Statistic
Table 13.4 vs. Table 13.5	2.2036
Table 13.6 vs. Table 13.7	2.6703

In this case study, it was appropriate to compare the results of the crown closure maps generated from 1981 aerial photo interpretation and 1990 satellite image processing. It was also appropriate to compare the cover type map derived from 1981 aerial photo interpretation with the 1990 cover type map created from satellite image processing. In both of these cases, the matrices (and therefore, the maps) were significantly different from each other. By examining the accuracy measures, it can be concluded that the 1990 maps generated from satellite imagery were significantly better than the maps created in 1981 from aerial photography.

ANALYSIS OF OFF-DIAGONAL SAMPLES

Following the statistical analysis of the matrices, the off-diagonal elements of the matrix need to be examined for possible:

1. Errors in the reference data
2. Sensitivity of the classification schemes to observer variability
3. Inappropriateness of the photo interpretation or satellite remote sensing for mapping hardwood rangeland crown closure and cover type
4. Mapping error

CROWN CLOSURE ANALYSIS

To learn whether the causes of differences in the matrix resulted from error or from variation in interpretation, two independent photo interpretations of the same accuracy assessment reference site were made for 173 sites: one in the office and one in the field. No two pairs of interpretations were made by the same photo interpreter. Table 13.12 compares these interpretations. In general, the class values of the paired interpretations fall along or very near the major diagonal, clearly illustrating the impacts of variation in interpretation.

The average difference in crown closure estimates between the office photo-interpreted and field photo-interpreted estimates was 9.31%. To compensate for the impacts of variation in human interpretation on crown closure map accuracy assessment, a ±9% variance in crown closure was implemented on all office-interpreted sites. For example, a field-interpreted estimate of 11% crown closure would

TABLE 13.12
Comparison of Office and Field Photo Interpretation of Crown Closure

		Field Photo Sites					
	Class	0%	<10%	10%–33%	34%–75%	76%–100%	Total
Office	0%	**4**	0	0	0	0	4
Photo	1%–9%	0	**3**	3	0	0	6
Sites	10%–33%	1	4	**41**	11	0	57
	34%–75%	1	1	10	**60**	3	75
	76%–100%	1	0	0	7	**23**	31
	Total	7	8	54	78	26	**173**

Producer's Accuracy		User's Accuracy	
Reference	Percent	Map	Percent
0%	57	0%	100
1%–9%	38	1%–9%	50
10%–33%	76	10%–33%	72
34%–75%	77	34%–75%	80
76%–100%	88	76%–100%	74

Overall agreement = 131/173 = 76%

TABLE 13.13
Crown Cover and Corresponding Acceptable Labels

If Field Site Equals	Acceptable Photo Labels
0%	0 or 1–9
1%–9%	0 or 1–9 or 10–33
10%–18%	1–9 or 10–33
19%–24%	11–33
25%–42%	11–33 or 34–75
43%–66%	34–75
67%–84%	34–75 or 76–100
85%–100%	76–100

be considered comparable to a photo-interpreted estimate of either the 1%–9% class (i.e. 11 − 9 = 2) or the 10%–33% class (i.e., 11 + 9 = 20). Table 13.13 illustrates how the variances were implemented across all crown closure classes.

Table 13.14 illustrates the implementation of the ranges on the matrix comparing the pairs of sites. A total of 16 sites fall outside of the allowable ranges. The photo interpretation of these 16 sites differs due to:

- *Photo interpretation error.* At two sites, the office photo interpreters mislabeled hardwoods for shrub.

- *Sensitivity of the classification system to observer variability.* One site differs in its labels of hardwood versus non-hardwood, because the site is mixed hardwood/conifer. A 9% variance on the estimates of hardwood or conifer would place the site in a different category. The photo interpreter in the field noted that a hardwood classification would have been acceptable.
- *Change.* Two sample sites had been harvested between the date of the photos (1981) and the date of the field visit (1991).

The remaining 11 sites represented variance in the photo interpretation estimates that are beyond those allowed in the adopted ranges. The 9% variance is an average value. By adopting the average (rather than the complete measured spread of variation), we are accepting that some of the differences in the matrices will be counted as map errors when they really are differences caused by variation in human interpretation.

The ±9% variance in crown closure estimates brings into question the appropriateness of photo interpretation for labeling crown closure. Will the labels be acceptable, given that each label could vary by as much as 9% in either direction? The answer to this question lies in the anticipated uses of the map. For over 80 years, land managers and regulators have accepted photo interpretation for crown closure with few investigations concerning the accuracy of the photo-interpreted maps. This history of acceptance indicates that the relative nature of crown closure estimates is "good enough" for many applications. However, it is important to be aware of the variance of crown closure estimates whether they are used to create a map or to assess another map.

CROWN CLOSURE MAP RESULTS

Table 13.15 presents the crown closure error matrices for the 1981 map. The matrix includes the ±9% variance, resulting in overall accuracy of 60%. In general, the 1981 map systematically underestimates crown closure. In addition, errors of omission are significant, with 18 of the 19 map non-hardwood sites actually containing more than 10% crown closure in hardwoods. Most of these errors of omission resulted from the hardwoods being misidentified as shrubs. Errors of commission of non-hardwoods to hardwoods were few and almost always included mixed hardwood–conifer stands where the map photo interpreter estimated that there was more conifer than hardwood on the site.

Table 13.16 presents the crown closure error matrix for the 1990 map. Overall agreement between the reference data and the map is 73%. As with the 1981 maps, there seems to be a systematic underestimation of crown closure. However, unlike the 1981 maps, the updated map suffers from the 10 year difference in date between the aerial photography used for reference data and the date of the imagery. Reductions in crown closure caused by harvesting, fires, and urban expansion will cause non-map error differences between the office photo interpretation of the 1981 photos and the 1990 maps.

Furthermore, it is assumed that the labeling procedures to produce a polygon label from the 1990 pixel-based map results in the "true" label for each

TABLE 13.14

Comparison of Office and Field Photo Interpretation Adjusted for Variation in Crown Closure

				Field Photo Sites					
Class	0%	1–9%	10–18%	19–24%	25–42%	43–66%	67–84%	85–100%	Total
Office 0%	4	0	0	0	0	0	0	0	4
Photo 1–9%	0	3	2	0	0	1	0	0	6
Sites 10–33%	1	4	12	10	24	5	1	0	57
34–75%	1	1	1	0	27	31	14	0	75
76–100%	1	0	0	0	0	4	6	20	31
Total	7	8	15	10	51	40	21	20	172

Producer's Accuracy

Reference	Percent
0%	57
1–9%	88
10–18%	93
19–24%	100
25–42%	98
43–66%	78
67–84%	95
85–100%	100

User's Accuracy

Map	Percent
0%	100
1–9%	83
10–33%	88
34–75%	96
76–100%	84

Overall agreement = 157/172 = 91%

TABLE 13.15
Crown Closure Error Matrix, 1981 Map

	Class	Reference Data								
		0%	1–9%	10–18%	19–24%	25–42%	43–66%	67–84%	85–100%	Total
1981	0%	0	1	2	2	8	4	1	1	19
Map	1–9%	4	17	40	20	33	15	5	2	136
	10–33%	1	2	18	13	59	32	10	3	138
	34–75%	2	1	4	4	36	54	18	9	128
	76–100%	3	0	0	1	11	20	13	11	59
	Total	10	21	64	40	147	125	47	26	480

Producer's Accuracy

Reference	Percent
0%	40
1–9%	95
10–18%	91
19–24%	33
25–42%	65
43–66%	43
67–84%	66
85–100%	42

User's Accuracy

Map	Percent
0%	5
1–9%	45
10–33%	67
34–75%	84
76–100%	41

Overall accuracy = 286/480 = 60%

TABLE 13.16
Crown Closure Error Matrix, 1990 Map

	Class	Reference Data								
		0%	1–9%	10–18%	19–24%	25–42%	43–66%	67–84%	85–100%	Total
1990 Map	0%	4	1	0	1	0	4	4	4	18
	1–9%	1	11	21	7	20	5	1	0	66
	10–33%	1	7	36	24	79	35	10	1	193
	34–75%	4	2	7	8	47	79	32	13	192
	76–100%	0	0	0	0	1	2	0	8	11
	Total	10	21	64	40	147	125	47	26	480

Producer's Accuracy

Reference	Percent
0%	50
1–9%	90
10–18%	89
19–24%	60
25–42%	86
43–66%	63
67–84%	68
85–100%	31

User's Accuracy

Map	Percent
0%	28
1–9%	50
10–33%	76
34–75%	82
76–100%	73

Overall accuracy = 350/480 = 73%

sample polygon. It is possible, however, to develop several different labeling algorithms that produce several different labels. The implemented labeling algorithm assumed that the midpoint of a class was a good estimate for class value. The use of the average percentage crown closure for each class increases accuracy by approximately 2%.

Errors of omission also exist on the 1990 maps. Nine of 13 samples mislabeled as non-hardwoods in the 1990 map were labeled as hardwoods in the 1981 photo-interpreted map. As with the 1981 map, errors of commission occur in mixed hardwood–conifer sites.

COVER TYPE ANALYSIS

To understand the sensitivity of the matrix to variation in crown cover estimates versus errors in photo interpretation, the 173 sites used in the crown closure analysis were also interpreted for cover type. Table 13.17 shows the results of comparing the office photo-interpreted labels with the field photo-interpreted labels of the same sample units. The off-diagonal elements are caused either by photo interpretation error (e.g., the misidentification of species) or by variation in interpretation. For example, of the 10 sites showing differences between BOW and BOGP labels, nine were recognized by the field photo interpreter as acceptable as either BOW or BOGP. All eight sites contained gray pine. At issue was the estimated crown closure percentage of gray pine.

TABLE 13.17
Comparison of Office versus Field Photo Interpretation of Cover Type

		Field Photo Sites						
	Class	NH	BOGP	BOW	COW	MH	VOW	Total
Office	NH	**4**	0	0	0	0	0	4
Photo	BOGP	0	**3**	1	0	1	0	5
Sites	BOW	1	9	**38**	0	4	2	54
	COW	0	2	10	**41**	7	5	65
	MH	1	4	4	1	**14**	1	25
	VOW	0	0	1	0	1	**3**	5
	Total	6	18	54	42	27	11	**158**

Producer's Accuracy		User's Accuracy	
Reference	Percent	Map	Percent
NH	67	NH	100
BOGP	17	BOGP	60
BOW	70	BOW	70
COW	98	COW	63
MH	52	MH	56
VOW	27	VOW	60

Overall agreement = 103/158 = 65%

To account for the variation, interpreters filled out a cover-type fuzzy logic matrix for every accuracy assessment sample unit. Each sample was evaluated for the likelihood of being identified as each of the six possible cover types (Gopal and Woodcock, 1992). "Likelihood" was measured using the terms "absolutely wrong," "probably wrong," "acceptable," "probably right," and "absolutely right." For example, if a site with 65% total tree crown closure consisted of 20% valley oak, 25% blue oak, 10% gray pine, and 10% interior live oak, then the following interpretation might occur:

Blue Oak Woodland => acceptable
Blue Oak Gray Pine => probably right
Valley Oak Woodland => acceptable
Coast Oak Woodland => probably wrong
Montane Hardwood => probably wrong

A site with 100% grass cover, on the other hand, would be interpreted as follows:

Blue Oak Woodland => absolutely wrong
Blue Oak Gray Pine => absolutely wrong
Valley Oak Woodland => absolutely wrong
Coast Oak Woodland => absolutely wrong
Montane Hardwood => absolutely wrong

Table 13.18 incorporates the fuzzy logic "acceptable" interpretations into the confusion matrix. The greater the heterogeneity of the site, the more likely it is to have more than one "acceptable" label. The increase in overall agreement from 65% to 80% shows that 15% of the disagreement was caused by variation in interpretation rather than photo interpreter error.

Sites where differences persist in the matrix occur primarily because of photo interpretation error. Hardwood species can be extremely difficult to distinguish in the field, still less from aerial photography:

- Almost all of the differences between BOW/BOGP and COW were caused by the office photo interpreter misidentifying coast live oak as blue oak.
- The differences between BOW/BOGP and MH are caused by misidentification of coast live oak as interior live oak and by an artifact of the dichotomous key that mistakenly allows mixed interior live oak–blue oak stands to be labeled montane hardwood instead of BOW.
- Differences between VOW and BOW or COW occur because of misidentification of valley oak as either blue oak or coast live oak.

COVER TYPE MAP RESULTS

Tables 13.19 and 13.20 present the error matrices for cover type for the 1981 and 1990 maps. Both tables include all "right," "probably right," and "acceptable" labels as matches on the diagonal of the matrix. Increases in overall accuracies of 7 to 9% over Tables 13.6 and 13.7 indicate the impact of variation in interpretation on the

TABLE 13.18

Comparison of Office versus Field Photo Interpretation of Cover Type using the Fuzzy Logic Approach

	Class	NH	BOGP	BOW	COW	MH	VOW	Total
Field	NH	**4**	0	0	0	0	0	4
Photo	BOGP	0	**4**	1	0	0	0	5
Sites	BOW	0	0	**49**	0	3	2	54
	COW	0	2	9	**48**	2	4	65
	MH	0	2	2	1	**19**	1	25
	VOW	0	0	1	0	1	**3**	5
	Total	4	8	62	49	25	10	**158**

Office Photo Sites

Producer's Accuracy		User's Accuracy	
Reference	Percent	Reference	Percent
NH	100	NH	100
BOGP	80	BOGP	50
BOW	91	BOW	79
COW	74	COW	98
MH	76	MH	76
VOW	60	VOW	30

Overall agreement = 127/158 = 80%

TABLE 13.19

Cover Type Error Matrix, 1981 Map

	Class	NH	BOGP	BOW	COW	MH	VOW	Total
1981	NH	**0**	1	8	5	4	2	20
Map	BOGP	1	**30**	33	4	19	4	91
	BOW	2	1	**77**	5	9	5	99
	COW	2	0	22	**59**	9	2	94
	MH	2	3	8	23	**88**	1	125
	VOW	1	0	7	14	7	**3**	32
	Total	8	35	155	110	136	17	**461**

Reference Data

Producer's Accuracy		User's Accuracy	
NH	0	NH	0
BOGP	86	BOGP	33
BOW	50	BOW	78
COW	54	COW	63
MH	65	MH	70
VOW	18	VOW	9

Overall accuracy = 257/461 = 56%

TABLE 13.20
Cover Type Error Matrix, 1990 Map

	Class	NH	BOGP	BOW	COW	MH	VOW	Total
1990	NH	**4**	0	1	2	9	0	16
Map	BOGP	1	**31**	17	3	15	3	70
	BOW	0	1	**111**	7	16	7	142
	COW	2	0	11	**77**	7	3	100
	MH	0	3	8	16	**82**	1	110
	VOW	1	0	7	5	7	**3**	23
	Total	8	35	155	110	136	17	**461**

Reference Data appears as the spanning header over NH, BOGP, BOW, COW, MH, VOW.

Producer's Accuracy		User's Accuracy	
Reference	Percent	Map	Percent
NH	50	NH	25
BOGP	89	BOGP	44
BOW	72	BOW	78
COW	70	COW	77
MH	60	MH	75
VOW	18	VOW	13

Overall accuracy = 308/461 = 67%

matrices. As with crown closure, ambiguity exists between class labels for sites on the margins of class boundaries.

The overall accuracies of the 1990 map exceed those of the 1981 map. The incorporation of ancillary data (including the 1981 maps) and editing from field notes significantly increased the 1990 map's cover type accuracy. Specifically, much of the 1981 confusion between BOGP and BOW was reduced, as was that between COW and BOW/BOGP. However, the results do *not* constitute a comparison of photo interpretation versus satellite image processing methods, because the 1981 map was an extremely important ancillary layer in the creation of the 1990 map.

The most significant confusion in both maps' error matrices persists between BOW with BOGP, BOGP/BOW with MH, MH with COW, and VOW with all other hardwood types. The following reviews the causes of the disparities:

- Differences caused by various estimates of the amounts of gray pine in blue oak stands continue to cause confusion between BOGP and BOW labels. Twelve of the 18 BOW–BOGP confused sites contained various percentages of gray pine. Because of its sparse canopy, gray pine is extremely difficult to see and estimate on both aerial photography and satellite imagery.
- Error in species identification continues to contribute to the differences. As Table 13.18 indicates, this confusion also occurs in the reference

data and is contributing to lower map accuracies. It is difficult to distinguish the live oaks from one another on the ground, still less from
remotely sensed data. Similarly, valley oak on slopes is difficult to
distinguish from blue oaks. The confusion between blue oak and the
live oaks could be aided through the use of multi-temporal imagery or
photography, because blue oak is deciduous, whereas the live oaks are
evergreen.

- Much of the apparent confusion may be a function of failures in the classification system. The system does not adequately address the classification
of hardwood types that are dominated by interior live oak or Oregon white
oak. For example, of the sites identified by the reference data as MH when
the maps listed BOW or BOGP, 20 of the 31 sites were pure interior live oak
or mixtures of only interior live oak and blue oak with some sites containing gray pine.

- Most MH indicator species—including interior live oak, black oak, tan
oak, Pacific madrone, and canyon live oak—in addition to many associate
species—particularly California bay laurel, valley oak, and coast live oak
itself—also commonly occur in COW stands.

EXTENT

The final error matrices assess the accuracy of the extent of the 1981 and 1990
maps. As Tables 13.21 and 13.22 show, the 1981 maps contain significant errors
of omission. Eighty-six of the 102 (84%) accuracy assessment sites labeled non-
hardwood on the 1981 map but hardwood on the 1990 map were also labeled as
hardwood by the reference data. The extent of the 1990 maps is more accurate than
that of the 1981 maps. This is especially significant given the fact that no editing
or quality control was performed on the 1990 maps in areas outside of the extent
of the 1981 map.

TABLE 13.21
Error Matrix of Extent, 1981 Map

	Class	Reference Data		
		Non-hardwood	Hardwood	Total
1981	Non-hardwood	**16**	86	102
Map	Hardwood	10	**451**	461
	Total	26	537	**563**

Producer's Accuracy		User's Accuracy	
Reference	**Percent**	**Map**	**Percent**
Non-hardwood	62	Non-hardwood	16
Hardwood	84	Hardwood	98

Overall accuracy = 467/563 = 83%

TABLE 13.22
Error Matrix of Extent, 1990 Map

	Class	Non-hardwood	Hardwood	Total
		Reference Data		
1990	Non-hardwood	**5**	16	21
Map	Hardwood	21	**521**	542
	Total	26	537	**563**

Producer's Accuracy		User's Accuracy	
Reference	Percent	Map	Percent
Non-hardwood	19	Non-hardwood	24
Hardwood	97	Hardwood	96

Overall accuracy = 526/563 = 93%

DISCUSSION

The 1990 mapping project resulted in the creation of GIS land cover information for over 32 million acres of land; approximately one-third of the State of California and three times the extent of the 1981 mapping project. Of these 32 million acres:

- 11.4 million are hardwood rangelands.
- 3.8 million are conifer lands.
- 3.9 million are shrublands.
- At least 7.9 million are grasslands.
- 0.5 million are urban.
- 0.6 million are water.
- 3.4 million are other lands (e.g., agricultural, etc.).

Substantially more acres of hardwood rangeland exist than reported in the 1981 photo-based mapping project. Most of the additional acreage occurs on the north coast and the northern part of the Sacramento Valley. Aside from the omitted hardwood rangelands in the photo-based maps, both the photo-based and the image-based maps provide valuable information about California's hardwood rangelands.

The accuracy assessment of the 1981 and the 1990 maps gave map users valuable information:

- Both maps consistently slightly underestimate crown closure.
- Because certain hardwood species are difficult to distinguish from one another, cover type confusion can exist in areas where indistinguishable species occur together. In areas where both coast live oak and interior live oak occur, confusion in cover type labels will also occur. Similarly, valley oak on slopes is difficult to distinguish from blue oaks. Valley oak is the most confused type, and it is unlikely that mixed valley oak stands on slopes can be adequately classified from either aerial photography or satellite imagery.

Finally, because of its sparse canopy, gray pine is extremely difficult to see on both aerial photography and satellite imagery. This lack of resolution in the remotely sensed data leads to confusion between blue oak woodland and blue oak–gray pine woodland.

- Errors of commission (labeling an area as hardwoods when it is not hardwoods) are rare in both maps and usually occur in mixed hardwood–conifer stands.
- Weakness in the classification system may create confusion, because the system did not adequately handle classification of hardwood types that are dominated by interior live oak or Oregon white oak. This issue has since been solved with more up-to-date classification schemes for vegetation in California.
- In addition, the assessment also showed that satellite image processing is a valuable tool for mapping and monitoring California's hardwood rangelands and can produce maps that exceed the accuracy of maps from photo interpretation. It also is flexible, allowing changes in classification systems, and cost-effective, allowing a tripling of the project area extent with no increase in project cost. Finally, it can provide increased detail of information and a richer database and facilitate the monitoring of land use and land cover change over time.

CONCLUSIONS

The accuracy assessment analysis presented in this chapter shows both the complexity and the subtle surprises that face any project. Several points are particularly important:

- The development and implementation of a robust classification system is critical to the success of any mapping or accuracy assessment project. As this chapter shows, the classification system plays a critical part in design, data collection, and analysis. Ambiguous systems without clear rules will result in a disastrous project.
- The analysis of accuracy assessment must go beyond the simple creation of an error matrix. The producer and user of the map need to understand why sites fall off the diagonal. Map users and producers must understand whether the confusion is inevitable and *acceptable*, such as that caused by variation in crown closure estimates; or inevitable and *unacceptable*, such as that resulting from the misidentification of hardwood species.
- Users and producers must also understand how much of the confusion between map and reference labels is caused by reference error, overlapping classification system boundaries, inappropriate use of remote sensing technologies, or map error. In this project:
  a. Reference error exists, especially in species identification. If photo interpretation of cover type is only 80% correct (Table 13.18), what is the impact on the map accuracy assessment, which relied heavily on photo interpretation for development of the reference data?

   b.   Fuzzy class boundaries and variation in estimation were significant. Is this acceptable? In this case, cost savings of using remote sensing more than made up for the loss in map label precision. In other applications, the loss may be unacceptable, resulting in abandonment of remote sensing in favor of ground mapping.

- Trade-offs between sample design and data collection rigor and practical considerations are inevitable in most accuracy assessments. In particular, field data collection frequently precludes the use of random site selection. However, in each and every case, it is important to remember that every method or procedure must be statistically valid as well as practically attainable. It is not acceptable to sacrifice either requirement; rather, it is necessary to find the best balance between the two.

14 Advanced Topics

There are a number of advanced topics that need to be covered beyond what has been presented in this book up to this point. This chapter begins with a discussion of change detection accuracy assessment. The complexities of conducting such an assessment are presented along with the formulation of the change detection error matrix. A key issue in any change detection accuracy assessment is the realization that change is a rare event, and sampling must occur to specifically deal with this issue. While the creation of a change detection error matrix is possible, it requires a tremendous amount of work and may not even be possible in many situations. Therefore, a compromise two-step method is proposed and demonstrated that may provide a more practical approach to assessing the accuracy of change. The chapter concludes with a short discussion of multi-layer accuracy assessment.

CHANGE DETECTION

An increasingly popular and extremely important application of remotely sensed data is for use in change detection. Change detection is the process of identifying differences in the state of an object or phenomenon by observing it at different times (Singh, 1989). Four aspects of change detection are important: (1) detecting that changes have occurred, (2) identifying the nature of the change, (3) measuring the areal extent of the change, and (4) assessing the spatial pattern of the change (Brothers and Fish, 1978; Malila, 1985; Singh, 1986). Techniques to perform change detection with digital imagery have become numerous because of an ever-increasing amount of imagery, improved versatility in manipulating digital data, better image analysis software, and significant developments in computing power. Just as assessing the accuracy of a single-date map is vital, change detection accuracy assessment is an important component of any change analysis project.

Assessing the accuracy of a single-date or one point in time (OPIT) thematic map generated from remotely sensed data as presented in this book is a complex, but attainable, endeavor. In addition to the complexities associated with a single-date accuracy assessment of remotely sensed data, change detection presents even more difficult and challenging issues to consider. The very nature of change detection makes quantitative analysis of the accuracy difficult. For example, how does one obtain reference data for images that were taken in the past? How does one sample enough areas that will change in the future to have a statistically valid assessment? Which change detection technique will work best for a given change in the environment? Positional accuracy also plays a big role in change detection. It is critical to determine whether an increase in the size of an area has actually occurred or whether the apparent change is simply due to a positional error. Figure 14.1 is a modification of the sources of error figure for a single-date assessment presented early on in this book (Figure 2.5) and shows how complicated the error sources get when

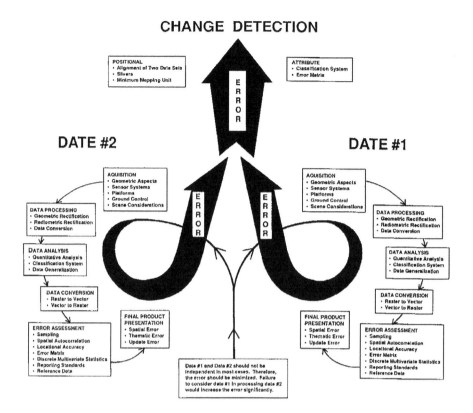

FIGURE 14.1 Sources of error in a change detection analysis from remotely sensed data. (Reproduced with permission from the American Society for Photogrammetry and Remote Sensing, from Congalton R.G. 1996. Accuracy assessment: A critical component of land cover mapping, in *Gap Analysis: A Landscape Approach to Biodiversity Planning*. A peer-reviewed proceedings of the ASPRS/GAP Symposium. Charlotte, NC, pp. 119–131.)

performing a change detection. Most of the studies on change detection conducted to this point do not present quantitative results of their work, which makes it difficult to determine which method should be applied to a future project.

The following section presents the topics to be considered when preparing to perform a change detection accuracy assessment. There are three critical components that must be considered in any change detection accuracy assessment. These are: (1) reference data, (2) sampling, and (3) the change detection error matrix.

Reference Data

The collection of valid reference data is central to any accuracy assessment, whether it is a single-date assessment or for evaluating a change detection. Let's imagine conducting a change detection project in 2018 by comparing a vegetation/land cover map generated from 1998 Landsat imagery (call this Time 1) with another map generated from 2018 Landsat imagery (call this Time 2). Let us further suppose that the classification scheme used for both maps is the same, since we created both maps. Reference data for evaluating the 2018 map could be collected on the ground in 2018 or even 2019 and still be considered valid. However, how can reference data for assessing the 1998 map and therefore, the change detection be obtained?

There are a few possible answers. The most probable answer is that there are no readily available reference data using the same classification scheme for 1998 of this area, and therefore, there really is no way to assess the change. Second, there might be some aerial photography of the area that was acquired around the same time as the 1998 Landsat imagery. Of course, scale is an issue here. If the photos are of such a small scale that sufficient detail cannot be accurately interpreted from them given the classification scheme used in the mapping project, then the photos cannot be used to provide reference data. Even if the scale is sufficient, photo interpretation is subject to error, and the reference data may not be appropriate here, depending on the complexity of the classification scheme. Third, there may be some ground inventory of the area in question that can be used as reference data. This third possibility is extremely slim, and even if it does exist, there is the issue of minimum mapping unit and sample unit size that must be considered. As previously discussed, many inventories conducted in the field lack sufficient size to be effective reference data sample units. Therefore, the lack of valid reference data is often a limiting factor when attempting to conduct a change detection accuracy assessment.

SAMPLING

There is one overriding issue that must be considered when sampling for change detection accuracy assessment that is beyond the sampling issues already presented in this book for a single-date assessment. Failure to consider this issue dooms the assessment to be a wasted effort. It must be remembered that *change is a rare event*. Under normal circumstances, it would be unusual for more than 10% of a given area to change in a 5–10 year period. More likely, the change would be closer to 5%. In extreme cases, high change rates such as 20% are possible. Of course, in certain catastrophic situations (e.g., wildfires, hurricanes, tornadoes, and the like), the percentage of an area that has changed may be even higher.

Now, consider sampling to find the change areas. Using a random sampling approach, even in a map with high change rates (20%), on average only two out of 10 samples will find any change. In the more usual case, it could take up to 20 samples before an area that has changed is found. Given the time and effort needed to collect samples for accuracy assessment, oversampling in the non-change areas must be avoided. Stratification of the area to prioritize sampling in the change areas should be employed. However, exactly how to delineate these strata is not always obvious. If all the change areas were known, then no new map of change would be needed. Fortunately, for many applications, logic or experience dictates the likely places for change to occur. For example, urban change occurs in areas around existing urban centers. It is extremely rare to find a new city built in the middle of nowhere. Sampling for urban change in a buffer zone around an urban center increases your chances of finding it when compared with a randomly placed sample. In this scenario, taking some portion of your sample in high-priority areas seems to make sense.

In another example, MacLeod and Congalton (1998) conducted a change detection accuracy assessment for monitoring eelgrass change in Great Bay, New Hampshire. Because change is such a rare event, it was necessary to proportionally allocate more sampling effort to areas where change was more likely to happen. In this example for mapping eelgrass, we know that it is very unlikely that eelgrass will grow in the channel (i.e., the deep water areas). Sampling should be limited in the channel. On the other hand, the eelgrass is more likely in two places: (1) around existing eelgrass beds and (2) in shallow areas where no eelgrass currently grows. The sampling effort should be increased in these areas. Therefore, we modified our sampling efforts in the following ways: (1) only 10% of our sampling effort occurred in the deep water areas, (2) 40% of our sampling effort was dedicated to a buffer area within one sample grid (i.e., pixel) of existing eelgrass, and (3) 50% of our sampling effort was dedicated to shallow areas where new eelgrass seedlings could occur. In this way, the sampling was designed to find the change areas (Congalton and Brennan, 1998).

There are many other factors to consider when sampling for change detection accuracy assessment. However, failure to note that change is a rare event influences all these other factors and must be considered first.

CHANGE DETECTION ERROR MATRIX

To apply established accuracy assessment techniques to change detection, the standard single-date classification error matrix needs to be adapted to a change detection error matrix as proposed by Congalton and Macleod (1994) and Macleod and Congalton (1998). This new matrix has the same characteristics as the single-date classification error matrix but also assesses errors in changes between two time periods (between Time 1 and Time 2) and not simply a single classification.

Table 14.1 reviews a single-date error matrix and the associated descriptive statistics, overall, producer's, and user's accuracies, that have already been presented in this book. This single-date error matrix is for three vegetation/land cover map classes (F = forest, U = urban, and W = water). The matrix is of dimension 3×3. The y-axis of the error matrix represents the three vegetation/land cover map classes as derived from the remotely sensed classification (i.e., the map), and the x-axis shows the three map classes identified in the reference data.

The major diagonal of this matrix is highlighted and indicates correct classification. In other words, when the classification indicates that the category was F and the reference data agree that it is F, then the [F, F] cell in the matrix is tallied. The same logic follows for the other map classes: U and W. Off-diagonal elements in the matrix indicate the different types of confusion (called *omission* and *commission* *error*) that exist in the classification. Omission error occurs when an area is omitted from the correct category. Commission error occurs when an area is placed in the wrong category. This information is helpful in guiding the user to where the major problems exist in the classification.

The top part of Figure 14.2 shows a change detection error matrix generated for the same three vegetation/land cover map classes (F, U, and W). Note, however, that the matrix is no longer of dimension 3×3 but rather, 9×9. This is because we are no longer looking at a single classification but rather, a change between two different

TABLE 14.1

Example of a Single-Date (One Point in Time) Error Matrix Showing Overall, User's, and Producer's Accuracies

| | Reference Data | | | Row | Land Cover Categories |
	F	U	W	Total	
F	40	9	8	57	F = Forest
U	1	15	5	21	U = Urban
W	1	1	20	22	W = Water
Column Total	42	25	33	100	

Classified Data

OVERALL ACCURACY
= 40 + 15 + 20
= 75/100 = 75%

PRODUCER'S ACCURACY	USER'S ACCURACY
F = 40/42 = 95%	F = 40/57 = 70%
U = 15/25 = 60%	U = 15/21 = 71%
W = 20/33 = 61%	W = 20/22 = 91%

maps generated at different times. Remember, in a single-date error matrix, there is one row and column for each map class. However, in assessing change detection, the error matrix is the size of the number of map classes squared. Therefore, the question of interest is, "What map class was this area at Time 1, and what is it at Time 2?" The answer has nine possible outcomes for each axis of the matrix (F at Time 1 and F at Time 2, U at Time 1 and U at Time 2, W at Time 1 and W at Time 2, F at Time 1 and U at Time 2, F at Time 1 and W at Time 2, U at Time 1 and F at Time 2, U at Time 1 and W at Time 2, W at Time 1 and F at Time 2, or W at Time 1 and U at Time 2), all of which are indicated along the rows and columns of the error matrix. It is then important to note what the remotely sensed data said about the change and compare it with what the reference data indicate. This comparison uses the exact same logic as for the single classification error matrix; it is just complicated by the two time periods (i.e., the change). Again, the major diagonal indicates correct classification, while the off-diagonal elements indicate the errors or confusion. The descriptive statistics (i.e., overall, user's, and producer's accuracies) can also be computed.

It is important to note that the change detection error matrix can also be simplified or collapsed into a 2×2 no change/change error matrix (bottom of Figure 14.2). The no change/change error matrix can be generated by summing the appropriate cells in the four sections of the complete change detection error matrix partitioned by the dotted lines. For example, to get the number of areas where both the classification and reference data determined that no change had occurred between the two dates, you would simply sum all nine cells in the upper left box (the areas that did not change in either the classification or reference data). To summarize or collapse the cells in which change occurred in both the classification and reference

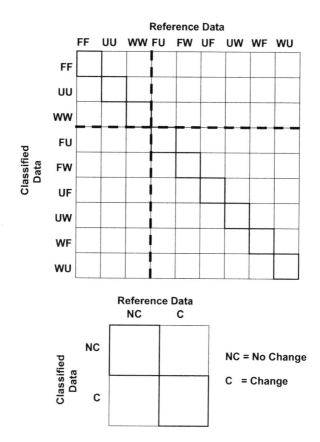

FIGURE 14.2 Change error matrix for the same three map classes (Forest, Urban, Water) as the single-date matrix and the collapsed no change/change matrix.

data, you would sum the 36 cells in the lower right box. The other two cells in the no change/change matrix would be determined in a similar manner. From this no change/change error matrix, the analysts can easily determine whether low accuracy was due to a poor change detection technique, misclassification, or both.

It should be obvious to the reader that performing a change detection accuracy assessment is a very complex undertaking. Simply scaling the single-date assessment methodology causes the size of the error matrix to increase, as well as the number of samples required for the assessment. In the example error matrix for a single-date, three-class map (Table 14.1), 150 samples (3 classes × 50 samples/class) are required. When a second time period is added, the number of samples grows to 450 (9 change classes × 50 samples/class). If a single-date mapping project had 10 classes, the required sample size would be 5000 (10 × 10 × 50 samples per class) samples. Since not all changes are logical/possible within a given period of time (e.g., one would not expect water to become forest in 5 years), that number would likely be smaller, but the number of required samples is still much greater than for a single-date accuracy assessment and probably not feasible under most time and budget conditions.

Therefore, while it may not be possible to perform a complete change detection accuracy assessment and generate a change detection error matrix for every change detection project, it is still relevant to try to answer the following two questions: 1) How accurately have the areas that have changed between Time 1 and Time 2 been mapped? and 2) How well was the change captured? To answer these questions, the change detection accuracy assessment process can be divided into two steps instead of using a single assessment and the change detection error matrix approach.

TWO-STEP APPROACH TO CHANGE DETECTION ACCURACY ASSESSMENT

If it is not possible to use the change detection error matrix approach to perform your change detection accuracy assessment, then you may wish to use this two-step approach. This method does not allow you to obtain the accuracy of all the change classes (e.g., the map was forest in Time 1 and is now residential in Time 2; from forest to residential), but it does provide for assessing the accuracy of the areas that changed in Time 2 and how well the overall changes were captured.

The first step in this process is to assess the accuracy of just the areas that changed between the two time periods in question. In other words, conduct a single-date accuracy assessment only on the areas that changed between Time 1 and Time 2. The sampling procedure is similar to that of a traditional single-date accuracy assessment with the requisite number of samples per land cover class selected using a chosen sampling strategy from the map area. However, in this case, only areas classified as change (i.e., the map class is different in Time 2 than it was in Time 1) are used to select the samples. The accuracy assessment only needs to be conducted for the change areas for Time 2, because the rest of the map has the same accuracy as the map did in Time 1 for all the areas that did not change.

The second step in this process is simply a change/no change validation. This step is similar to collapsing the change detection error matrix to the change/no change (2×2) matrix presented in the bottom of Figure 14.2. The difference here is that instead of having to sample to fill in the entire change detection error matrix, the sampling is performed only to assess the change/no change. Treating the map as a binary or two-class scheme, change/no change, requires a simpler sampling technique than the multinomial situation of a complete change detection error matrix. Since we are working with a two-case situation, where we only wish to know whether the classification is change or no change, we can use the binomial distribution to calculate the sample size. Ginevan (1979) introduced this sampling method to the remote sensing community and concluded that:

- The method should have a low probability of accepting a map of low accuracy.
- It should have a high probability of accepting a map of high accuracy.
- It should require a minimum number of samples.

Computing the sample size for the binomial approach requires the use of a look-up table that presents the required sample size for a given minimum error and a desired level of confidence. For example, a map at with a chosen accuracy of 90%

(10% error) and using a 95% confidence level (at 95%, we run the risk of a 1 in 20 chance that we reject a map that is actually correct), the minimum number of samples required for the assessment is 298. Given this sample size, the map is rejected as inaccurate if more than 21 are misclassified.

Therefore, this two-step approach is quite effective. While not producing a complete change detection error matrix or assessing the accuracy of each change (to–from) class, it does provide a means of assessing the accuracy of the labeling (thematic accuracy) of the areas that changed between the two time frames. In addition, an assessment of whether or not the change is accurately captured can be generated using the binomial change/no change approach. These two steps are considerably easier and require significantly less time, money, and resources than using the change detection error matrix approach. However, if the required resources are available, the change detection error matrix provides the most information about the change analysis and is the recommended approach to use.

CASE STUDY

This case study details the change detection accuracy assessment for the Kentucky Landscape Census (KLC) National Land Cover Dataset (NLCD) update from 2001 to 2005. Appendix 14.1 presents a list of the land cover classes and a brief description of each. It was not possible in this project, due to limited time and resources, to collect enough data to generate a change detection error matrix. Instead, the goal for this change analysis was to assess the accuracy of the change classification (the areas that were changed between 2001 and 2005) and to determine how well change, in general, was captured between the two dates.

To accomplish this task, the accuracy assessment was completed in two steps. First, the 2005 change areas were assessed as a single-date land cover map. Reference data samples were collected by interpreting high-resolution imagery collected within a year of the 2005 Landsat imagery (NLCD classification). Generating an error matrix with at least 30 samples per class, the overall map accuracy was computed as well as the omission and commission error rates for each individual thematic class within the map. Second, a change mask was assessed as a binary change/no change map. Samples were collected using a stratified random selection approach within Kentucky. To limit the selection area to areas of likely change, various strata layers were created to prioritize the selection of the samples. By conducting the assessment in these two separate steps, the two questions of "How accurate is the 2005 change map?" and "How well was land cover change captured?" were answered.

Step 1: Accuracy of the Change Areas

The first step in the change detection accuracy assessment was to assess the accuracy of the areas that changed as a separate, single-date map. The sampling procedure is similar to that of any traditional accuracy assessment with between 30 and 50 samples per land cover class randomly selected from the mapping area. However, in this case, only areas classified as change between 2001 and 2005 were used to draw samples, and only the 2005 classification was assessed. The reference data for this time period were 1 meter color imagery from the National Agricultural Inventory Program (NAIP). Complete coverage for all of Kentucky was available from NAIP.

The results of the accuracy assessment of the 2005 change areas are presented in error matrix form in Table 14.2. Inspection of the error matrix shows that not all classes were assessed for accuracy and included in the error matrix. While all classes in the United States Geological Survey (USGS) classification scheme (see Appendix 14.1) were classified, the bulk of the change occurred in only some of the land cover classes. Changes to map classes such as wetland features or forest regrowth classes did not occur in sufficient amounts, and therefore, too few samples were available with which to assess the accuracy of these classes.

The overall deterministic accuracy for the 2005 change areas is 58.8%, and the fuzzy accuracy is 79.9%. The 21.1% difference between deterministic and fuzzy accuracies can be attributed to two similar effects. First, much of the increase in fuzzy accuracy is related to class confusion within the developed classes, and second, there is a separate but similar confusion between grassland and shrub land. The four developed classes are defined by the percentage of impervious surface in each class:

- Developed, Open: 0%–25%
- Developed, Low Intensity: 26%–50%
- Developed, Medium Intensity: 51%–75%
- Developed, High Intensity: 76%–100%

While this division results in well-defined class boundaries, there is a degree of uncertainty associated with the percent impervious map that translates to the final classification. As the pixels in the percent impervious map are assigned through a statistical regression analysis technique, a degree of error is associated with each estimated value, generally ±10%. This results in pixels within less than 10% of the class boundaries potentially being in two developed map classes. For example, a pixel with a value of 55% would be categorized as Developed, Medium Intensity; however, by factoring in the degree of uncertainty with the estimate, it could also be categorized as Developed, Low Intensity. For the purposes of this accuracy assessment, a developed accuracy sample unit was given a fuzzy interpretation if, by factoring in the degree of uncertainty, it satisfied the categorization criteria for more than one developed class. The predominance of the developed map classes in the final map and their inherent uncertainty contribute to the variance between the deterministic and fuzzy accuracy estimates.

Similar confusion between grass and shrub is the second major contributor to the difference between the deterministic and fuzzy accuracies. These classes are rarely found naturally in Kentucky. Instead, the two classes more often represent a transition or succession of vegetation growth after a disturbance related to forestry or mining. Determining the amount of shrub to grass vegetation based on the hard class breaks leads to fuzziness in some accuracy calls.

TABLE 14.2
Error Matrix Shows the Accuracy of the 2005 Change Areas

| | REFERENCE | | | | | | | | User's Accuracies | | | |
LABELS (MAP)	Water	Developed Open Space	Developed Low Intensity	Developed Medium Intensity	Developed High Intensity	Bare Land	Shrub	Grassland	Deterministic Totals	Deterministic Accuracies	Fuzzy Totals	Fuzzy Accuracies
Water	15	0,0	0,0	0,0	0,0	0,4	0,0	0,0	15/19	78.9%	15/19	78.9%
Developed Open Space	0,0	9	21,1	4,0	0,0	2,5	0,0	1,2	9/45	20.0%	37/45	82.2%
Developed Low Intensity	0,0	4,0	24	3,0	1,1	1,3	0,0	0,1	24/38	63.2%	33/38	86.8%
Developed Medium Intensity	0,0	0,0	4,0	5	1,0	0,3	0,0	0,0	5/13	38.5%	10/13	76.9%
Developed High Intensity	0,0	0,0	2,0	0,0	11	0,0	0,0	0,0	11/13	84.6%	13/13	100.0%
Bare Land	0,0	0,0	1,2	0,1	0,0	49	0,0	1,0	49/54	90.7%	51/54	94.4%
Shrub	0,0	0,1	0,1	0,0	0,0	0,6	31,0	19,12	31/61	50.8%	50/61	82.0%
Grassland	0,0	1,0	0,0	0,0	0,0	9,18	0,2	40,0	40/70	57.1%	50/70	71.4%
Producer's Accuracies												
Deterministic Totals	15/15	9/15	24/56	5/13	11/14	49/100	31/33	40/76				
Deterministic Accuracies	100.0%	60.0%	42.9%	38.5%	78.6%	49.0%	91.7%	52.6%				
Fuzzy Totals	15/15	14/15	52/56	12/13	13/14	61/100	31/33	61/76				
Fuzzy Accuracies	100.0%	93.3%	92.9%	92.3%	92.9%	61.0%	91.7%	80.3%				

Overall Accuracies

Deterministic		Fuzzy	
184/313	58.8%	250/313	79.9%

Step 2: Change/No Change Assessment

Treating the map as a binary scheme, change/no change, requires a simpler sampling technique than generating a complete change detection error matrix with all the "from" and "to" classes. We can use a binomial distribution to calculate the sample size (Ginevan, 1979). A simple look-up table can be used to determine the required sample size for a given minimum error and a desired level of confidence. For a map accuracy of 90% and using a 95% confidence level (at 95%, we run the risk of a 1 in 20 chance that we reject a map that is actually correct), the minimum number of samples required is 298, with the map being rejected as not meeting the accuracy standard if more than 21 are misclassified.

To compensate for the rarity of change within the landscape, an approach was designed employing five strata layers to increase the sampling to areas of likely change. The first stratum is called the *change mask* and incorporates all the areas indicated to have changed by the image analysis change methodology used in this project; 30% of the sampling was performed within the change mask. The second area sampled was a buffer surrounding the change mask. It is expected that change will occur near change, so it follows that sampling should occur around the change areas, and 25% of the samples were taken in this buffer area around the change mask. A third stratum used for another 25% of the sampling included those areas indicated by spectral analysis of the two images as changed. Fourth, 10% of the samples were allocated to those map classes that had the highest amounts of change. In other words, sampling was increased for those map classes that had significant change occur between 2001 and 2005. Finally, the last 10% of the sampling was allocated to the rest of the map. Table 14.3 presents a summary of the sampling allocation by strata along with the number of samples taken in each stratum.

The overall accuracy for the change/no change assessment was 96% (Table 14.4). Seven samples were labeled change on the map but were not change on the reference data, while six samples were labeled as no change on the map but actually did change. Only 13 total errors were found. Given the binomial sampling selected with a desired map accuracy of 90% and a 95% confidence level, 21 errors were permitted. Therefore, this map was accurate at the 90% level, and the error matrix shows the true accuracy to be 96%.

TABLE 14.3
Sampling Breakdown Based on Strata Layer

Strata Layer	Percentage of Total Samples	Number of Samples
Change Mask	30	88
Distance from Change Mask	25	75
Spectral Magnitude	25	75
Probability of Change	10	30
Remaining Unsampled Area	10	30

TABLE 14.4
Final Change/No Change Matrix

		Reference		Producer's Accuracies	
				Totals	Accuracies
		Change	**No Change**		
M	Change	**75**	7	75/82	92%
A	No Change	6	**210**	210/216	97%
P					

	User's Accuracies		Overall Accuracy	
Totals	75/81	210/217		
Accuracies	93%	97%	285/298	96.0%

Determining the accuracy of the KLC change map was a critical component of this project. The process demonstrated by this case study was designed to assess the accuracy of the change areas on the 2005 map and evaluate how well change was captured between 2001 and 2005. It was not possible, in this project, to conduct a full change detection accuracy assessment and generate a change detection error matrix. This two-step approach is an effective compromise when the time and resources are not available to conduct a full assessment. These results show that change was captured with a success rate of 96%.

While the deterministic accuracy assessment is low at 58.8%, the fuzzy assessment of the classification shows a favorable overall classification accuracy of 79.9%.

MULTI-LAYER ASSESSMENTS

Everything that has been presented in the book up to this point, with the exception of the last section on change detection, has dealt with the accuracy of a single map layer. However, it is important to at least mention multi-layer assessments. Figure 14.3 demonstrates a scenario in which four different map layers are combined to produce a map of wildlife habitat suitability. In this scenario, accuracy assessments have been performed on each of the map layers; each layer is 90% accurate. The question is: how accurate is the wildlife suitability map?

If the four map layers are independent (i.e., the errors in each map are not correlated), then probability tells us that the accuracy would be computed by multiplying the accuracies of the layers together. Therefore, the accuracy of the final map is 90% × 90% × 90% × 90% = 66%. However, if the four map layers are not independent but rather, completely correlated with each other (i.e., the errors are in the exact same place in all four layers), then the accuracy of the final map is 90%. In reality, neither of these cases is very likely. There is usually some correlation between the map layers. For instance, vegetation is certainly related to proximity to a stream and also to elevation. Therefore, the actual accuracy of the final map could only be determined by performing another accuracy assessment on this layer. We do know that this accuracy will be between 66% and 90% and will probably be closer to 90% than to 66%.

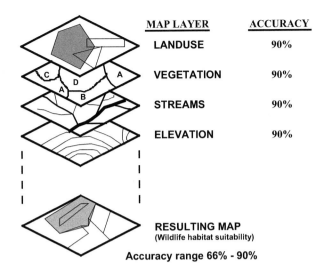

FIGURE 14.3 The range of accuracies for a decision made from combining multiple layers of spatial data.

One final observation should be mentioned here. It is quite eye-opening that using four map layers, all with very high accuracies, could result in a final map with an accuracy of only 66%. In fact, if we added a fifth map layer and it also had an accuracy of 90%, we see that the accuracy of our wildlife map could drop below 60%. On the other hand, we have been using these types of maps for a long time without any knowledge of their accuracy. Therefore, do not despair; certainly, this knowledge can only help us improve our ability to effectively use spatial data.

APPENDIX 14.1
Class Descriptions of the 2005 NLCD Land Cover

Class	Description
Open Water	All areas of open water, generally with less than 25% cover of vegetation or soil.
Developed, Open Space	Includes areas with a mixture of some constructed materials, but mostly vegetation in the form of lawn grasses. Impervious surfaces account for less than 20% of total cover. These areas most commonly include large-lot single-family housing units, parks, golf courses, and vegetation planted in developed settings for recreation, erosion control, or aesthetic purposes.
Developed, Low Intensity	Includes areas with a mixture of constructed materials and vegetation. Impervious surfaces account for 20–49% of total cover. These areas most commonly include single-family housing units.
Developed, Medium Intensity	Includes areas with a mixture of constructed materials and vegetation. Impervious surfaces account for 50–79% of the total cover. These areas most commonly include single-family housing units.
Developed, High Intensity	Includes highly developed areas where people reside or work in high numbers. Examples include apartment complexes, row houses, and commercial/industrial. Impervious surfaces account for 80 to100% of the total cover.

(Continued)

APPENDIX 14.1 (CONTINUED)
Class Descriptions of the 2005 NLCD Land Cover

Class	Description
Bare Land	Barren areas of bedrock, scarps, talus, slides, volcanic material, glacial debris, sand dunes, strip mines, gravel pits, and other accumulations of earthen material. Generally, vegetation accounts for less than 15% of total cover.
Deciduous Forest	Areas dominated by trees generally greater than 5 m tall, and greater than 20% of total vegetation cover. More than 75% of the tree species shed foliage simultaneously in response to seasonal change.
Evergreen Forest	Areas dominated by trees generally greater than 5 m tall, and greater than 20% of total vegetation cover. More than 75% of the tree species maintain their leaves all year. Canopy is never without green foliage.
Mixed Forest	Areas dominated by trees generally greater than 5 m tall, and greater than 20% of total vegetation cover. Neither deciduous nor evergreen species are greater than 75% of total tree cover.
Scrub Shrub	Areas dominated by shrubs less than 5 m tall with shrub canopy typically greater than 20% of total vegetation. This class includes true shrubs, young trees in an early successional stage, or trees stunted from environmental conditions.
Grassland Herbaceous	Areas dominated by graminoid or herbaceous vegetation, generally greater than 80% of total vegetation. These areas are not subject to intensive management such as tilling but can be used for grazing.
Pasture Hay	Areas of grasses, legumes, or grass–legume mixtures planted for livestock grazing or the production of seed or hay crops, typically on a perennial cycle. Pasture/hay vegetation accounts for greater than 20% of total vegetation.
Cultivated Crop	Areas used for the production of annual crops, such as corn, soybeans, vegetables, tobacco, and cotton, and also perennial woody crops such as orchards and vineyards. Crop vegetation accounts for greater than 20% of total vegetation. This class also includes all land being actively tilled.
Woody Wetland	Areas where forest or shrubland vegetation accounts for 25–100% of the cover, and the soil or substrate is periodically saturated with or covered with water.
Emergent Herbaceous Wetland	Areas where perennial herbaceous vegetation accounts for 75–100% of the cover, and the soil or substrate is periodically saturated with or covered with water.

15 Summary and Conclusions

This book provides you with the principles and practices needed to conduct an effective accuracy assessment of a map generated from remotely sensed data. It was not written to review every academic exercise ever published on this topic. Instead, it was written to be a practical guide providing the geospatial analyst with the principles necessary to conduct the assessment while at the same time demonstrating the considerations and limitations that occur in practice. You now know about the history and development of these tools and understand that it is not just a matter of following a simple recipe. Rather, there are many considerations, trade-offs, and decisions that must be made along the way to be successful. Positional accuracy and thematic accuracy must be considered together. It does you no good to label the area correctly if you are in the wrong place; nor is it valuable to be in the correct place but label it wrongly. A large amount of planning and thought must go into each accuracy assessment if it is going to be cost effective and yet statistically sound. The collection of valid reference data to compare with the map is essential regardless of whether you are conducting a positional assessment, a thematic one, or both. Planning this collection as effectively and efficiently as possible is a critical component determining the success of the project.

This book has provided you with the knowledge that you will need to conduct an effective and valid accuracy assessment. However, there are some nuggets of wisdom that should be emphasized when performing such an assessment. These nuggets are presented in the following list in bullet form with a brief explanation of each. It is our hope that this summary will remind you of the many considerations, trade-offs, and decisions that you must know about in your quest to perform the most appropriate accuracy assessment for your specific project:

- *Accuracy assessment is a complex process that must be carefully considered and planned at the beginning of a mapping project. THINK ABOUT IT.* It is human nature to just jump into things without the careful planning that is required to be successful. Performing an accuracy assessment without careful consideration and planning will doom it to being significantly more expensive than it need be in the best-case scenario and result in a flawed or invalid assessment in the worst case. Use the flow charts and other materials in this book to really *think about it* at the beginning of the project, and you will be successful.
- *Positional accuracy and thematic accuracy are not independent and must be planned for together. PLAN FOR BOTH.* It is important to *plan* for both positional accuracy and thematic accuracy, since one impacts the other.

You need to be in the correct place with the correct map label for the map to be deemed accurate in that location.

- *The classification scheme must be well defined, and the same scheme must be used for making the map and in collecting the reference data. DEFINE IT.* It seems a simple concept to make sure that the classification scheme is *defined*, such that it is clear to everyone. However, this is not always the case. It is critical that the scheme has both labels and some method of defining those labels; that is, definitions or a key of some type. It is also important that the scheme be mutually exclusive, totally exhaustive, and hierarchical. Finally, a minimum mapping unit (mmu) should be determined.

- *Effective collection of the reference data is the most important component of any accuracy assessment. BE CONSCIENTIOUS.* Whether you are conducting a positional or a thematic accuracy assessment, the collection of the necessary reference data is the most expensive and time-consuming component. Therefore, the success of the entire assessment hinges on the *conscientious* collection of these data. The next few bullets provide details about the considerations for an effective reference data collection.

  - *The sample unit size must be based on positional accuracy considerations and the pixel size of the imagery. CONSIDER IT.* The size of the sample unit must be such that the analyst has strong confidence that the map class label given to the area for the reference data corresponds to the same location on the map. In other words, any positional error is removed, and the reference data are indicative of only the thematic accuracy being assessed. In almost every situation when dealing with moderate to high–spatial resolution imagery, the sample unit should not be a single pixel. Therefore, the size of the sample unit should be carefully *considered*.

  - *The number of samples used to assess accuracy will influence the statistical validity and cost of the assessment. DETERMINE IT.* It is important that sufficient samples be collected for each map class, so that a statistically valid assessment can be conducted. There are rules of thumb and equations to aid in *determining* the appropriate number of samples. If a sufficient number of samples are not collected, then the assessment is not valid, and the effort is wasted.

  - *The choice of the appropriate sampling scheme allows the collection of the reference data to be as efficient as possible. SELECT IT.* There are trade-offs between random sampling and cost/effort effectiveness, especially if the reference data are to be collected in the field. In most cases, a stratified sampling approach works well to allow sufficient samples to be collected for each map class.

  - *Failure to account for spatial autocorrelation results in wasted effort in the collection of the reference data, as the samples are not independent of each other and therefore, violate a key statistical assumption of the sampling process. ACCOUNT FOR IT.* Avoid spatial autocorrelation, so that the reference samples are independent from one another and the training data, by setting a minimum distance between all the samples used in the mapping project (both for training and for accuracy assessment).

- *Pick the appropriate error matrix for your project. CHOOSE IT.* There are a number of possible ways to represent the error matrix for assessing the accuracy of a mapping project. These include the traditional tally-based matrix, the area-based matrix when using an object-based classification approach, and the fuzzy error matrix. Each of these is discussed in great detail in this book. In some cases, the analyst may *choose* to use more than one method to represent the error matrix. Care must be taken at the beginning of the project to *choose* the appropriate representations for your specific project.
- *Calculate descriptive statistics from the error matrix. COMPUTE THEM.* Once the error matrix has been generated, it is useful to *compute* overall, producer's, and user's accuracies that provide more information about the map being assessed.
- *Run basic analysis techniques on the error matrix.* RUN THEM. There are many analysis techniques that can be run to further investigate the results of the accuracy assessment indicated by the error matrix. It is up to the analyst and the map user to *run* those techniques they deem most appropriate for their specific needs.
- *Both the map user and the map producer will learn more from the analysis of the off-diagonal cells of the error matrix than from the diagonal cells. STUDY IT.* Knowing which map classes are confused with one another allows the map user to more effectively use the map. *Studying* the error matrix to understand why map classes are confused with one another allows the map producer to develop new methods by focusing on correcting the confusion.
- *Humans will always vary in their estimates of vegetative cover. ACCEPT IT.* There is no absolute right or wrong in many accuracy assessments, only shades of gray. The more complex the classification scheme, the higher the impact of human variation in labeling the reference samples. *Accept* that the variation exists, and use the techniques presented in this book to account for it.
- *Accuracy assessment is expensive. BUDGET FOR IT.* Too often, accuracy assessment is an afterthought, added to a project at the end, when not enough funds are available to perform a credible assessment. Understand that accuracy assessment is a fundamental requirement of any mapping project that should be planned from the beginning of the project, and make sure that you have a *budget* for it.
- *The entire accuracy assessment process must be fully described. DOCUMENT IT.* Given the many considerations, trade-offs, and decisions that must be made when conducting an accuracy assessment, it is

imperative that the analyst fully describes and *documents* the process used for their particular project. In this way, everyone will understand exactly how the assessment was performed and will be able to judge for themselves whether the map is useful for their purposes. Failure to *document* the process leaves the assessment unfinished and the map less useful.

Appendix 1: ASPRS Positional Accuracy Standards for Digital Geospatial Data

(EDITION 1, VERSION 1.0 - NOVEMBER 2014)

CONTENTS

Foreword

The goal of American Society for Photogrammetry and Remote Sensing (ASPRS) is to advance the science of photogrammetry and remote sensing: to educate individuals in the science of photogrammetry and remote sensing; to foster the exchange of information pertaining to the science of photogrammetry and remote sensing; to develop, place into practice and maintain standards and ethics applicable to aspects of the science; to provide a means for the exchange of ideas among those interested in the sciences; and to encourage, publish and distribute books, periodicals, treatises, and other scholarly and practical works to further the science of photogrammetry and remote sensing.

This standard was developed by the ASPRS Map Accuracy Standards Working Group, a joint committee under the Photogrammetric Applications Division, Primary Data Acquisition Division and Lidar Division, which was formed for the purpose of reviewing and updating ASPRS map accuracy standards to reflect current technologies. A subcommittee of this group, consisting of Dr. Qassim Abdullah of Woolpert, Inc., Dr. David Maune of Dewberry Consultants, Doug Smith of David C. Smith and Associates, Inc., and Hans Karl Heidemann of the U.S. Geological Survey, was responsible for drafting the document.

ASPRS Positional Accuracy Standards for Digital Geospatial Data

1 PURPOSE

The objective of the *ASPRS Positional Accuracy Standards for Digital Geospatial Data* is to replace the existing *ASPRS Accuracy Standards for Large-Scale Maps* (1990), and the *ASPRS Guidelines, Vertical Accuracy Reporting for Lidar Data* (2004) to better address current technologies.

This standard includes positional accuracy standards for digital orthoimagery, digital planimetric data and digital elevation data. Accuracy classes, based on RMSE values, have been revised and upgraded from the 1990 standard to address the higher accuracies achievable with newer technologies. The standard also includes additional accuracy measures, such as orthoimagery seam lines, aerial triangulation accuracy, lidar relative swath-to-swath accuracy, recommended minimum nominal pulse density (NPD), horizontal accuracy of elevation data, delineation of low confidence areas for vertical data, and the required number and spatial distribution of checkpoints based on project area.

1.1 Scope and Applicability

This standard addresses geo-location accuracies of geospatial products and it is not intended to cover classification accuracy of thematic maps. Further, the standard does not specify the best practices or methodologies needed to meet the accuracy thresholds stated herein. Specific requirements for the testing methodologies are specified as are some of the key elemental steps that are critical to the development of data if they are to meet these standards. However, it is the responsibility of the data provider to establish all final project design parameters, implementation steps and quality control procedures necessary to ensure the data meet final accuracy requirements.

The standard is intended to be used by geospatial data providers and users to specify the positional accuracy requirements for final geospatial products.

1.2 Limitations

This standard is limited in scope to addressing accuracy thresholds and testing methodologies for the most common mapping applications and to meet immediate shortcomings in the outdated 1990 and 2004 standards referenced above. While the standard is intended to be technology independent and broad based, there are several specific accuracy assessment needs that were identified but are not addressed herein at this time, including:

1. Methodologies for accuracy assessment of linear features (as opposed to well-defined points);
2. Rigorous total propagated uncertainty (TPU) modeling (as opposed to – or in addition to – ground truthing against independent data sources);
3. Robust statistics for data sets that do not meet the criteria for normally distributed data and therefore cannot be rigorously assessed using the statistical methods specified herein;
4. Image quality factors, such as edge definition and other characteristics;
5. Robust assessment of checkpoint distribution and density;
6. Alternate methodologies to TIN interpolation for vertical accuracy assessment.

This standard is intended to be the initial component upon which future work can build. Additional supplemental standards or modules should be pursued and added by subject matter experts in these fields as they are developed and approved by the ASPRS.

At this time this standard does not reference existing international standards. International standards could be addressed in future modules or versions of this standard if needed.

1.3 STRUCTURE AND FORMAT

The standard is structured as follows: The primary terms and definitions, references and requirements are stated within the main body of the standard, according to the ASPRS standards template and without extensive explanation or justification. Detailed supporting guidelines and background information are attached as Annexes A through D. Annex A provides a background summary of other standards, specifications and/or guidelines relevant to ASPRS but which do not satisfy current requirements for digital geospatial data. Annex B provides accuracy/quality examples and overall guidelines for implementing the standard. Annex C provides guidelines for accuracy testing and reporting. Annex D provides guidelines for statistical assessment and examples for computing vertical accuracy in vegetated and non-vegetated terrain.

2   CONFORMANCE

No conformance requirements are established for this standard.

3   REFERENCES

American Society for Photogrammetry and Remote Sensing (ASPRS), 2013. ASPRS Accuracy Standards for Digital Geospatial Data (DRAFT), PE&RS, December 2013, pp. 1073–1085.

American Society for Photogrammetry and Remote Sensing (ASPRS), 1990. ASPRS Accuracy Standards for Large-Scale Maps, URL: http://www.asprs.org/a/society/committees/standards/1990_jul_1068–1070.pdf (last date accessed: 22 January 2015)

American Society for Photogrammetry and Remote Sensing (ASPRS), 2004. ASPRS Guidelines, Vertical Accuracy Reporting for Lidar Data, URL: http://www.asprs.org/a/society/committees/standards/Vertical_Accuracy_Reporting_for_Lidar_Data.pdf (last date accessed: 22 January 2015)

Dieck, R.H., 2007. *Measurement Uncertainty: Methods and Applications*, Instrument Society of America, Research Triangle Park, North Carolina, 277.

Federal Geographic Data Committee, 1998. FGDC-STD-007.2-1998, Geospatial Positioning Accuracy Standards, Part 2: Standards for Geodetic Networks, FGDC, c/o U.S. Geological Survey, URL: https://www.fgdc.gov/standards/projects/FGDC-standards-projects/accuracy/part2/chapter2 (last date accessed: 22 January 2015)

Federal Geographic Data Committee, 1998. FGDC-STD-007.3-1998, Geospatial Positioning Accuracy Standards, Part 3: National Standard for Spatial Data Accuracy (NSSDA), FGDC, c/o U.S. Geological Survey, URL: https://www.fgdc.gov/standards/projects/FGDC-standards-projects/accuracy/part3/chapter3 (last date accessed: 22 January 2015)

National Digital Elevation Program (NDEP), 2004. *NDEP Guidelines for Digital Elevation Data*, URL: http://www.ndep.gov/NDEP_Elevation_Guidelines_Ver1_10May2004.pdf (last date accessed: 22 January 2015)

National Geodetic Survey (NGS), 1997. NOAA Technical Memorandum NOS NGS-58, V. 4.3: Guidelines for Establishing GPS-Derived Ellipsoid Heights (Standards: 2 cm and 5 cm), URL: https://www.ngs.noaa.gov/PUBS_LIB/NGS-58.html (last date accessed: 22 January 2015)

National Geodetic Survey (NGS), 2008. NOAA Technical Memorandum NOS NGS-59, V1.5: Guidelines for Establishing GPS-Derived Orthometric Heights, URL: http://www.ngs.noaa.gov/PUBS_LIB/NGS592008069FINAL2.pdf (last date accessed: 22 January 2015)

Additional informative references for other relevant and related guidelines and specifications are included in Annex A.

4   AUTHORITY

The responsible organization for preparing, maintaining, and coordinating work on this guideline is the American Society for Photogrammetry and Remote Sensing (ASPRS), Map Accuracy Standards Working Group, a joint committee formed by the Photogrammetric Applications Division, Primary Data Acquisition Division and the Lidar Division. For further information, contact the Division Directors using the contact information posted on the APSRS website, http://www.asprs.org.

5   TERMS AND DEFINITIONS

absolute accuracy – A measure that accounts for all systematic and random errors in a data set.

accuracy – The closeness of an estimated value (for example, measured or computed) to a standard or accepted (true) value of a particular quantity. Not to be confused with *precision*.

bias – A systematic error inherent in measurements due to some deficiency in the measurement process or subsequent processing.

blunder – A mistake resulting from carelessness or negligence.

confidence level – The percentage of points within a data set that are estimated to meet the stated accuracy; e.g., accuracy reported at the 95% confidence level means that 95% of the positions in the data set will have an error with respect to true ground position that are equal to or smaller than the reported accuracy value.

consolidated vertical accuracy (CVA) – Replaced by the term Vegetated Vertical Accuracy (VVA) in this standard, CVA is the term used by the NDEP guidelines for vertical accuracy at the 95th percentile in all land cover categories combined.

fundamental vertical accuracy (FVA) – Replaced by the term Non-vegetated Vertical Accuracy (NVA), in this standard, FVA is the term used by the NDEP guidelines for vertical accuracy at the 95% confidence level in open terrain only where errors should approximate a normal error distribution.

ground sample distance (GSD) – The linear dimension of a sample pixel's footprint on the ground. Within this document GSD is used when referring to the collection GSD of the raw image, assuming near-vertical imagery. The actual GSD of each pixel is not uniform throughout the raw image and varies significantly with terrain height and other factors. Within this document, GSD is assumed to be the value computed using the calibrated camera focal length and camera height above average horizontal terrain.

horizontal accuracy – The horizontal (radial) component of the positional accuracy of a data set with respect to a horizontal datum, at a specified confidence level.

inertial measurement unit (IMU) – The primary component of an INS. Measures 3 components of acceleration and 3 components of rotation using orthogonal triads of accelerometers and gyros.

inertial navigation system (INS) – A self-contained navigation system, comprised of several subsystems – IMU, navigation computer, power supply, interface, etc. Uses measured accelerations and rotations to estimate velocity, position and orientation. An unaided INS loses accuracy over time, due to gyro drift.

kurtosis – The measure of relative "peakedness" or flatness of a distribution compared with a normally distributed data set. Positive kurtosis indicates a relatively peaked distribution near the mean while negative kurtosis indicates a flat distribution near the mean.

local accuracy – The uncertainty in the coordinates of points with respect to coordinates of other directly connected, adjacent points at the 95% confidence level.

mean error – The average positional error in a set of values for one dimension (x, y, or z); obtained by adding all errors in a single dimension together and then dividing by the total number of errors for that dimension.

network accuracy – The uncertainty in the coordinates of mapped points with respect to the geodetic datum at the 95% confidence level.

non-vegetated vertical accuracy (NVA) – The vertical accuracy at the 95% confidence level in non-vegetated open terrain, where errors should approximate a normal distribution.

percentile – A measure used in statistics indicating the value below which a given percentage of observations in a group of observations fall. For example, the 95th percentile is the value (or score) below which 95 percent of the observations may be found. For accuracy testing, percentile calculations are based on the absolute values of the errors, as it is the magnitude of the errors, not the sign that is of concern.

pixel resolution or pixel size – As used within this document, pixel size is the ground size of a pixel in a digital orthoimage, after all rectifications and resampling procedures.

positional error – The difference between data set coordinate values and coordinate values from an independent source of higher accuracy for identical points.

positional accuracy – The accuracy of the position of features, including horizontal and vertical positions, with respect to horizontal and vertical datums.

precision (repeatability) – The closeness with which measurements agree with each other, even though they may all contain a systematic bias.

relative accuracy – A measure of variation in point-to-point accuracy in a data set.

resolution – The smallest unit a sensor can detect or the smallest unit an orthoimage depicts. The degree of fineness to which a measurement can be made.

root-mean-square error (RMSE) – The square root of the average of the set of squared differences between data set coordinate values and coordinate values from an independent source of higher accuracy for identical points.

skew – A measure of symmetry or asymmetry within a data set. Symmetric data will have skewness toward zero.

standard deviation – A measure of spread or dispersion of a sample of errors around the sample mean error. It is a measure of precision, rather than accuracy; the standard deviation does not account for uncorrected systematic errors.

supplemental vertical accuracy (SVA) – Merged into the Vegetated Vertical Accuracy (VVA) in this standard, SVA is the NDEP guidelines term for reporting the vertical accuracy at the 95th percentile in each separate land cover category where vertical errors may not follow a normal error distribution.

systematic error – An error whose algebraic sign and, to some extent, magnitude bears a fixed relation to some condition or set of conditions. Systematic errors follow some fixed pattern and are introduced by data collection procedures, processing or given datum.

uncertainty (of measurement) – a parameter that characterizes the dispersion of measured values, or the range in which the "true" value most likely lies. It can also be defined as an estimate of the limits of the error in a measurement (where "error" is defined as the difference between the theoretically-unknowable "true" value of a parameter and its measured value). Standard uncertainty refers to uncertainty expressed as a standard deviation.

vegetated vertical accuracy (VVA) – An estimate of the vertical accuracy, based on the 95th percentile, in vegetated terrain where errors do not necessarily approximate a normal distribution.

vertical accuracy – The measure of the positional accuracy of a data set with respect to a specified vertical datum, at a specified confidence level or percentile.

For additional terms and more comprehensive definitions of the terms above, reference is made to the *Glossary of Mapping Sciences; Manual of Photogrammetry,* 6th edition; *Digital Elevation Model Technologies and Applications: The DEM Users Manual,* 2nd edition; and/or the *Manual of Airborne Topographic Lidar,* all published by ASPRS.

6 SYMBOLS, ABBREVIATED TERMS, AND NOTATIONS

ACC$_r$ the horizontal (radial) accuracy at the 95% confidence level
ACC$_z$ the vertical linear accuracy at the 95% confidence level
ASPRS American Society for Photogrammetry and Remote Sensing
CVA   Consolidated Vertical Accuracy
DEM   Digital Elevation Model
DTM   Digital Terrain Model
FVA   Fundamental Vertical Accuracy
GSD   Ground Sample Distance
GNSS   Global Navigation Satellite System
GPS   Global Positioning System
IMU   Inertial Measurement Unit
INS   Inertial Navigation System
NGPS   Nominal Ground Point Spacing
NPD   Nominal Pulse Density
NMAS   National Map Accuracy Standard
NPS   Nominal Pulse Spacing
NSSDA National Standard for Spatial Data Accuracy
NVA   Non-vegetated Vertical Accuracy
RMSE$_r$ the horizontal linear RMSE in the radial direction that includes both x- and y-coordinate errors.
RMSE$_x$ the horizontal linear RMSE in the X direction (Easting)
RMSE$_y$ the horizontal linear RMSE in the Y direction (Northing)
RMSE$_z$ the vertical linear RMSE in the Z direction (Elevation)
RMSE   Root Mean Square Error
RMSD$_z$ root-mean-square-difference in elevation (z)
SVA   Supplemental Vertical Accuracy
TIN   Triangulated Irregular Network
VVA   Vegetated Vertical Accuracy
$\bar{x}$ sample mean error, for x
S sample standard deviation
γ_1 sample skewness
γ_2 sample kurtosis

7 SPECIFIC REQUIREMENTS

This standard defines accuracy classes based on RMSE thresholds for digital ortho-imagery, digital planimetric data, and digital elevation data.

Testing is always recommended but may not be required for all data sets; specific requirements must be addressed in the project specifications.

When testing is required, horizontal accuracy shall be tested by comparing the planimetric coordinates of well-defined points in the data set with coordinates determined from an independent source of higher accuracy. Vertical accuracy shall be tested by comparing the elevations of the surface represented by the data set with elevations determined from an independent source of higher accuracy. This is done by comparing the elevations of the checkpoints with elevations interpolated from the data set at the same x/y coordinates. See Annex C, Section C.11 for detailed guidance on interpolation methods.

All accuracies are assumed to be relative to the published datum and ground control network used for the data set and as specified in the metadata. Ground control and checkpoint accuracies and processes should be established based on project requirements. Unless specified to the contrary, it is expected that all ground control and checkpoints should normally follow the guidelines for network accuracy as detailed in the Geospatial Positioning Accuracy Standards, Part 2: Standards for Geodetic Networks, Federal Geodetic Control Subcommittee, Federal Geographic Data Committee (FGDC-STD-007.2-1998). When local control is needed to meet specific accuracies or project needs, it must be clearly identified both in the project specifications and the metadata.

7.1 STATISTICAL ASSESSMENT OF HORIZONTAL AND VERTICAL ACCURACIES

Horizontal accuracy is to be assessed using root-mean-square-error (RMSE) statistics in the horizontal plane, i.e., $RMSE_x$, $RMSE_y$ and $RMSE_r$. Vertical accuracy is to be assessed in the z dimension only. For vertical accuracy testing, different methods are used in non-vegetated terrain (where errors typically follow a normal distribution suitable for RMSE statistical analyses) and vegetated terrain (where errors do not necessarily follow a normal distribution). When errors cannot be represented by a normal distribution, the 95th percentile value more fairly estimates accuracy at a 95% confidence level. For these reasons vertical accuracy is to be assessed using $RMSE_z$ statistics in non-vegetated terrain and 95th percentile statistics in vegetated terrain. Elevation data sets shall also be assessed for horizontal accuracy where possible, as outlined in Section 7.5.

With the exception of vertical data in vegetated terrain, error thresholds stated in this standard are presented in terms of the acceptable RMSE value. Corresponding estimates of accuracy at the 95% confidence level values are computed using *National Standard for Spatial Data Accuracy* (NSSDA) methodologies according to the assumptions and methods outlined in Annex D, Accuracy Statistics and Examples.

7.2 Assumptions Regarding Systematic Errors and Acceptable Mean Error

With the exception of vertical data in vegetated terrain, the assessment methods outlined in this standard, and in particular those related to computing NSSDA 95% confidence level estimates, assume that the data set errors are normally distributed and that any significant systematic errors or biases have been removed. It is the responsibility of the data provider to test and verify that the data meet those requirements including an evaluation of statistical parameters such as the kurtosis, skew and mean error, as well as removal of systematic errors or biases in order to achieve an acceptable mean error prior to delivery.

The exact specification of an acceptable value for mean error may vary by project and should be negotiated between the data provider and the client. As a general rule, these standards recommend that the mean error be less than 25% of the specified RMSE value for the project. If a larger mean error is negotiated as acceptable, this should be documented in the metadata. In any case, mean errors that are greater than 25% of the target RMSE, whether identified pre-delivery or post-delivery, should be investigated to determine the cause of the error and to determine what actions, if any, should be taken. These findings should be clearly documented in the metadata.

Where RMSE testing is performed, discrepancies between the x, y, or z coordinates of the ground point check survey and the data set that exceed three times the specified RMSE error threshold shall be interpreted as blunders and should be investigated and either corrected or explained before the data are considered to meet this standard. Blunders may not be discarded without proper investigation and explanation in the metadata.

7.3 Horizontal Accuracy Standards for Geospatial Data

Table 7.1 specifies the primary horizontal accuracy standard for digital data, including digital orthoimagery, digital planimetric data, and scaled planimetric maps. This standard defines horizontal accuracy classes in terms of their $RMSE_x$ and $RMSE_y$ values. While prior ASPRS standards used numerical ranks for discrete accuracy classes tied directly to map scale (i.e., Class 1, Class 2, etc.), many modern applications require more flexibility than these classes allowed. Furthermore, many applications of horizontal accuracy cannot be tied directly to compilation scale, resolution of the source imagery, or final pixel resolution.

TABLE 7.1

Horizontal Accuracy Standards for Geospatial Data

Horizontal Accuracy Class	Absolute Accuracy			Orthoimagery Mosaic Seamline Mismatch (cm)
	RMSE$_x$ and RMSE$_y$ (cm)	RMSEr (cm)	Horizontal Accuracy at 95% Confidence Level (cm)	
X-cm	$\leq X$	$\leq 1.414*X$	$\leq 2.448*X$	$\leq 2*X$

A Scope of Work, for example, can specify that digital orthoimagery, digital planimetric data, or scaled maps must be produced to meet ASPRS Accuracy Standards for 7.5 cm $RMSE_x$ and $RMSE_y$ Horizontal Accuracy Class.

Annex B includes extensive examples that relate accuracy classes of this standard to their equivalent classes according to legacy standards. $RMSE_x$ and $RMSE_y$ recommendations for digital orthoimagery of various pixel sizes are presented in Table B.5. Relationships to prior map accuracy standards are presented in Table B.6. Table B.6 lists $RMSE_x$ and $RMSE_y$ recommendations for digital planimetric data produced from digital imagery at various GSDs and their equivalent map scales according to the legacy standards of ASPRS 1990 and NMAS of 1947. The recommended associations of $RMSE_x$ and $RMSE_y$, pixel size, and GSD that are presented in the above mentioned tables of Annex B are based on current status of mapping technologies and best practices. Such associations may change in the future as mapping technologies continue to advance and evolve.

7.4   VERTICAL ACCURACY STANDARDS FOR ELEVATION DATA

Vertical accuracy is computed using RMSE statistics in non-vegetated terrain and 95th percentile statistics in vegetated terrain. The naming convention for each vertical accuracy class is directly associated with the RMSE expected from the product. Table 7.2 provides the vertical accuracy classes naming convention for any digital elevation data. Horizontal accuracy requirements for elevation data are specified and reported independent of the vertical accuracy requirements. Section 7.5 outlines the horizontal accuracy requirements for elevation data.

Annex B includes examples on typical vertical accuracy values for digital elevation data and examples on relating the vertical accuracy of this standard to the legacy map standards. Table B.7 of Annex B lists 10 common vertical accuracy classes and their corresponding accuracy values and other quality measures according to this standard. Table B.8 of Annex B provides the equivalent vertical accuracy measures for the same ten classes according to the legacy standards of ASPRS 1990 and NMAS of 1947. Table B.9 provides examples on vertical accuracy and the recommended Lidar points density for digital elevation data according to the new ASPRS 2014 standard.

The Non-vegetated Vertical Accuracy at the 95% confidence level in non-vegetated terrain (NVA) is approximated by multiplying the accuracy value of the Vertical Accuracy Class (or $RMSE_z$) by 1.9600. This calculation includes survey checkpoints located in traditional open terrain (bare soil, sand, rocks, and short grass) and urban terrain (asphalt and concrete surfaces). The NVA, based on an $RMSE_z$ multiplier, should be used only in non-vegetated terrain where elevation errors typically follow a normal error distribution. $RMSE_z$-based statistics should not be used to estimate vertical accuracy in vegetated terrain or where elevation errors often do not follow a normal distribution.

The Vegetated Vertical Accuracy at the 95% confidence level in vegetated terrain (VVA) is computed as the 95th percentile of the absolute value of vertical errors in all vegetated land cover categories combined, including tall weeds and crops, brush lands, and fully forested areas. For all vertical accuracy classes, the VVA standard is 3.0 times the accuracy value of the Vertical Accuracy Class.

TABLE 7.2
Vertical Accuracy Standards for Digital Elevation Data

	Absolute Accuracy			Relative Accuracy (where applicable)		
Vertical Accuracy Class	RMSE$_z$ Non-Vegetated (cm)	NVA[1] at 95% Confidence Level (cm)	VVA[2] at 95th Percentile (cm)	Within-Swath Hard Surface Repeatability (Max Diff) (cm)	Swath-to-Swath Non-Vegetated Terrain (RMSD$_z$) (cm)	Swath-to-Swath Non-Vegetated Terrain (Max Diff) (cm)
X-cm	$\leq X$	$\leq 1.96*X$	$\leq 3.00*X$	$\leq 0.60*X$	$\leq 0.80*X$	$\leq 1.60*X$

Both the RMSE$_z$ and 95th percentile methodologies specified above are currently widely accepted in standard practice and have been proven to work well for typical elevation data sets derived from current technologies. However, both methodologies have limitations, particularly when the number of checkpoints is small. As more robust statistical methods are developed and accepted, they will be added as new Annexes to supplement and/or supersede these existing methodologies.

7.5   HORIZONTAL ACCURACY REQUIREMENTS FOR ELEVATION DATA

This standard specifies horizontal accuracy thresholds for two types of digital elevation data with different horizontal accuracy requirements:

- **Photogrammetric elevation data:** For elevation data derived using stereo photogrammetry, the horizontal accuracy equates to the horizontal accuracy class that would apply to planimetric data or digital orthoimagery produced from the same source imagery, using the same aerial triangulation/INS solution.

---

[1] Statistically, in non-vegetated terrain and elsewhere when elevation errors follow a normal distribution, 68.27% of errors are within one standard deviation (s) of the mean error, 95.45% of errors are within ($2 * s$) of the mean error, and 99.73% of errors are within ($3 * s$) of the mean error. The equation ($1.9600 * s$) is used to approximate the maximum error either side of the mean that applies to 95% of the values. Standard deviations do not account for systematic errors in the data set that remain in the mean error. Because the mean error rarely equals zero, this must be accounted for. Based on empirical results, if the mean error is small, the sample size sufficiently large and the data are normally distributed, $1.9600 * RMSE_z$ is often used as a simplified approximation to compute the NVA at a 95% confidence level. This approximation tends to overestimate the error range as the mean error increases. A precise estimate requires a more robust statistical computation based on the standard deviation and mean error. ASPRS encourages standard deviation, mean error, skew, kurtosis and RMSE to all be computed in error analyses in order to more fully evaluate the magnitude and distribution of the estimated error.

[2] VVA standards do not apply to areas previously defined as low confidence areas and delineated with a low confidence polygon (see Annex C). If VVA accuracy is required for the full data set, supplemental field survey data may be required within low confidence areas where VVA accuracies cannot be achieved by the remote sensing method being used for the primary data set.

- **Lidar elevation data:** Horizontal error in lidar derived elevation data is largely a function of positional error as derived from the Global Navigation Satellite System (GNSS), attitude (angular orientation) error (as derived from the INS) and flying altitude; and can be estimated based on these parameters. The following equation[3] provides an estimate for the horizontal accuracy for the lidar-derived dataset assuming that the positional accuracy of the GNSS, the attitude accuracy of the Inertial Measurement Unit(IMU) and the flying altitude are known:

$$\text{Lidar Horizontal Error}\left(\text{RMSE}_r\right)$$

$$= \sqrt{\left(\text{GNSS positional error}\right)^2 + \left(\frac{\tan\left(\text{IMU error}\right)}{0.55894170} \times \text{flying altitude}\right)^2}$$

The above equation considers flying altitude (in meters), GNSS errors (radial, in cm), IMU errors (in decimal degrees), and other factors such as ranging and timing errors (which is estimated to be equal to 25% of the orientation errors). In the above equation, the values for the "GNSS positional error" and the "IMU error" can be derived from published manufacturer specifications for both the GNSS receiver and the IMU.

If the desired horizontal accuracy figure for lidar data is agreed upon, then the following equation can be used to estimate the flying altitude:

$$\text{Flying Altitude} \approx$$

$$\frac{0.55894170}{\tan\left(\text{IMU error}\right)} \sqrt{\left(\text{Lidar Horizontal Error}\left(\text{RMSEr}\right)\right)^2 - \left(\text{GNSS positional error}\right)^2}$$

Table B.10 can be used as a guide to estimate the horizontal errors to be expected from lidar data at various flying altitudes, based on estimated GNSS and IMU errors.

Guidelines for testing the horizontal accuracy of elevation data sets derived from lidar are outlined in Annex C.

Horizontal accuracies at the 95% confidence level, using NSSDA reporting methods for either "produced to meet" or "tested to meet" specifications should be reported for all elevation data sets.

For technologies or project requirements other than as specified above for photogrammetry and airborne lidar, appropriate horizontal accuracies should be negotiated between the data provider and the client. Specific error thresholds, accuracy thresholds or methods for testing will depend on the technology used and project design. The data provider has the responsibility to establish appropriate methodologies, applicable to the technologies used, to verify that horizontal accuracies meet the stated project requirements.

---

[3] The method presented here is one approach; there other methods for estimating the horizontal accuracy of lidar data sets, which are not presented herein. Abdullah, Q., 2014, unpublished data.

7.6 LOW CONFIDENCE AREAS FOR ELEVATION DATA

If the VVA standard cannot be met, low confidence area polygons shall be developed and explained in the metadata. For elevation data derived from imagery, the low confidence areas would include vegetated areas where the ground is not visible in stereo. For elevation data derived from lidar, the low confidence areas would include dense cornfields, mangrove or similar impenetrable vegetation. The low confidence area polygons are the digital equivalent to using dashed contours in past standards and practice. Annex C, Accuracy Testing and Reporting Guidelines, outlines specific guidelines for implementing low confidence area polygons.

7.7 ACCURACY REQUIREMENTS FOR AERIAL TRIANGULATION AND INS-BASED SENSOR ORIENTATION OF DIGITAL IMAGERY

The quality and accuracy of the aerial triangulation (if performed) and/or the Inertial Navigation System-based (INS-based) sensor orientations (if used for direct orientation of the camera) play a key role in determining the final accuracy of imagery derived mapping products.

For photogrammetric data sets, the aerial triangulation and/or INS-based direct orientation accuracies must be of higher accuracy than is needed for the final, derived products.

For INS-based direct orientation, image orientation angles quality shall be evaluated by comparing checkpoint coordinates read from the imagery (using stereo photogrammetric measurements or other appropriate method) to the coordinates of the checkpoint as determined from higher accuracy source data.

Aerial triangulation accuracies shall be evaluated using one of the following methods:

1. By comparing the values of the coordinates of the checkpoints as computed in the aerial triangulation solution to the coordinates of the checkpoints as determined from higher accuracy source data;
2. By comparing the values of the coordinates read from the imagery (using stereo photogrammetric measurements or other appropriate method) to the coordinates of the checkpoint as determined from higher accuracy source data.

For projects providing deliverables that are only required to meet accuracies in x and y (orthoimagery or two-dimensional vector data), aerial triangulation errors in z have a smaller impact on the horizontal error budget than errors in x and y. In such cases, the aerial triangulation requirements for $RMSE_z$ can be relaxed. For this reason the standard recognizes two different criteria for aerial triangulation accuracy:

- Accuracy of aerial triangulation designed for digital planimetric data (orthoimagery and/or digital planimetric map) **only**:
  $RMSE_{x(AT)}$ or $RMSE_{y(AT)} = \frac{1}{2} * RMSE_{x(Map)}$ or $RMSE_{y(Map)}$

$RMSE_{z(AT)} = RMSE_{x(Map)}$ or $RMSE_{y(Map)}$ of orthoimagery

- Note: The exact contribution of aerial triangulation errors in z to the overall horizontal error budget for the products depends on ground point location in the image and other factors. The relationship stated here for an $RMSE_z$ (AT) of twice the allowable RMSE in x or y is a conservative estimate that accommodates the typical range of common camera geometries and provides allowance for many other factors that impact the horizontal error budget.
- Accuracy of aerial triangulation designed for elevation data, or planimetric data (orthoimagery and/or digital planimetric map) and elevation data production:
  - $RMSE_{x(AT)}$, $RMSE_{y(AT)}$ or $RMSE_{z(AT)} = \frac{1}{2} * RMSE_{x(Map)}$, $RMSE_{y(Map)}$ or $RMSE_{z(DEM)}$

Annex B, Data Accuracy and Quality Examples, provides practical examples of these requirements.

7.8 ACCURACY REQUIREMENTS FOR GROUND CONTROL USED FOR AERIAL TRIANGULATION

Ground control points used for aerial triangulation should have higher accuracy than the expected accuracy of derived products according to the following two categories:

- Accuracy of ground control designed for planimetric data (orthoimagery and/or digital planimetric map) production **only**:
  $RMSE_x$ or $RMSE_y = \frac{1}{4} * RMSE_{x(Map)}$ or $RMSE_{y(Map)}$,
  $RMSE_z = \frac{1}{2} * RMSE_{x(Map)}$ or $RMSE_{y(Map)}$
- Accuracy of ground control designed for elevation data, or planimetric data **and** elevation data production:

  $RMSE_x$, $RMSE_y$ or $RMSE_z = \frac{1}{4} * RMSE_{x(Map)}$, $RMSE_{y(Map)}$ or $RMSE_{z(DEM)}$

Annex B, Data Accuracy and Quality Examples, provides practical examples of these requirements.

7.9 CHECK POINT ACCURACY AND PLACEMENT REQUIREMENTS

The independent source of higher accuracy for checkpoints shall be at least three times more accurate than the required accuracy of the geospatial data set being tested.

Horizontal checkpoints shall be established at well-defined points. A well-defined point represents a feature for which the horizontal position can be measured to a high degree of accuracy and position with respect to the geodetic datum. For the purpose of accuracy testing, well-defined points must be easily visible or identifiable on the ground, on the independent source of higher accuracy, and on the product itself. For testing orthoimagery, well-defined points shall not be selected on features elevated with respect to the elevation model used to rectify the imagery.

Unlike horizontal checkpoints, vertical checkpoints are not necessarily required to be clearly defined or readily identifiable point features.

Vertical checkpoints shall be established at locations that minimize interpolation errors when comparing elevations interpolated from the data set to the elevations of the checkpoints. Vertical checkpoints shall be surveyed on flat or uniformly sloped open terrain and with slopes of 10% or less and should avoid vertical artifacts or abrupt changes in elevation.

7.10   CHECK POINT DENSITY AND DISTRIBUTION

When testing is to be performed, the distribution of the checkpoints will be project specific and must be determined by mutual agreement between the data provider and the end user. In no case shall an NVA, digital orthoimagery accuracy or planimetric data accuracy be based on less than 20 checkpoints.

A methodology to provide quantitative characterization and specification of the spatial distribution of checkpoints across the project extents, accounting for land cover type and project shape, is both realistic and necessary. But until such a methodology is developed and accepted, checkpoint density and distribution will be based primarily on empirical results and simplified area-based methods.

Annex C, Accuracy Testing and Reporting Guidelines, provides details on the recommended checkpoint density and distribution. The requirements in Annex C may be superseded and updated as newer methods for determining the appropriate distribution of checkpoints are established and approved.

7.11   RELATIVE ACCURACY OF LIDAR AND IFSAR DATA

Relative accuracy assessment characterizes the internal geometric quality of an elevation data set without regard to surveyed ground control. The assessment includes two aspects of data quality: within-swath accuracy (smooth surface repeatability), and swath-to-swath accuracy. Within-swath accuracy is usually only associated with lidar collections. The requirements for relative accuracy are more stringent than those for absolute accuracy. Acceptable limits for relative accuracy are stated in Table 7.2.

For lidar collections, within-swath relative accuracy is a measure of the repeatability of the system when detecting flat, hard surfaces. Within-swath relative accuracy also indicates the internal stability of the instrument. Within-swath accuracy is evaluated against single swath data by differencing two raster elevation surfaces generated from the minimum and maximum point elevations in each cell (pixel), taken over small test areas of relatively flat, hard surfaces. The raster cell size should be twice the NPS of the lidar data. Suitable test areas will have produced only single return lidar points and will not include abrupt changes in reflectivity (e.g., large paint stripes, shifts between black asphalt and white concrete, etc.), as these may induce elevation shifts that could skew the assessment. The use of a difference test normalizes for the actual elevation changes in the surfaces. Acceptable thresholds for each accuracy class are based on the maximum difference between minimum and maximum values within each pixel.

For lidar and IFSAR collections, relative accuracy between swaths (swath-to-swath) in overlap areas is a measure of the quality of the system calibration/boresighting and airborne GNSS trajectories.

Swath-to-swath relative accuracy is assessed by comparing the elevations of overlapping swaths. As with within-swath accuracy assessment, the comparisons are performed in areas producing only single return lidar points. Elevations are extracted at checkpoint locations from each of the overlapping swaths and computing the root-mean-square-difference ($RMSD_z$) of the residuals. Because neither swath represents an independent source of higher accuracy, as used in $RMSE_z$ calculations, the comparison is made using the RMS differences rather than RMS errors. Alternatively, the so called "delta-z" raster file representing the differences in elevations can be generated from the subtraction of the two raster files created for each swath over the entire surface and it can be used to calculate the $RMSD_z$. This approach has the advantages of a more comprehensive assessment, and provides the user with a visual representation of the error distribution.

Annex C, Accuracy Testing and Reporting Guidelines, outlines specific criteria for selecting checkpoint locations for swath-to-swath accuracies. The requirements in the annex may be superseded and updated as newer methods for determining the swath-to-swath accuracies are established and approved.

7.12   REPORTING

Horizontal and vertical accuracies shall be reported in terms of compliance with the RMSE thresholds and other quality and accuracy criteria outlined in this standard. In addition to the reporting stated below, ASPRS endorses and encourages additional reporting statements stating the estimated accuracy at a 95% confidence level in accordance with the FGDC NSSDA standard referenced in Section 3. Formulas for relating the RMSE thresholds in this standard to the NSSDA standard are provided in Annexes B and D.

If testing is performed, accuracy statements should specify that the data are "tested to meet" the stated accuracy.

If testing is not performed, accuracy statements should specify that the data are "produced to meet" the stated accuracy. This "produced to meet" statement is equivalent to the "compiled to meet" statement used by prior standards when referring to cartographic maps. The "produced to meet" method is appropriate for mature or established technologies where established procedures for project design, quality control and the evaluation of relative and absolute accuracies compared to ground control have been shown to produce repeatable and reliable results. Detailed specifications for testing and reporting to meet these requirements are outlined in Annex C.

The horizontal accuracy of digital orthoimagery, planimetric data, and elevation data sets shall be documented in the metadata in one of the following manners:

- "This data set was tested to meet ASPRS Positional Accuracy Standards for Digital Geospatial Data (2014) for a ___ (cm) $RMSE_x/RMSE_y$ Horizontal Accuracy Class. Actual positional accuracy was found to be $RMSE_x = $ ___ (cm) and $RMSE_y = $___cm which equates to Positional Horizontal Accuracy = +/– ___ at 95% confidence level."[4]
- "This data set was produced to meet ASPRS Positional Accuracy Standards for Digital Geospatial Data (2014) for a ___ (cm) $RMSE_x/RMSE_y$ Horizontal Accuracy Class which equates to Positional Horizontal Accuracy = +/– ___ cm at a 95% confidence level."[5]

The vertical accuracy of elevation data sets shall be documented in the metadata in one of the following manners:

- "This data set was tested to meet ASPRS Positional Accuracy Standards for Digital Geospatial Data (2014) for a ___ (cm) $RMSE_z$ Vertical Accuracy Class. Actual NVA accuracy was found to be $RMSE_z = $ ___ (cm), equating to +/– ___ (cm) at 95% confidence level. Actual VVA accuracy was found to be +/– ___ (cm) at the 95th percentile."[4]
- "This data set was produced to meet ASPRS Positional Accuracy Standards for Digital Geospatial Data (2014) for a ___ (cm) $RMSE_z$ Vertical Accuracy Class equating to NVA = +/– ___ (cm) at 95% confidence level and VVA = +/– ___ (cm) at the 95th percentile."[5]

---

[4] "Tested to meet" is to be used only if the data accuracies were verified by testing against independent checkpoints of higher accuracy.

[5] "Produced to meet" should be used by the data provider to assert that the data meet the specified accuracies, based on established processes that produce known results, but that independent testing against checkpoints of higher accuracy was not performed.

Annex A
Background and Justifications (Informative)

A.1   LEGACY STANDARDS AND GUIDELINES

Accuracy standards for geospatial data have broad applications nationally and/or internationally, whereas specifications provide technical requirements/acceptance criteria that a geospatial product must conform to in order to be considered acceptable for a specific intended use. Guidelines provide recommendations for acquiring, processing and/or analyzing geospatial data, normally intended to promote consistency and industry best practices.

The following is a summary of standards, specifications and guidelines relevant to ASPRS but which do not fully satisfy current requirements for accuracy standards for digital geospatial data:

- The *National Map Accuracy Standard* (NMAS) of 1947 established horizontal accuracy thresholds for the *Circular Map Accuracy Standard* (CMAS) as a function of map scale, and vertical accuracy thresholds for the *Vertical Map Accuracy Standard* (VMAS) as a function of contour interval – both reported at the 90% confidence level. Because NMAS accuracy thresholds are a function of the map scale and/or contour interval of a printed map, they are inappropriate for digital geospatial data where scale and contour interval are changed with a push of a button while not changing the underlying horizontal and/or vertical accuracy.
- The *ASPRS 1990 Accuracy Standards for Large-Scale Maps* established horizontal and vertical accuracy thresholds in terms of RMSE values in X, Y, and Z at ground scale. However, because the RMSE thresholds for Class 1, Class 2 and Class 3 products pertain to printed maps with published map scales and contour intervals, these ASPRS standards from 1990 are similarly inappropriate for digital geospatial data.
- The *National Standard for Spatial Data Accuracy* (NSSDA), published by the Federal Geographic Data Committee (FGDC) in 1998, was developed to report accuracy of digital geospatial data at the 95% confidence level as a function of RMSE values in X, Y, and Z at ground scale, unconstrained by map scale or contour interval. The NSSDA states, "The reporting standard in the horizontal component is the radius of a circle of uncertainty,

such that the true or theoretical location of the point falls within that circle 95% of the time. The reporting standard in the vertical component is a linear uncertainty value, such that the true or theoretical location of the point falls within +/– of that linear uncertainty value 95% of the time. The reporting accuracy standard should be defined in metric (International System of Units, SI) units. However, accuracy will be reported in English units (inches and feet) where point coordinates or elevations are reported in English units ... The NSSDA uses root-mean-square error (RMSE) to estimate positional accuracy ... Accuracy reported at the 95% confidence level means that 95% of the positions in the data set will have an error with respect to true ground position that is equal to or smaller than the reported accuracy value." The NSSDA does not define threshold accuracy values, stating, "Agencies are encouraged to establish thresholds for their product specifications and applications and for contracting purposes." In its Appendix 3-A, the NSSDA provides equations for converting RMSE values in X, Y, and Z into horizontal and vertical accuracies at the 95% confidence levels. The NSSDA assumes normal error distributions with systematic errors eliminated as best as possible.

- The National Digital Elevation Program (NDEP) published the *NDEP Guidelines for Digital Elevation Data* in 2004, recognizing that lidar errors of Digital Terrain Models (DTMs) do not necessarily follow a normal distribution in vegetated terrain. The NDEP developed Fundamental Vertical Accuracy (FVA), Supplemental Vertical Accuracy (SVA) and Consolidated Vertical Accuracy (CVA). The FVA is computed in non-vegetated, open terrain only, based on the NSSDA's $RMSE_z * 1.9600$ because elevation errors in open terrain do tend to follow a normal distribution, especially with a large number of checkpoints. SVA is computed in individual land cover categories, and CVA is computed in all land cover categories combined — both based on 95th percentile errors (instead of RMSE multipliers) because errors in DTMs in other land cover categories, especially vegetated/forested areas, do not necessarily follow a normal distribution. The NDEP Guidelines, while establishing alternative procedures for testing and reporting the vertical accuracy of elevation data sets when errors are not normally distributed, also do not provide accuracy thresholds or quality levels.

- The *ASPRS Guidelines: Vertical Accuracy Reporting for Lidar Data*, published in 2004, essentially endorsed the NDEP Guidelines, to include FVA, SVA and CVA reporting. Similarly, the ASPRS 2004 Guidelines, while endorsing the NDEP Guidelines when elevation errors are not normally distributed, also do not provide accuracy thresholds or quality levels.

- Between 1998 and 2010, the Federal Emergency Management Agency (FEMA) published *Guidelines and Specifications for Flood Hazard Mapping Partners* that included $RMSE_z$ thresholds and requirements for testing and reporting the vertical accuracy separately for all major land cover categories within floodplains being mapped for the National Flood Insurance Program (NFIP). With its Procedure Memorandum No. 61 — Standards for Lidar and Other High-Quality Digital Topography, dated 27 September

2010, FEMA endorsed the USGS Draft Lidar Base Specifications V13, relevant to floodplain mapping in areas of highest flood risk only, with poorer accuracy and point density in areas of lesser flood risks. USGS' draft V13 specification subsequently became the USGS Lidar Base Specification V1.0 specification summarized below. FEMA's Guidelines and Procedures only address requirements for flood risk mapping and do not represent accuracy standards that are universally applicable.

- In 2012, USGS published its Lidar Base Specification, Version 1.0, which is based on $RMSE_z$ of 12.5 cm in open terrain and elevation post spacing no greater than 1 to 2 meters. FVA, SVA, and CVA values are also specified. This document is not a standard but a specification for lidar data used to populate the National Elevation Dataset (NED) at 1/9th arc-second post spacing (~3 meters) for gridded Digital Elevation Models (DEMs).
- In 2012, USGS also published the final report of the *National Enhanced Elevation Assessment* (NEEA), which considered five Quality Levels of enhanced elevation data to satisfy nationwide requirements; each Quality Level having different $RMSE_z$ and point density thresholds. With support from the National Geospatial Advisory Committee (NGAC), USGS subsequently developed its new 3D Elevation Program (3DEP) based on lidar Quality Level 2 data with 1′ equivalent contour accuracy ($RMSE_z < 10$ cm) and point density of 2 points per square meter for all states except Alaska in which IFSAR Quality Level 5 data are specified with $RMSE_z$ between 1 and 2 meters and with 5 meter post spacing. The 3DEP lidar data are expected to be high-resolution data capable of supporting DEMs at 1 meter resolution. The 3DEP Quality Level 2 and Quality Level 5 products are expected to become industry standards for digital elevation data, respectively replacing the older elevation data from the USGS' National Elevation Dataset.
- In 2014, the latest USGS Lidar Base Specification Version 1.2 was published to accommodate lidar Quality Levels 0, 1, 2 and 3.

A.2   NEW STANDARD FOR A NEW ERA

The current standard was developed in response to the pressing need of the GIS and mapping community for a new standard that embraces the digital nature of current geospatial technologies. The following are some of the justifications for the development of the new standard:

- Legacy map accuracy standards, such as the ASPRS 1990 standard and the NMAS of 1947, are outdated. Many of the data acquisition and mapping technologies that these standards were based on are no longer used. More recent advances in mapping technologies can now produce better quality and higher accuracy geospatial products and maps. New standards are needed to reflect these advances.
- Legacy map accuracy standards were designed to deal with plotted or drawn maps as the only medium to represent geospatial data. The concept

of hardcopy map scale dominated the mapping industry for decades. Digital mapping products need different measures (besides scale) that are suitable for the digital medium that users now utilize.

• Within the past two decades (during the transition period between the hardcopy and softcopy mapping environments), most standard measures for relating GSD and map scale to the final mapping accuracy were inherited from photogrammetric practices using scanned film. New mapping processes and methodologies have become much more sophisticated with advances in technology and advances in our knowledge of mapping processes and mathematical modeling. Mapping accuracy can no longer be associated with the camera geometry and flying altitude alone. Many other factors now influence the accuracy of geospatial mapping products. Such factors include the quality of camera calibration parameters, quality and size of a Charged Coupled Device (CCD) used in the digital camera CCD array, amount of imagery overlap, quality of parallax determination or photo measurements, quality of the GPS signal, quality and density of ground control, quality of the aerial triangulation solution, capability of the processing software to handle GPS drift and shift and camera self-calibration, and the digital terrain model used for the production of orthoimagery. These factors can vary widely from project to project, depending on the sensor used and specific methodology. For these reasons, existing accuracy measures based on map scale, film scale, GSD, c-factor, and scanning resolution no longer apply to current geospatial mapping practices.

• Elevation products from the new technologies and active sensors such as lidar and IFSAR are not considered by the legacy mapping standards. New accuracy standards are needed to address elevation products derived from these technologies.

A.2.1 MAPPING PRACTICES DURING THE FILM-BASED ERA

Since the early history of photogrammetric mapping, film was the only medium to record an aerial photographic session. During that period, film scale, film-to-map enlargement ratio, and c-factor were used to define final map scale and map accuracy. A film-to-map enlargement ratio value of 6 and a c-factor value of 1800 to 2000 were widely accepted and used during this early stage of photogrammetric mapping. C-factor is used to determine the flying height based on the desired contour interval from the following formula:

$$\text{c-factor} = \frac{\text{flying altitude}}{\text{contour interval}}$$

Values in Table A.1 were historically utilized by the mapping community for photogrammetric mapping from film.

TABLE A.1

Common Photography Scales Using Camera with 9″ Film Format and 6″ Lens

Film Scale	$1'' = 300'$	$1'' = 600'$	$1'' = 1200'$	$1'' = 2400'$	$1'' = 3333'$
	1:3,600	1:7,200	1:14,400	1:28,800	1:40,000
Flying Altitude	1,800′/550 m	3,600′/1,100 m	7,200′/2,200 m	14,400′/4,400 m	20,000′/6,100 m
Map Scale	$1'' = 50'$	$1'' = 100'$	$1'' = 200'$	$1'' = 400'$	$1'' = 1000'$
	1:600	1:1,200	1:2,400	1:4,800	1:12,000

A.2.2 MAPPING PRACTICES DURING THE SOFTCOPY PHOTOGRAMMETRY ERA

When the softcopy photogrammetric mapping approach was first introduced to the mapping industry in the early 1990s, large format film scanners were used to convert the aerial film to digital imagery. The mapping community needed guidelines for relating the scanning resolution of the film to the supported map scale and contour interval used by legacy standards to specify map accuracies. Table A.2 relates the resulting GSD of the scanned film and the supported map scale and contour interval derived from film-based cameras at different flying altitudes. Table A.2 assumes a scan resolution of 21 microns as that was in common use for many years. The values in Table A.2 are derived based on the commonly used film-to-map enlargement ratio of 6 and a c-factor of 1800. Such values were endorsed and widely used by both map users and data providers during and after the transition period from film to the softcopy environment.

A.2.3 MAPPING PRACTICES DURING THE DIGITAL SENSORS PHOTOGRAMMETRY ERA

Since first introduced to the mapping community in 2000, digital large format metric mapping cameras have become the main aerial imagery acquisition system utilized for geospatial mapping. The latest generation of digital metric mapping cameras have enhanced optics quality, extended radiometric resolution through a higher dynamic range, finer CCD resolution, rigid body construction, and precise electronics. These new camera technologies, coupled with advances in the airborne GPS and mathematical modeling performed by current photogrammetric processing software, make it possible to extend the limits on the flying altitude and still achieve higher-quality mapping products, of equal or greater accuracy, than what could be achieved with older technologies.

Many of the rules that have influenced photogrammetric practices for the last six or seven decades (such as those outlined in Sections A.2.1 and A.2.2 above) are based on the capabilities of outdated technologies and techniques. For instance, standard guidelines like using a film-to-map enlargement ratio value of 6 and a c-factor between 1,800 and 2,000 are based on the limitations of optical-mechanical photogrammetric

TABLE A.2

Relationship between Film Scale and Derived Map Scale

		Common Photography Scales (with 9″ film format camera and 6″ lens)				Scanning Resolution (μm)
Photo Scale		1″ = 300′	1″ = 600′	1″ = 1200′	1″ = 2400′	
		1:3,600	1:7,200	1:14,400	1:28,800	
Flying Altitude		1,800′/550 m	3,600′/1,100 m	7,200′/2,200 m	14,400′/4,400 m	
Approximate Ground Sampling Distance (GSD) of Scan		0.25′/7.5 cm	0.50′/0.15 m	1.0′/0.3 m	2.0′/0.6 m	21

	Supported Map/Orthoimagery Scales and Contour Intervals			
GSD	3″/7.5 cm	6″/15 cm	1.0′/30 cm	2.0′/60 cm
C.I.	1.0′/30 cm	2.0′/60 cm	4′/1.2 m	8′/2.4 m
Map Scale	1″ = 50′	1″ = 100′	1″ = 200′	1″ = 400′
	1:600	1:1,200	1:2,400	1:4,800

plotters and aerial film resolution. These legacy rules no longer apply to mapping processes utilizing digital mapping cameras and current technologies.

Unfortunately, due to a lack of clear guidelines, outdated practices and guidelines from previous eras are commonly misapplied to newer technologies. The majority of users and data providers still utilize the figures given in Table A.2 for associating the imagery GSD to a supported map scale and associated accuracy, even though these associations are based on scanned film and do not apply to current digital sensors. New relationships between imagery GSD and product accuracy are needed to account for the full range factors that influence the accuracy of mapping products derived from digital sensors.

Annex B
Data Accuracy and Quality Examples (Normative)

B.1   AERIAL TRIANGULATION AND GROUND CONTROL ACCURACY EXAMPLES

Sections 7.7 and 7.8 describe the accuracy requirements for aerial triangulation, IMU, and ground control points relative to product accuracies. These requirements differ depending on whether the products include elevation data. Tables B.1 and B.2 provide an example of how these requirements are applied in practice for a typical product with $RMSE_x$ and $RMSE_y$ of 50 cm.

B.2   DIGITAL ORTHOIMAGERY HORIZONTAL ACCURACY CLASSES

This standard does not associate product accuracy with the GSD of the source imagery, pixel size of the orthoimagery, or map scale for scaled maps.

The relationship between the recommended $RMSE_x$ and $RMSE_y$ accuracy class and the orthoimagery pixel size varies depending on the imaging sensor characteristics and the specific mapping processes used. The appropriate horizontal accuracy class must be negotiated and agreed upon between the end user and the data provider, based on specific project needs and design criteria. This section provides some general guidance to assist in making that decision.

Example tables are provided to show the following: The general application of the standard as outlined in Section 7.3 (Table B.3); A cross reference to typical past associations between pixel size, map scale and the 1990 ASPRS legacy standard (Table B.4); and, Typical values associated with different levels of accuracy using current technologies (Table B.5).

Table B.3 presents examples of 24 horizontal accuracy classes and associated quality criteria as related to orthoimagery according to the formula and general requirements stated in Section 7.3.

As outlined in Annex A, in the transition between hardcopy and softcopy mapping environments, user's and the mapping community established generally accepted associations between orthoimagery pixel size, final map scale and the ASPRS 1990 map accuracy classes. These associations are based primarily on relationships for scanned film, older technologies and legacy standards. While they may not directly apply to digital geospatial data produced with newer technologies, these practices

TABLE B.1

Aerial Triangulation and Ground Control Accuracy Requirements, Orthoimagery and/or Planimetric Data Only

Product Accuracy (RMSE$_x$, RMSE$_y$) (cm)	A/T Accuracy		Ground Control Accuracy	
	RMSE$_x$ and RMSE$_y$ (cm)	RMSE$_z$ (cm)	RMSE$_x$ and RMSE$_y$ (cm)	RMSE$_z$ (cm)
50	25	50	12.5	25

TABLE B.2

Aerial Triangulation and Ground Control Accuracy Requirements, Orthoimagery and/or Planimetric Data and Elevation Data

Product Accuracy (RMSE$_x$, RMSE$_y$) (cm)	A/T Accuracy		Ground Control Accuracy	
	RMSE$_x$ and RMSE$_y$ (cm)	RMSE$_z$ (cm)	RMSE$_x$ and RMSE$_y$ (cm)	RMSE$_z$ (cm)
50	25	25	12.5	12.5

have been in widespread use for many years and many existing data sets are based on these associations. As such, it is useful to have a cross reference relating these legacy specifications to their corresponding RMSEx and RMSEy accuracy classes in the new standard.

Table B.4 lists the most common associations that have been established (based on user's interpretation and past technologies) to relate orthoimagery pixel size to map scale and the ASPRS 1990 legacy standard map accuracy classes.

Given current sensor and processing technologies for large and medium format metric cameras, an orthoimagery accuracy of 1-pixel RMSE$_x$ and RMSE$_y$ is considered achievable, assuming proper project design and best practices implementation. This level of accuracy is more stringent by a factor of two than orthoimagery accuracies typically associated with the ASPRS 1990 Class 1 accuracies presented in Table B.4.

Achieving the highest level of accuracy requires specialized consideration related to sensor type, ground control density, ground control accuracies, and overall project design. In many cases, this results in higher cost. As such, the highest achievable accuracies may not be appropriate for all projects. Many geospatial mapping projects require high-resolution and high-quality imagery, but do not require the highest level of positional accuracy. This fact is particularly true for update or similar projects where the intent is to upgrade the image resolution, but still leverage existing elevation model data and ground control data that may originally have been developed to a lower accuracy standard.

TABLE B.3
Common Horizontal Accuracy Classes According to the New Standard[6]

Horizontal Accuracy Class RMSE$_x$ and RMSE$_y$ (cm)	RMSE$_r$ (cm)	Orthoimage Mosaic Seamline Maximum Mismatch (cm)	Horizontal Accuracy at the 95% Confidence Level (cm)
0.63	0.9	1.3	1.5
1.25	1.8	2.5	3.1
2.50	3.5	5.0	6.1
5.00	7.1	10.0	12.2
7.50	10.6	15.0	18.4
10.00	14.1	20.0	24.5
12.50	17.7	25.0	30.6
15.00	21.2	30.0	36.7
17.50	24.7	35.0	42.8
20.00	28.3	40.0	49.0
22.50	31.8	45.0	55.1
25.00	35.4	50.0	61.2
27.50	38.9	55.0	67.3
30.00	42.4	60.0	73.4
45.00	63.6	90.0	110.1
60.00	84.9	120.0	146.9
75.00	106.1	150.0	183.6
100.00	141.4	200.0	244.8
150.00	212.1	300.0	367.2
200.00	282.8	400.0	489.5
250.00	353.6	500.0	611.9
300.00	424.3	600.0	734.3
500.00	707.1	1000.0	1223.9
1000.00	1414.2	2000.0	2447.7

Table B.5 provides a general guideline to determine the appropriate orthoimagery accuracy class for three different levels of geospatial accuracy values listed as "Highest accuracy work" specify an RMSEx and RMSEy accuracy class of 1-pixel (or better) and are considered to reflect the highest tier accuracy for the specified resolution given current technologies. This accuracy class is appropriate when geospatial accuracies are of higher importance and when the higher accuracies are supported by sufficient sensor, ground control and digital terrain model accuracies. Values listed as "Standard Mapping and GIS work" specify a 2-pixel RMSE$_x$ and RMSE$_y$ accuracy class. This accuracy is appropriate for a standard level of high-quality and high-accuracy geospatial mapping applications. It is equivalent to ASPRS 1990 Class 1 accuracies, as interpreted by users as industry standard and presented in

---

6   For Tables B.3 through B.8, values were rounded to the nearest mm after full calculations were performed with all decimal places.

TABLE B.4

Examples on Horizontal Accuracy for Digital Orthoimagery Interpreted from ASPRS 1990 Legacy Standard

Common Orthoimagery Pixel Sizes	Associated Map Scale	ASPRS 1990 Accuracy Class	Associated Horizontal Accuracy According to Legacy ASPRS 1990 Standard	
			$RMSE_x$ and $RMSE_y$ (cm)	$RMSE_x$ and $RMSE_y$ in Terms of Pixels
0.625 cm	1:50	1	1.3	2-pixels
		2	2.5	4-pixels
		3	3.8	6-pixels
1.25 cm	1:100	1	2.5	2-pixels
		2	5.0	4-pixels
		3	7.5	6-pixels
2.5 cm	1:200	1	5.0	2-pixels
		2	10.0	4-pixels
		3	15.0	6-pixels
5 cm	1:400	1	10.0	2-pixels
		2	20.0	4-pixels
		3	30.0	6-pixels
7.5 cm	1:600	1	15.0	2-pixels
		2	30.0	4-pixels
		3	45.0	6-pixels
15 cm	1:1,200	1	30.0	2-pixels
		2	60.0	4-pixels
		3	90.0	6-pixels
30 cm	1:2,400	1	60.0	2-pixels
		2	120.0	4-pixels
		3	180.0	6-pixels
60 cm	1:4,800	1	120.0	2-pixels
		2	240.0	4-pixels
		3	360.0	6-pixels
1 meter	1:12,000	1	200.0	2-pixels
		2	400.0	4-pixels
		3	600.0	6-pixels
2 meter	1:24,000	1	400.0	2-pixels
		2	800.0	4-pixels
		3	1,200.0	6-pixels
5 meter	1:60,000	1	1,000.0	2-pixels
		2	2,000.0	4-pixels
		3	3,000.0	6-pixels

TABLE B.5
Digital Orthoimagery Accuracy Examples for Current Large and Medium Format Metric Cameras

Common Orthoimagery Pixel Sizes	Recommended Horizontal Accuracy Class RMSE$_x$ and RMSE$_y$ (cm)	Orthoimage RMSE$_x$ and RMSE$_y$ in Terms of Pixels	Recommended Use[7]
1.25 cm	≤1.3	≤1-pixel	Highest accuracy work
	2.5	2-pixels	Standard Mapping and GIS work
	≥3.8	≥3-pixels	Visualization and less accurate work
2.5 cm	≤2.5	≤1-pixel	Highest accuracy work
	5.0	2-pixels	Standard Mapping and GIS work
	≥7.5	≥3-pixels	Visualization and less accurate work
5 cm	≤5.0	≤1-pixel	Highest accuracy work
	10.0	2-pixels	Standard Mapping and GIS work
	≥15.0	≥3-pixels	Visualization and less accurate work
7.5 cm	≤7.5	≤1-pixel	Highest accuracy work
	15.0	2-pixels	Standard Mapping and GIS work
	≥22.5	≥3-pixels	Visualization and less accurate work
15 cm	≤15.0	≤1-pixel	Highest accuracy work
	30.0	2-pixels	Standard Mapping and GIS work
	≥45.0	≥3-pixels	Visualization and less accurate work
30 cm	≤30.0	≤1-pixel	Highest accuracy work
	60.0	2-pixels	Standard Mapping and GIS work
	≥90.0	≥3-pixels	Visualization and less accurate work
60 cm	≤60.0	≤1-pixel	Highest accuracy work
	120.0	2-pixels	Standard Mapping and GIS work
	≥180.0	≥3-pixels	Visualization and less accurate work
1 meter	≤100.0	≤1-pixel	Highest accuracy work
	200.0	2-pixels	Standard Mapping and GIS work
	≥300.0	≥3-pixels	Visualization and less accurate work

(Continued)

TABLE B.5 (CONTINUED)

Digital Orthoimagery Accuracy Examples for Current Large and Medium Format Metric Cameras

Common Orthoimagery Pixel Sizes	Recommended Horizontal Accuracy Class RMSE$_x$ and RMSE$_y$ (cm)	Orthoimage RMSE$_x$ and RMSE$_y$ in Terms of Pixels	Recommended Use[7]
2 meter	≤200.0	≤1-pixel	Highest accuracy work
	400.0	2-pixels	Standard Mapping and GIS work
	≥600.0	≥3-pixels	Visualization and less accurate work
5 meter	≤500.0	≤1-pixel	Highest accuracy work
	1,000.0	2-pixels	Standard Mapping and GIS work
	≥1,500.0	≥3-pixels	Visualization and less accurate work

Table B.4. This level of accuracy is typical of a large majority of existing projects designed to legacy standards. RMSE$_x$ and RMSE$_y$ accuracies of 3 or more pixels would be considered appropriate for "Visualization and less accurate work" when higher accuracies are not needed.

Users should be aware that the use of the symbol ≥ in Table B.5 is intended to infer that users can specify larger threshold values for RMSE$_x$ and RMSE$_y$. The symbol ≤ in Table B.5 indicates that users can specify lower thresholds at such time as they may be supported by current or future technologies.

The orthoimagery pixel sizes and associated RMSEx and RMSEy accuracy classes presented in Table B.5 are largely based on experience with current sensor technologies and primarily apply to large and medium format metric cameras. The table is only provided as a guideline for users during the transition period to the new standard. These associations may change in the future as mapping technologies continue to advance and evolve.

It should be noted that in Tables B.4 and B.5, it is the pixel size of the final digital orthoimagery that is used to associate the horizontal accuracy class, not the Ground Sample Distance (GSD) of the raw image. When producing digital orthoimagery, the GSD as acquired by the sensor (and as computed at mean average terrain) should not be more than 95% of the final orthoimage pixel size. In extremely steep terrain, additional consideration may need to be given to the variation of the GSD across low

---

[7] "Highest accuracy work" in Table B.5 refers only to the highest level of achievable accuracies relative to that specific resolution; it does not indicate "highest accuracy work" in any general sense. The final choice of both image resolution and final product accuracy class depends on specific project requirements and is the sole responsibility of the end user; this should be negotiated with the data provider and agreed upon in advance.

lying areas in order to ensure that the variation in GSD across the entire image does not significantly exceed the target pixel size.

B.3   DIGITAL PLANIMETRIC DATA HORIZONTAL ACCURACY CLASSES

Table B.6 presents 24 common horizontal accuracy classes for digital planimetric data, approximate GSD of source imagery for high accuracy planimetric data, and equivalent map scales per legacy NMAS and ASPRS 1990 accuracy standards. In Table B.6, the values for the approximate GSD of source imagery only apply to imagery derived from common large and medium format metric cameras. The range

TABLE B.6
Horizontal Accuracy/Quality Examples for High Accuracy Digital Planimetric Data

ASPRS 2014				Equivalent to Map Scale in		
Horizontal Accuracy Class RMSE$_x$ and RMSE$_y$ (cm)	RMSE$_r$ (cm)	Horizontal Accuracy at the 95% Confidence Level (cm)	Approximate GSD of Source Imagery (cm)	ASPRS 1990 Class 1	ASPRS 1990 Class 2	Equivalent to Map Scale in NMAS
0.63	0.9	1.5	0.31 to 0.63	1:25	1:12.5	1:16
1.25	1.8	3.1	0.63 to 1.25	1:50	1:25	1:32
2.5	3.5	6.1	1.25 to 2.5	1:100	1:50	1:63
5.0	7.1	12.2	2.5 to 5.0	1:200	1:100	1:127
7.5	10.6	18.4	3.8 to 7.5	1:300	1:150	1:190
10.0	14.1	24.5	5.0 to 10.0	1:400	1:200	1:253
12.5	17.7	30.6	6.3 to12.5	1:500	1:250	1:317
15.0	21.2	36.7	7.5 to 15.0	1:600	1:300	1:380
17.5	24.7	42.8	8.8 to 17.5	1:700	1:350	1:444
20.0	28.3	49.0	10.0 to 20.0	1:800	1:400	1:507
22.5	31.8	55.1	11.3 to 22.5	1:900	1:450	1:570
25.0	35.4	61.2	12.5 to 25.0	1:1000	1:500	1:634
27.5	38.9	67.3	13.8 to 27.5	1:1100	1:550	1:697
30.0	42.4	73.4	15.0 to 30.0	1:1200	1:600	1:760
45.0	63.6	110.1	22.5 to 45.0	1:1800	1:900	1:1,141
60.0	84.9	146.9	30.0 to 60.0	1:2400	1:1200	1:1,521
75.0	106.1	183.6	37.5 to 75.0	1:3000	1:1500	1:1,901
100.0	141.4	244.8	50.0 to 100.0	1:4000	1:2000	1:2,535
150.0	212.1	367.2	75.0 to 150.0	1:6000	1:3000	1:3,802
200.0	282.8	489.5	100.0 to 200.0	1:8000	1:4000	1:5,069
250.0	353.6	611.9	125.0 to 250.0	1:10000	1:5000	1:6,337
300.0	424.3	734.3	150.0 to 300.0	1:12000	1:6000	1:7,604
500.0	707.1	1223.9	250.0 to 500.0	1:20000	1:10000	1:21,122
1000.0	1414.2	2447.7	500.0 to 1000.0	1:40000	1:20000	1:42,244

of the approximate GSD of source imagery is only provided as a general recommendation, based on the current state of sensor technologies and mapping practices. Different ranges may be considered in the future depending on future advances of such technologies and mapping practices.

B.4   DIGITAL ELEVATION DATA VERTICAL ACCURACY CLASSES

Table B.7 provides vertical accuracy examples and other quality criteria for ten common vertical accuracy classes. Table B.8 compares the ten vertical accuracy classes with contours intervals from legacy ASPRS 1990 and NMAS 1947 standards. Table B.9 provides ten vertical accuracy classes with the recommended lidar point density suitable for each of them.

B.5   CONVERTING ASPRS 2014 ACCURACY VALUES
TO LEGACY ASPRS 1990 ACCURACY VALUES

In this section easy methods and examples will be provided for users who are faced with the issue of relating the standard (ASPRS 2014) to the legacy ASPRS 1990 Accuracy Standards for Large-Scale Maps. A major advantage of the new standard is it indicates accuracy based on RMSE at the ground scale. Although both the new 2014 standard and the legacy ASPRS map standard of 1990 are using the same measure of RMSE, they are different on the concept of representing the accuracy classes. The legacy ASPRS map standard of 1990 uses Class 1 for higher accuracy

TABLE B.7
Vertical Accuracy/Quality Examples for Digital Elevation Data

Vertical Accuracy Class (cm)	Absolute Accuracy			Relative Accuracy (Where Applicable)		
	RMSE$_z$ Non-Vegetated (cm)	NVA at 95% Confidence Level (cm)	VVA at 95th Percentile (cm)	Within-Swath Hard Surface Repeatability (Max Diff) (cm)	Swath-to-Swath Non-Veg Terrain (RMSD$_z$) (cm)	Swath-to-Swath Non-Veg Terrain (Max Diff) (cm)
1	1.0	2.0	3	0.6	0.8	1.6
2.5	2.5	4.9	7.5	1.5	2	4
5	5.0	9.8	15	3	4	8
10	10.0	19.6	30	6	8	16
15	15.0	29.4	45	9	12	24
20	20.0	39.2	60	12	16	32
33.3	33.3	65.3	100	20	26.7	53.3
66.7	66.7	130.7	200	40	53.3	106.7
100	100.0	196.0	300	60	80	160
333.3	333.3	653.3	1000	200	266.7	533.3

TABLE B.8

Vertical Accuracy of the New ASPRS 2014 Standard Compared with Legacy Standards

Vertical Accuracy Class (cm)	RMSE$_z$ Non-Vegetated (cm)	Equivalent Class 1 Contour Interval per ASPRS 1990 (cm)	Equivalent Class 2 Contour Interval per ASPRS 1990 (cm)	Equivalent Contour Interval per NMAS (cm)
1	1.0	3.0	1.5	3.29
2.5	2.5	7.5	3.8	8.22
5	5.0	15.0	7.5	16.45
10	10.0	30.0	15.0	32.90
15	15.0	45.0	22.5	49.35
20	20.0	60.0	30.0	65.80
33.3	33.3	99.9	50.0	109.55
66.7	66.7	200.1	100.1	219.43
100	100.0	300.0	150.0	328.98
333.3	333.3	999.9	500.0	1096.49

and Classes 2 and 3 for data with lower accuracy while the new 2014 standard refers to the map accuracy by the value of RMSE without limiting it to any class. The following examples illustrate the procedures users can follow to relate horizontal and vertical accuracies values between the new ASPRS standard of 2014 and the legacy ASPRS 1990 Accuracy Standards for Large-Scale Maps.

Example 1: Converting the horizontal accuracy of a map or orthoimagery from the new 2014 standard to the legacy ASPRS map standard of 1990

Given a map or orthoimagery with an accuracy of $RMSE_x = RMSE_y = 15$ cm according to new 2014 standard, compute the equivalent accuracy and map scale according to the legacy ASPRS map standard of 1990, for the given map or orthoimagery.

SOLUTION

1. Because both standards utilize the same RMSE measure, then the accuracy of the map according to the legacy ASPRS map standard of 1990 is $RMSE_x = RMSE_y = 15$ cm
2. To find the equivalent map scale according to the legacy ASPRS map standard of 1990, follow the following steps:
   a. Multiply the $RMSE_x$ and $RMSE_y$ value in centimeters by 40 to compute the map scale factor (MSF) for a Class 1 map, therefore:

TABLE B.9

Examples on Vertical Accuracy and Recommended Lidar Point Density for Digital Elevation Data according to the New ASPRS 2014 Standard

Vertical Accuracy Class (cm)	Absolute Accuracy		Recommended Minimum NPD[a] (pls/m²)	Recommended Maximum NPS[8] (m)
	RMSE_z Non-Vegetated (cm)	NVA at 95% Confidence Level (cm)		
1	1.0	2.0	≥ 20	≤ 0.22
2.5	2.5	4.9	16	0.25
5	5.0	9.8	8	0.35
10	10.0	19.6	2	0.71
15	15.0	29.4	1	1.0
20	20.0	39.2	0.5	1.4
33.3	33.3	65.3	0.25	2.0
66.7	66.7	130.7	0.1	3.2
100	100.0	196.0	0.05	4.5
333.3	333.3	653.3	0.01	10.0

$MSF = 15 \ (cm) \times 40 = 600$

b. The map scale according to the legacy ASPRS map standard of 1990 is equal to:

i. Scale = 1:MSF or 1:600 Class 1;

ii. The accuracy value of $RMSE_x = RMSE_y = 15$ cm is also equivalent to Class 2 accuracy for a map with a scale of 1:300.

Example 2: Converting the vertical accuracy of an elevation dataset from the new standard to the legacy ASPRS map standard of 1990

Given an elevation data set with a vertical accuracy of $RMSE_z = 10$ cm according to the new standard, compute the equivalent contour interval according to the legacy ASPRS map standard of 1990, for the given dataset.

---

[8] Nominal Pulse Density (NPD) and Nominal Pulse Spacing (NPS) are geometrically inverse methods to measure the pulse density or spacing of a lidar collection. NPD is a ratio of the number of points to the area in which they are contained, and is typically expressed as pulses per square meter (ppsm or pts/m²). NPS is a linear measure of the typical distance between points, and is most often expressed in meters. Although either expression can be used for any data set, NPD is usually used for lidar collections with NPS <1, and NPS is used for those with NPS ≥1. Both measures are based on all 1st (or last)-return lidar point data as these return types each reflect the number of pulses. Conversion between NPD and NPS is accomplished using the equation $NPS = 1 / \sqrt{NPD}$ and $NPD = 1 / NPS^2$. Although typical point densities are listed for specified vertical accuracies, users may select higher or lower point densities to best fit project requirements and complexity of surfaces to be modeled.

SOLUTION

The legacy ASPRS map standard of 1990 states that:

"The limiting rms error in elevation is set by the standard at one-third the indicated contour interval for well-defined points only. Spot heights shall be shown on the map within a limiting rms error of one-sixth of the contour interval."

1. Because both standards utilize the same RMSE measure to express the vertical accuracy, then the accuracy of the elevation dataset according to the legacy ASPRS map standard of 1990 is also equal to the given $RMSE_z = 10$ cm

2. Using the legacy ASPRS map standard of 1990 accuracy measure of $RMSE_z = 1/3 \times$ contour interval (CI), the equivalent contour interval is computed according to the legacy ASPRS map standard of 1990 using the following formula:

   $CI = 3 \times RMSE_z = 3 \times 10$ cm $= 30$ cm with Class 1,

   Or $CI = 15$ cm with Class 2 accuracy

   However, if the user is interested in evaluating the spot height requirement according to the ASPRS 1990 standard, then the results will differ from the one obtained above. The accuracy for spot heights is required to be twice the accuracy of the contours (one-sixth versus one-third for the contours) or:

   For a 30 cm CI, the required spot height accuracy, $RMSE_z = 1/6 \times 30$ cm $= 5$ cm

   Since our data is $RMSEz = 10$ cm, it would only support Class 2 accuracy spot elevations for this contour interval.

B.6 CONVERTING ASPRS 2014 ACCURACY VALUES TO LEGACY NMAS 1947 ACCURACY VALUES

In this section easy methods and examples will be provided for users who are faced with the issue of relating the new standard (ASPRS 2014) to the legacy National Map Accuracy Standard (NMAS) of 1947. In regard to the horizontal accuracy measure, the NMAS of 1947 states that:

"Horizontal Accuracy: For maps on publication scales larger than 1:20,000, not more than 10 percent of the points tested shall be in error by more than 1/30 inch, measured on the publication scale; for maps on publication scales of 1:20,000 or smaller, 1/50 inch." This is known as the Circular Map Accuracy Standard (CMAS) or Circular Error at the 90% confidence level (CE90).

Therefore, the standard uses two accuracy measures based on the map scale with the figure of "1/30 inch" for map scales larger than 1:20,000 and "1/50 inch" for maps with a scale of 1:20,000 or smaller. As for the vertical accuracy measure, the standard states:

*"Vertical Accuracy, as applied to contour maps on all publication scales, shall
be such that not more than 10 percent of the elevations tested shall be in
error more than one-half the contour interval."* This is known as the Vertical
Map Accuracy Standard (VMAS) or Linear Error at the 90% confidence
level (LE90).

The following examples illustrate the procedures users can follow to relate horizontal and vertical accuracy values between the new ASPRS standard of 2014 and the
legacy National Map Accuracy Standard (NMAS) of 1947.

Example 3: Converting the horizontal accuracy of a map or orthoimagery from the new ASPRS 2014 standard to the legacy National Map Accuracy Standard (NMAS) of 1947

Given a map or orthoimagery with an accuracy of $RMSE_x = RMSE_y = 15$ cm according to the new 2014 standard, compute the equivalent accuracy and map scale
according to the legacy National Map Accuracy Standard (NMAS) of 1947, for the
given map or orthoimagery.

SOLUTION

1. Because the accuracy figure of $RMSE_x = RMSE_y = 15$ cm is relatively small,
   it is safe to assume that such accuracy value is derived for a map with a
   scale larger than 1:20,000. Therefore, we can use the factor "1/30 inch."
   Use the formula CMAS $(CE90) = 2.1460 \times RMSE_x = 2.1460 \times RMSE_y$
   CE 90% $= 2.1460 \times 15$ cm $= 32.19$ cm
2. Convert the CE 90% to feet
   32.19 cm $= 1.0561$ foot
3. Use the NMAS accuracy relation of $CE90\% = 1/30$ inch on the map,
   compute the map scale
   CE 90% $= 1/30 \times$ (ground distance covered by an inch of the map), or
   ground distance covered by an inch of the map $=$ CE 90% $\times 30 = 1.0561$
   foot $\times 30 = 31.68$ feet
4. The equivalent map scale according to NMAS is equal to $1" = 31.68'$ or
   1:380

Example 4: Converting the vertical accuracy of an elevation dataset from the new ASPRS 2014 standard to the legacy National Map Accuracy Standard (NMAS) of 1947

Given an elevation data set with a vertical accuracy of $RMSE_z = 10$ cm according to
the new ASPRS 2014 standard, compute the equivalent contour interval according to
the legacy National Map Accuracy Standard (NMAS) of 1947, for the given dataset.

SOLUTION

As mentioned earlier, the legacy ASPRS map standard of 1990 states that:

"Vertical Accuracy, as applied to contour maps on all publication scales, shall be such that not more than 10 percent of the elevations tested shall be in error more than one-half the contour interval."

1. Use the following formula to compute the 90% vertical error:
   VMAS (LE90) $= 1.6449 \times RMSE_z = 1.6449 \times 10$ cm $= 16.449$ cm
2. Compute the contour interval (CI) using the following criteria set by the NMAS standard:
   VMAS (LE90) $= \frac{1}{2}$ CI, or
   CI $= 2 \times LE90 = 2 \times 16.449$ cm $= 32.9$ cm

B.7 EXPRESSING THE ASPRS 2014 ACCURACY VALUES ACCORDING TO THE FGDC NATIONAL STANDARD FOR SPATIAL DATA ACCURACY (NSSDA)

In this section easy methods and examples will be provided for users who are faced with the issue of relating the new standard (ASPRS 2014) to the FGDC National Standard for Spatial Data Accuracy (NSSDA).

Example 5: Converting the horizontal accuracy of a map or orthoimagery from the new 2014 standard to the FGDC National Standard for Spatial Data Accuracy (NSSDA)

Given a map or orthoimagery with an accuracy of $RMSE_x = RMSE_y = 15$ cm according to new 2014 standard, express the equivalent accuracy according to the FGDC National Standard for Spatial Data Accuracy (NSSDA), for the given map or orthoimagery.

SOLUTION

According to NSSDA, the horizontal positional accuracy is estimated at 95% confidence level from the following formula:

Accuracy at 95% or Accuracy$_r = 2.4477 \times RMSE_x = 2.4477 \times RMSE_y$
If we assume that:

$RMSE_x = RMSE_y$ and $RMSE_r = \sqrt{RMSE_x^2 + RMSE_y^2}$,

then $RMSE_r = \sqrt{2RMSE_x^2} = \sqrt{2RMSE_y^2} = 1.4142 \times RMSE_x = 1.4142 \times RMSE_y =$

   $1.4142 \times 15 = 21.21$ cm
also

$RMSE_x$ or $RMSE_y = \dfrac{RMSE_r}{1.4142}$
Then,

Accuracy$_r = 2.4477 \left(\dfrac{RMSE_r}{1.4142} \right) = 1.7308 (RMSE_r) = 1.7308 (21.21$ cm$) = 36.71$ cm

Example 6: Converting the vertical accuracy of an elevation dataset from the new ASPRS 2014 standard to the FGDC National Standard for Spatial Data Accuracy (NSSDA)

Given an elevation data set with a vertical accuracy of $RMSE_z = 10$ cm according to the new ASPRS 2014 standard, express the equivalent accuracy according to the FGDC National Standard for Spatial Data Accuracy (NSSDA), for the given dataset.

SOLUTION

According to NSSDA, the vertical accuracy of an elevation dataset is estimated at 95% confidence level according to the following formula:

$$\text{Vertical Accuracy at 95\% Confidence Level} = 1.9600\left(RMSE_z\right) = 1.9600(10) = 19.6 \text{ cm}$$

B.8   HORIZONTAL ACCURACY EXAMPLES FOR LIDAR DATA

As described in Section 7.5, the horizontal errors in lidar data are largely a function of GNSS positional error, INS angular error, and flying altitude. Therefore for a given project, if the radial horizontal positional error of the GNSS is assumed to be equal to 0.11314 m (based on 0.08 m in either X or Y), and the IMU error is 0.00427 degree in roll, pitch, and heading, the following table can be used to estimate the horizontal accuracy of lidar derived elevation data.

Table B.10 provides estimated horizontal errors, in terms of $RMSE_r$, in lidar elevation data as computed by the equation in section 7.5 for different flying altitudes above mean terrain.

Different lidar systems in the market have different specifications for the GNSS and IMU and therefore, the values in Table B.10 should be modified according to the equation in section 7.5.

TABLE B.10
Expected Horizontal Errors ($RMSE_r$) for Lidar Data in Terms of Flying Altitude

Altitude (m)	Positional $RMSE_r$ (cm)		Altitude (m)	Positional $RMSE_r$ (cm)
500	13.1		3,000	41.6
1,000	17.5		3,500	48.0
1,500	23.0		4,000	54.5
2,000	29.0		4,500	61.1
2,500	35.2		5,000	67.6

B.9 ELEVATION DATA ACCURACY VERSUS ELEVATION DATA QUALITY

In aerial photography and photogrammetry, the accuracy of the individual points in a data set is largely dependent on the scale and resolution of the source imagery. Larger-scale imagery, flown at a lower altitude, produces smaller GSDs and higher measurement accuracies (both vertical and horizontal). Users have quite naturally come to equate higher density imagery (smaller GSD or smaller pixel sizes) with higher accuracies and higher quality.

In airborne topographic lidar, this is not entirely the case. For many typical lidar collections, the maximum accuracy attainable, theoretically, is now limited by physical error budgets of the different components of the lidar system such as laser ranging, the GNSS, the IMU, and the encoder systems. Increasing the density of points does not change those factors. Beyond the physical error budget limitations, all data must also be properly controlled, calibrated, boresighted, and processed. Errors introduced during any of these steps will affect the accuracy of the data, regardless of how dense the data are. That said, high density lidar data are usually of higher *quality* than low density data, and the increased quality can manifest as *apparently* higher accuracy.

In order to accurately represent a complex surface, denser data are necessary to capture the surface details for accurate mapping of small linear features such as curbs and micro drainage features, for example. The use of denser data for complex surface representation does not make the individual lidar measurements any more accurate, but does improve the accuracy of the derived surface at locations between the lidar measurements (as each reach between points is shorter).

In vegetated areas, where many lidar pulses are fully reflected before reaching the ground, a higher density data set tends to be more accurate because more points will penetrate through vegetation to the ground. More ground points will result in less interpolation between points and improved surface definition because more characteristics of the actual ground surface are being measured, not interpolated. The use of more ground points is more critical in variable or complex surfaces, such as mountainous terrain, where generalized interpolation between points would not accurately model all of the changes in the surface.

Increased density may not improve the accuracy in flat, open terrain where interpolation between points would still adequately represent the ground surface. However, in areas where denser data may not be necessary to improve the vertical accuracy of data, a higher density data set may still improve the *quality* of the data by adding additional detail to the final surface model, by better detection of edges for breaklines, and by increasing the confidence of the relative accuracy in swath overlap areas through the reduction of interpolation existing within the data set. When lidar intensity is to be used in product derivation or algorithms, high collection density is always useful.

Annex C
Accuracy Testing and Reporting Guidelines (Normative)

When errors are normally distributed, accuracy testing can be performed with RMSE values, standard deviations, mean errors, maximum and minimum errors, and unitless skew and kurtosis values. When errors are not normally distributed, alternative methods must be used. If the number of test points (checkpoints) is sufficient, testing and reporting can be performed using 95th percentile errors. A percentile rank is the percentage of errors that fall at or below a given value. Errors are visualized with histograms that show the pattern of errors relative to a normal error distribution.

The ability of RMSE, 95th percentile, or any other statistic to estimate accuracy at the 95% confidence level is largely dependent on the number and accuracy of the checkpoints used to test the accuracy of a dataset being evaluated. Whereas 100 or more is a desirable number of checkpoints, that number of checkpoints may be impractical and unaffordable for many projects, especially small project areas.

C.1 CHECKPOINT REQUIREMENTS

Both the total number of points and spatial distribution of checkpoints play an important role in the accuracy evaluation of any geospatial data. Prior guidelines and accuracy standards typically specify the required number of checkpoints and, in some cases, the land-cover types, but defining and/or characterizing the spatial distribution of the points was not required. While characterizing the point distribution is not a simple process and no practical method is available at this time, characterizing the point distribution by some measure and, consequently, providing a quality number is undoubtedly both realistic and necessary. ASPRS encourages research into this topic, peer reviewed and published in Photogrammetric Engineering & Remote Sensing for public testing and comment.

Until a quantitative characterization and specification of the spatial distribution of checkpoints across a project is developed, more general methods of determining an appropriate checkpoint distribution must be implemented. In the interim, this Annex provides general recommendations and guidelines related to the number of checkpoints, distribution across land cover types, and spatial distribution.

C.2 NUMBER OF CHECKPOINTS REQUIRED

Table C.1 lists ASPRS recommendations for the number of checkpoints to be used for vertical and horizontal accuracy testing of elevation data sets and for horizontal accuracy testing of digital orthoimagery and planimetric data sets.

TABLE C.1

Recommended Number of Checkpoints Based on Area

Project Area (Square Kilometers)	Horizontal Accuracy Testing of Orthoimagery and Planimetrics	Vertical and Horizontal Accuracy Testing of Elevation Data sets		
	Total Number of Static 2D/3D Checkpoints (clearly-defined points)	Number of Static 3D Checkpoints in NVA[9]	Number of Static 3D Checkpoints in VVA	Total Number of Static 3D Checkpoints
≤500	20	20	5	25
501–750	25	20	10	30
751–1000	30	25	15	40
1001–1250	35	30	20	50
1251–1500	40	35	25	60
1501–1750	45	40	30	70
1751–2000	50	45	35	80
2001–2250	55	50	40	90
2251–2500	60	55	45	100

Using metric units, ASPRS recommends 100 static vertical checkpoints for the first 2500 square kilometer area within the project, which provides a statistically defensible number of samples on which to base a valid vertical accuracy assessment.

For horizontal testing of areas >2500 km², clients should determine the number of additional horizontal checkpoints, if any, based on criteria such as resolution of imagery and extent of urbanization.

For vertical testing of areas >2500 km², add 5 additional vertical checkpoints for each additional 500 km² area. Each additional set of 5 vertical checkpoints for 500 km² would include 3 checkpoints for NVA and 2 for VVA. The recommended number and distribution of NVA and VVA checkpoints may vary depending on the importance of different land cover categories and client requirements.

C.3   DISTRIBUTION OF VERTICAL CHECKPOINTS ACROSS LAND COVER TYPES

In contrast to the recommendations in Table C.1, both the 2003 and the current FEMA guidelines reference the five general land cover types, and specify a minimum of 20 checkpoints in each of three to five land cover categories as they exist within the project area, for a total of 60–100 checkpoints. Under the current FEMA guidelines, this quantity applies to each 5,180 square kilometer (2000 square mile) area, or partial area, within the project.

---

[9] Although vertical checkpoints are normally not well defined, where feasible, the horizontal accuracy of lidar data sets should be tested by surveying approximately half of all NVA checkpoints at the ends of paint stripes or other point features that are visible and can be measured on lidar intensity returns.

ASPRS recognizes that some project areas are primarily non-vegetated, whereas other areas are primarily vegetated. For these reasons, the distribution of checkpoints can vary based on the general proportion of vegetated and non-vegetated area in the project. Checkpoints should be distributed generally proportionally among the various vegetated land cover types in the project.

C.4   NSSDA METHODOLOGY FOR CHECKPOINT DISTRIBUTION (HORIZONTAL AND VERTICAL TESTING)

The NSSDA offers a method that can be applied to projects that are generally rectangular in shape and are largely non-vegetated. These methods do not apply to the irregular shapes of many projects or to most vegetated land cover types. The NSSDA specifies the following:

> "Due to the diversity of user requirements for digital geospatial data and maps, it is not realistic to include statements in this standard that specify the spatial distribution of checkpoints. Data and/or map producers must determine checkpoint locations.
>
> Checkpoints may be distributed more densely in the vicinity of important features and more sparsely in areas that are of little or no interest. When data exist for only a portion of the data set, confine test points to that areasss. When the distribution of error is likely to be nonrandom, it may be desirable to locate checkpoints to correspond to the error distribution.
>
> For a data set covering a rectangular area that is believed to have uniform positional accuracy, checkpoints may be distributed so that points are spaced at intervals of at least 10% of the diagonal distance across the data set and at least 20% of the points are located in each quadrant of the data set.
>
> **(FGDC, 1998)"**[10]

ASPRS recommends that, where appropriate and to the highest degree possible, the NSSDA method be applied to the project and incorporated land cover type areas. In some areas, access restrictions may prevent the desired spatial distribution of checkpoints across land cover types; difficult terrain and transportation limitations may make some land cover type areas practically inaccessible. Where it is not geometrically or practically applicable to strictly apply the NSSDA method, data vendors should use their best professional judgment to apply the spirit of that method in selecting locations for checkpoints.

Clearly, the recommendations in sections C.1 through C.3 offer a good deal of discretion in the location and distribution of checkpoints, and this is intentional. It would not be worthwhile to locate 50 vegetated checkpoints in a fully urbanized county such as Orange County, California; 80 non-vegetated checkpoints might be more appropriate. Likewise, projects in areas that are overwhelmingly forested with only a few small towns might support only 20 non-vegetated checkpoints.

---

[10] Federal Geographic Data Committee. (1998). FGDC-STD-007.3-1998, *Geospatial Positioning Accuracy Standards, Part 3: National Standard for Spatial Data Accuracy*, FGDC, c/o U.S. Geological Survey, www.fgdc.fgdc.gov/standards/documents/standards/accuracy/chapter3

The general location and distribution of checkpoints should be discussed between and agreed upon by the vendor and customer as part of the project plan.

C.5   VERTICAL CHECKPOINT ACCURACY

Vertical checkpoints need not be clearly-defined point features. Kinematic checkpoints (surveyed from a moving platform), which are less accurate than static checkpoints, can be used in any quantity as supplemental data, but the core accuracy assessment must be based on static surveys, consistent with NOAA Technical Memorandum NOS NGS-58, *Guidelines for Establishing GPS-Derived Ellipsoid Heights (Standards: 2 cm and 5 cm)*, or equivalent. NGS-58 establishes ellipsoid height accuracies of 5 cm at the 95% confidence level for network accuracies relative to the geodetic network, as well as ellipsoid height accuracies of 2 cm and 5 cm at the 95% confidence level for accuracies relative to local control.

As with horizontal accuracy testing, vertical checkpoints should be three times more accurate than the required accuracy of the elevation dataset being tested.

C.6   TESTING AND REPORTING OF HORIZONTAL ACCURACIES

When errors are normally distributed and the mean is small, ASPRS endorses the NSSDA procedures for testing and reporting the horizontal accuracy of digital geospatial data. The NSSDA methodology applies to most digital orthoimagery and planimetric data sets where systematic errors and bias have been appropriately removed. Accuracy statistics and examples are outlined in more detail in Annex D.

Elevation data sets do not always contain the type of well-defined points that are required for horizontal testing to NSSDA specifications. Specific methods for testing and verifying horizontal accuracies of elevation data sets depend on technology used and project design.

For horizontal accuracy testing of lidar data sets, at least half of the NVA vertical checkpoints should be located at the ends of paint stripes or other point features visible on the lidar intensity image, allowing them to double as horizontal checkpoints. The ends of paint stripes on concrete or asphalt surfaces are normally visible on lidar intensity images, as are 90-degree corners of different reflectivity, e.g., a sidewalk corner adjoining a grass surface. The data provider has the responsibility to establish appropriate methodologies, applicable to the technologies used, to verify that horizontal accuracies meet the stated requirements.

The specific testing methodology used should be identified in the metadata.

C.7   TESTING AND REPORTING OF VERTICAL ACCURACIES

For testing and reporting the vertical accuracy of digital elevation data, ASPRS endorses the *NDEP Guidelines for Digital Elevation Data*, with slight modifications from FVA, SVA and CVA procedures. This ASPRS standard reports the Nonvegetated Vertical Accuracy (NVA) at the 95% confidence level in all non-vegetated land cover categories combined and reports the Vegetated Vertical Accuracy (VVA) at the 95th percentile in all vegetated land cover categories combined.

If the vertical errors are normally distributed, the sample size sufficiently large, and the mean error is sufficiently small, ASPRS endorses NSSDA and NDEP methodologies for approximating vertical accuracies at the 95% confidence level, which applies to NVA checkpoints in all open terrain (bare soil, sand, rocks, and short grass) as well as urban terrain (asphalt and concrete surfaces) land cover categories.

In contrast, VVA is computed by using the 95th percentile of the absolute value of all elevation errors in all vegetated land cover categories combined, to include tall weeds and crops, brush lands, and lightly to fully-forested land cover categories. By testing and reporting the VVA separate from the NVA, ASPRS draws a clear distinction between non-vegetated terrain where errors typically follow a normal distribution suitable for RMSE statistical analyses, and vegetated terrain where errors do not necessarily follow a normal distribution and where the 95th percentile value more fairly estimates vertical accuracy at a 95% confidence level.

C.8   LOW CONFIDENCE AREAS

For stereo-compiled elevation datasets, photogrammetrists should capture two-dimensional closed polygons for "low confidence areas" where the bare-earth DTM may not meet the overall data accuracy requirements. Because photogrammetrists cannot see the ground in stereo beneath dense vegetation, in deep shadows or where the imagery is otherwise obscured, reliable data cannot be collected in those areas. Traditionally, contours within these obscured areas would be published as dashed contour lines. A compiler should make the determination as to whether the data being digitized is within NVA and VVA accuracies or not; areas not delineated by an obscure area polygon are presumed to meet accuracy standards. The extent of photogrammetrically derived obscure area polygons and any assumptions regarding how NVA and VVA accuracies apply to the photogrammetric data set must be clearly documented in the metadata.

Low confidence areas also occur with lidar and IFSAR where heavy vegetation causes poor penetration of the lidar pulse or radar signal. Although costs will be slightly higher, ASPRS recommends that "low confidence areas" for lidar be required and delivered as two-dimensional (2D) polygons based on the following four criteria:

1. Nominal ground point density (NGPD);
2. Cell size for the raster analysis;
3. Search radius to determine average ground point densities; and
4. Minimum size area appropriate to aggregate ground point densities and show a generalized low confidence area (minimum mapping unit).

This approach describes a raster-based analysis where the raster cell size is equal to the Search Radius listed for each Vertical Data Accuracy Class. Raster results are to be converted into polygons for delivery.

This section describes possible methods for the collection or delineation of low confidence areas in elevation datasets being created using two common paradigms. Other methodologies currently exist, and additional techniques will certainly emerge in the future. The data producer may use any method they deem suitable provided the detailed technique is clearly documented in the metadata.

Table C.2 lists the values for the above low confidence area criteria that apply to each vertical accuracy class.

Low confidence criteria and the values in Table C.2 are based on the following assumptions:

- Ground Point Density: Areas with ground point densities less than or equal to 1/4 of the recommended nominal pulse density (pulse per square meter) or twice the nominal pulse spacing are candidates for Low Confidence Areas. For example: a specification requires an NPS of 1 meter (or an NPD of 1 ppsm) but the elevation data in some areas resulted in a nominal ground point density of 0.25 point per square meter (nominal ground point spacing of 2 meters). Such areas are good candidate for "low confidence" areas.
- Raster Analysis Cell Size: Because the analysis of ground point density will most likely be raster based, the cell size at which the analysis will be performed needs to be specified. The recommendation is that the cell size equals the search radius.
- Search Radius for Computing Point Densities: Because point data are being assessed, an area must be specified in order to compute the average point density within this area. The standards recommend a search area with a radius equal to 3 * NPS (*not the Low Confidence NGPS*).This distance is small enough to allow good definition of low density areas while not being so small as to cause the project to look worse than it really is.

TABLE C.2
Low Confidence Areas

Vertical Accuracy Class	Recommended Project Min NPD (pts/m²) (Max NPS (m))	Recommended Low Confidence Min NGPD (pts/m²) (Max NGPS (m))	Search Radius and Cell Size for Computing NGPD (m)	Low Confidence Polygons Min Area (acres (m²))
1 cm	≥20 (≤0.22)	≥5 (≤0.45)	0.67	0.5 (2,000)
2.5	16 (0.25)	4 (0.50)	0.75	1 (4,000)
5	8 (0.35)	2 (0.71)	1.06	2 (8,000)
10	2 (0.71)	0.5 (1.41)	2.12	5 (20,000)
15	1 (1.0)	0.25 (2.0)	3.00	5 (20,000)
20	0.5 (1.4)	0.125 (2.8)	4.24	5 (20,000)
33.3	0.25 (2.0)	0.0625 (4.0)	6.0	10 (40,000)
66.7	0.1 (3.2)	0.025 (6.3)	9.5	15 (60,000)
100	0.05 (4.5)	0.0125 (8.9)	13.4	20 (80,000)
333.3	0.01 (10.0)	0.0025 (20.0)	30.0	25 (100,000)

- Minimum Size for Low Confidence Polygons: The areas computed with low densities should be aggregated together. Unless specifically requested by clients, structures/buildings and water should be removed from the aggregated low density polygons as these features are not true Low Confidence.

Aggregated polygons greater than or equal to the stated minimum size as provided in Table C.2 should be kept and defined as Low Confidence Polygons. In certain cases, too small an area will "checker board" the Low Confidence Areas; in other cases too large an area will not adequately define Low Confidence Area polygons. These determinations should be a function of the topography, land cover, and final use of the maps.

Acres should be used as the unit of measurement for the Low Confidence Area polygons as many agencies (USGS, NOAA, USACE, etc.) use acres as the mapping unit for required polygon collection. Approximate square meter equivalents are provided for those whose work is exclusively in the metric system. Smoothing algorithms could be applied to the Low Confidence Polygons, if desired.

There are two distinctly different types of low confidence areas:

- The first types of low confidence areas are identified by the data producer – *in advance* – where passable identification of the bare earth is expected to be unlikely or impossible. These are areas where no control or checkpoints should be located and where contours, if produced, should be dashed. They are exempt from accuracy assessment. Mangroves, swamps, and inundated wetland marshes are prime candidates for such advance delineation.
- The second types of low confidence areas are valid VVA areas, normally forests that should also be depicted with dashed contours, but where checkpoints *should* be surveyed and accuracy assessment *should* be performed. Such low confidence areas are delineated subsequent to classification and would usually be identifiable by the notably reduced density of bare-earth points.

Providing Low Confidence Area polygons allows lidar data providers to protect themselves from unusable/unfair checkpoints in swamps and protects the customer from data providers who might try to alter their data.

If reliable elevation data in low confidence areas is critical to a project, it is common practice to supplement the remote sensing data with field surveys.

C.9   ERRONEOUS CHECKPOINTS

Occasionally, a checkpoint may be erroneous or inappropriate for use at no fault of the lidar survey. Such a point may be removed from the accuracy assessment calculation:

- if it is demonstrated, with pictures and descriptions, that the checkpoint was improperly located, such as when a vertical checkpoint is on steep terrain or within a few meters of a significant breakline that redefines the slope of the area being interpolated surrounding the checkpoint;

- if it is demonstrated and documented that the topography has changed significantly between the time the elevation data were acquired and the time the checkpoint was surveyed; or
- if (a) the point is included in the survey and accuracy reports, but not the assessment calculation, with pictures and descriptions; (b) reasonable efforts to correct the discrepancy are documented, e.g., rechecked airborne GNSS and IMU data, rechecked point classifications in the area, rechecked the ground checkpoints; and (c) a defensible explanation is provided in the accuracy report for discarding the point.
- An explanation that the error exceeds three times the standard deviation ($>3 *s$) is NOT a defensible explanation.

C.10 RELATIVE ACCURACY COMPARISON POINT LOCATION AND CRITERIA FOR LIDAR SWATH-TO-SWATH ACCURACY ASSESSMENT

To the greatest degree possible, relative accuracy testing locations should meet the following criteria:

1. include all overlap areas (sidelap, endlap, and cross flights);
2. be evenly distributed throughout the full width and length of each overlap area;
3. be located in non-vegetated areas (clear and open terrain and urban areas);
4. be at least 3 meters away from any vertical artifact or abrupt change in elevation;
5. be on uniform slopes; and,
6. be within the geometrically reliable portion of both swaths (excluding the extreme edge points of the swaths). For lidar sensors with zigzag scanning patterns from oscillating mirrors, the geometrically reliable portion excludes about 5% (2.5% on either side); lidar sensors with circular or elliptical scanning patterns are generally reliable throughout.

While the $RMSD_z$ value may be calculated from a set of specific test location points, the Maximum Difference requirement is not limited to these locations; it applies to all locations within the entire data set that meet the above criteria.

C.11 INTERPOLATION OF ELEVATION REPRESENTED SURFACE FOR CHECKPOINT COMPARISONS

The represented surface of an elevation data set is normally a TIN (Figure C.1) or a raster DEM (Figure C.2).

Vertical accuracy testing is accomplished by comparing the elevation of the represented surface of the elevation dataset to elevations of checkpoints at the horizontal (x/y) coordinates of the checkpoints. The data set surface is most commonly represented by a TIN or raster DEM.

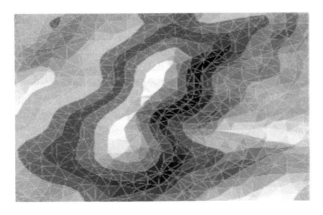

FIGURE C.1   Topographic surface represented as a TIN.

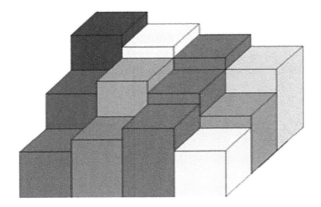

FIGURE C.2   Topographic surface represented as a raster DEM.

Vertical accuracy of point based elevation datasets should be tested by creating a TIN from the point based elevation dataset and comparing the TIN elevations to the checkpoint elevations. TINs should be used to test the vertical accuracy of point based elevation datasets because it is unlikely a checkpoint will be located at the location of a discrete elevation point. The TIN methodology is the most commonly used method used for interpolating elevations from irregularly spaced point data. Other potentially more accurate methods of interpolation exist and could be addressed by future versions of this standard as they become more commonly used and accepted.

Vertical accuracy of raster DEMs should be tested by comparing the elevation of the DEM, which is already a continuous surface, to the checkpoint elevations. For most DEM datasets, it is recommended that the elevation of the DEM is determined by extracting the elevation of the pixel that contains the x/y coordinates of the checkpoint. However, in some instances, such as when the DEM being tested is at a lower resolution typical of global datasets or when the truth data has an area footprint associated with it rather than a single x/y coordinate, it may be better to use interpolation methods to determine the elevation of the DEM dataset. Vendors should seek approval from clients if methods other than extraction are to be used to determine elevation values of the DEM dataset. Vertical accuracy testing methods listed in metadata and reports should state if elevation values were extracted from the tested dataset at the x/y location of the checkpoints or if further interpolation was used after the creation of the tested surface (TIN or raster) to determine the elevation of the tested dataset. If further interpolation was used, the interpolation method and full process used should be detailed accordingly.

Annex D
Accuracy Statistics and Example (Normative)

D.1 NSSDA REPORTING ACCURACY STATISTICS

The National Standard for Spatial Data Accuracy (NSSDA) documents the equations for computation of $RMSE_x$, $RMSE_y$, $RMSE_r$ and $RMSE_z$, as well as horizontal (radial) and vertical accuracies at the 95% confidence levels, $Accuracy_r$ and $Accuracy_z$, respectively. These statistics assume that errors approximate a normal error distribution and that the mean error is small relative to the target accuracy.

EXAMPLE ON THE NSSDA ACCURACY COMPUTATIONS

For the purposes of demonstration, suppose you have five checkpoints to verify the final horizontal and vertical accuracy for a data set (normally a minimum of 20 points would be needed). Table D.1 provides the map-derived coordinates and the surveyed coordinated for the five points. The table also shows the computed accuracy and other necessary statistics. In this abbreviated example, the data are intended to meet a horizontal accuracy class with a maximum $RMSE_x$ and $RMSE_y$ of 15 cm and the 10 cm vertical accuracy class.

COMPUTATION OF MEAN ERRORS IN X/Y/Z

$$\bar{x} = \frac{1}{(n)} \sum_{i=1}^{n} x_i$$

where:
   x_i is the ith error in the specified direction,
   n is the number of checkpoints tested,
   i is an integer ranging from 1 to n.

Mean error in Easting:

$$\bar{x} = \frac{-0.140 - 0.100 + 0.017 - 0.070 + 0.130}{5} = -0.033\,\text{m}$$

Mean error in Northing:

$$\bar{y} = \frac{-0.070 - 0.100 - 0.070 + 0.150 + 0.120}{5} = 0.006\,\text{m}$$

TABLE D.1
NSSDA Accuracy Statistics for Example Data Set with 3D Coordinates

Point ID	Map-Derived Values Easting (E)	Northing (N)	Elevation (H)	Surveyed Checkpoint Values Easting (E)	Northing (N)	Elevation (H)	Residuals (Errors) Δx (Easting)	Δy (Northing)	Δz (Elevation)
	meters	meters	meters	meters	meters	meters	meters	meters	meters
GCP1	359584.394	5142449.934	477.127	359584.534	5142450.004	477.198	−0.140	−0.070	−0.071
GCP2	359372.190	5147939.180	412.406	359872.290	5147939.280	412.396	−0.100	−0.100	0.010
GCP3	359393.089	5136979.824	487.292	359893.072	5136979.894	487.190	0.017	−0.070	0.102
GCP4	359927.194	5151084.129	393.591	359927.264	5151083.979	393.691	−0.070	0.150	−0.100
GCP5	372737.074	5151675.999	451.305	372736.944	5151675.879	451.218	0.130	0.120	0.087
					Number of checkpoints		5	5	5
					Mean Error (m)		−0.033	0.006	0.006
					Standard Deviation (m)		0.108	0.119	0.091
					RMSE (m)		0.102	0.106	0.081
					RMSEr (m)		0.147	$= \mathrm{SQRT}\left(\mathrm{RMSE}_x^2 + \mathrm{RMSE}_y^2\right)$	
				NSSDA Horizontal Accuracyr (ACCr) at 95% Confidence Level			0.255	$= \mathrm{RMSEr} \times 1.7308$	
				NSSDA Vertical Accuracyz (ACCz) at 95% Confidence Level			0.160	$= \mathrm{RMSEz} \times 1.9600$	

Mean error in Elevation:

$$\bar{z} = \frac{-0.070 + 0.010 + 0.102 - 0.100 + 0.087}{5} = 0.006\,\text{m}$$

COMPUTATION OF SAMPLE STANDARD DEVIATION

$$s_x = \sqrt{\frac{1}{(n-1)} \sum_{i=1}^{n} (x_i - \bar{x})^2}$$

where:

x_i is the *ith* error in the specified direction,

$\bar{x}$ is the mean error in the specified direction,

n is the number of checkpoints tested,

i is an integer ranging from 1 to n.

Sample Standard Deviation in Easting:

$$s_x = \sqrt{\frac{\begin{array}{l}\left(-0.140-(-0.033)\right)^2 + \left(-0.100-(-0.033)\right)^2 + \left(0.017-(-0.033)\right)^2 \\ +\left(-0.070-(-0.033)\right)^2 + \left(0.130-(-0.033)\right)^2\end{array}}{(5-1)}}$$

$$= 0.108\,\text{m}$$

Sample Standard Deviation in Northing:

$$s_y = \sqrt{\frac{\begin{array}{l}\left(-0.070-0.006\right)^2 + \left(-0.100-0.006\right)^2 + \left(-0.070-0.006\right)^2 \\ +\left(0.150-0.006\right)^2 + \left(0.120-0.006\right)^2\end{array}}{(5-1)}}$$

$$= 0.119\,\text{m}$$

Sample Standard Deviation in Elevation:

$$s_z = \sqrt{\frac{\begin{array}{l}\left(-0.071-0.006\right)^2 + (0.010-0.006)^2 + \left(0.102-0.006\right)^2 \\ +(-0.100-0.006)^2 + (0.087-0.006)^2\end{array}}{(5-1)}}$$

$$= 0.091\,\text{m}$$

COMPUTATION OF ROOT MEAN SQUARES ERROR

$$\text{RMSE}_x = \sqrt{\frac{1}{n}\sum_{i=1}^{n}\left(x_{i(\text{map})} - x_{i(\text{surveyed})}\right)^2}$$

where:

$x_{i(\text{map})}$ is the coordinate in the specified direction of the *ith* checkpoint in the data set,

$x_{i(\text{surveyed})}$ is the coordinate in the specified direction of the *ith* checkpoint in the independent source of higher accuracy,

n is the number of checkpoints tested,

i is an integer ranging from 1 to n.

$$\text{RMSE}_x = \sqrt{\frac{(-0.140)^2 + (-0.100)^2 + (0.017)^2 + (-0.070)^2 + (0.130)^2}{5}} = 0.102\,\text{m}$$

$$\text{RMSE}_y = \sqrt{\frac{(-0.070)^2 + (-0.100)^2 + (-0.070)^2 + (0.150)^2 + (0.120)^2}{5}} = 0.107\,\text{m}$$

$$\text{RMSE}_z = \sqrt{\frac{(-0.071)^2 + (0.010)^2 + (0.102)^2 + (-0.100)^2 + (0.087)^2}{5}} = 0.081\,\text{m}$$

$$\text{RMSE}_r = \sqrt{\text{RMSE}_x^2 + \text{RMSE}_y^2}$$

$$\text{RMSE}_r = \sqrt{(0.102)^2 + (0.107)^2} = 0.147\,\text{m}$$

COMPUTATION OF NSSDA ACCURACY AT 95% CONFIDENCE LEVEL

(Note: There are no significant systematic biases in the measurements. The mean errors are all smaller than 25% of the specified RMSE in Northing, Easting, and Elevation.)

Positional Horizontal Accuracy at 95% Confidence Level

$$= 2.4477\left(\frac{\text{RMSE}_r}{1.4142}\right) = 1.7308\left(\text{RMSE}_r\right)$$

$$= 1.7308\left(0.147\right) = \mathbf{0.255\,m}$$

$$\text{Vertical Accuracy at 95\% Confidence Level}$$
$$= 1.9600\left(\text{RMSE}_z\right) = 1.9600\left(0.081\right) = \mathbf{0.160\,m}$$

D.2   COMPARISON WITH NDEP VERTICAL ACCURACY STATISTICS

Whereas the NSSDA assumes that systematic errors have been eliminated as best as possible and that all remaining errors are random errors that follow a normal distribution, the ASPRS standard recognizes that elevation errors, especially in dense vegetation, do not necessarily follow a normal error distribution, as demonstrated by the error histogram of 100 checkpoints at Figure D.1 used as an example elevation data set for this Annex.

In vegetated land cover categories, the ASPRS standard (based on NDEP vertical accuracy statistics) uses the 95th percentile errors because a single outlier, when squared in the RMSE calculation, will unfairly distort the tested vertical accuracy statistic at the 95% confidence level. Unless errors can be found in the surveyed checkpoint, or the location of the checkpoint does not comply with ASPRS guidelines for location of vertical checkpoints, such outliers should not be discarded. Instead, such outliers should be included in the calculation of the 95th percentile because: (a) the outliers help identify legitimate issues in mapping the bare-earth terrain in dense vegetation, and (b) the 95th percentile, by definition, identifies that 95% of errors in the data set have errors with respect to true ground elevation that are equal to or smaller than the 95th percentile – the goal of the NSSDA.

EXAMPLE ELEVATION DATA SET

Figure D.1, plus Tables D.2 and D.3, refer to an actual elevation data set tested by prior methods compared to the current ASPRS standard.

Figure D.1 shows an actual error histogram resulting from 100 checkpoints, 20 each in five land cover categories: (1) open terrain, (2) urban terrain, concrete and asphalt, (3) tall weeds and crops, (4) brush lands and trees, and (5) fully forested. In this lidar example, the smaller outlier of 49 cm is in tall weeds and crops, and the larger outlier of 70 cm is in the fully forested land cover category. The remaining 98 elevation error values appear to approximate a normal error distribution with a mean error close to zero; therefore, the sample standard deviation and RMSE values are nearly identical. When mean errors are not close to zero, the sample standard deviation values will normally be smaller than the RMSE values.

Without considering the 95th percentile errors, traditional accuracy statistics, which preceded these *ASPRS Positional Accuracy Standards for Digital Geospatial Data*, would be as shown in Table D.2. Note that the maximum error, skewness (γ_1),kurtosis (γ_2), standard deviation and RMSE$_z$ values are somewhat higher for weeds and crops because of the 49 cm outlier, and they are much higher for the fully forested land cover category because of the 70 cm outlier.

The ASPRS standards listed in Table 7.5 define two new terms: Non-vegetated Vertical Accuracy (NVA) based on RMSE$_z$ statistics and Vegetated Vertical Accuracy (VVA) based on 95th percentile statistics. The NVA consolidates the NDEP's

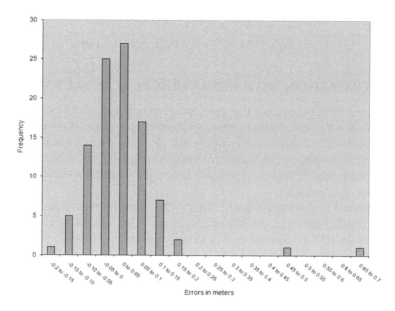

FIGURE D.1   Error histogram of typical elevation data set, showing two outliers in vegetated areas.

TABLE D.2
Traditional Error Statistics for Example Elevation Data Set

Land Cover Category	# of Check points	Min (m)	Max (m)	Mean (m)	Mean Absolute (m)	Median (m)	γ_1	γ_2	s (m)	RMSE$_z$ (m)
Open Terrain	20	−0.10	0.08	−0.02	0.04	0.00	−0.19	−0.64	0.05	0.05
Urban Terrain	20	−0.15	0.11	0.01	0.06	0.02	−0.84	0.22	0.07	0.07
Weeds & Crops	20	−0.13	0.49	0.02	0.08	−0.01	2.68	9.43	0.13	0.13
Brush Lands	20	−0.10	0.17	0.04	0.06	0.04	−0.18	−0.31	0.07	0.08
Fully Forested	20	−0.13	0.70	0.03	0.10	0.00	3.08	11.46	0.18	0.17
Consolidated	100	−0.15	0.70	0.02	0.07	0.01	3.18	17.12	0.11	0.11

non-vegetated land cover categories (open terrain and urban terrain, in this example), whereas the VVA consolidates the NDEP's vegetated land cover categories (weeds and crops, brush lands, and fully forested, in this example). Table D.3 shows ASPRS statistics and reporting methods compared to both NSSDA and NDEP.

TABLE D.3

Comparison of NSSDA, NDEP and ASPRS Statistics for Example Elevation Data Set

Land Cover Category	NSSDA Accuracy$_z$ at 95% Confidence Level Based on RMSE$_z$ * 1.9600 (m)	NDEP FVA, plus SVAs and CVA Based on the 95th Percentile (m)	NDEP Accuracy Term	ASPRS Vertical Accuracy (m)	ASPRS Accuracy Term
Open Terrain	0.10	0.10	FVA	0.12	NVA
Urban Terrain	0.14	0.13	SVA		
Weeds & Crops	0.25	0.15	SVA	0.167	VVA
Brush Lands	0.16	0.14	SVA		
Fully Forested	0.33	0.21	SVA		
Consolidated	0.22	0.13	CVA	N/A	N/A

D.3 COMPUTATION OF PERCENTILE

There are different approaches to determining percentile ranks and associated values. This standard recommends the use of the following equations for computing percentile rank and percentile as the most appropriate for estimating the Vegetated Vertical Accuracy.

Note that percentile calculations are based on the absolute values of the errors, as it is the magnitude of the errors, not the sign that is of concern.

The percentile rank (*n*) is first calculated for the desired percentile using the following equation:

$$n = \left(\left(\left(\frac{P}{100} \right) * (N-1) \right) + 1 \right)$$

where:
  n is the rank of the observation that contains the *Pth* percentile,
  P is the proportion (of 100) at which the percentile is desired (e.g., 95 for 95th percentile), and
  N is the number of observations in the sample data set.

Once the rank of the observation is determined, the percentile (Q_p) can then be interpolated from the upper and lower observations using the following equation:

$$Q_p = \left(A[n_w] + \left(n_d * \left(A[n_w + 1] - A[n_w] \right) \right) \right)$$

where:

Q_p is the *Pth* percentile; the value at rank *n*,

A is an array of the absolute values of the samples, indexed in ascending order from *1* to *N*,

$A[i]$ is the sample value of array *A* at index *i* (e.g., n_w or n_d). *i* must be an integer between *1* and *N*,

n is the rank of the observation that contains the *Pth* percentile,

n_w is the whole number component of *n* (e.g., 3 of 3.14), and

n_d is the decimal component of *n* (e.g., 0.14 of 3.14).

Example:

Given a sample data set $\{X_1, X_2 \ldots X_N\} = \{7, -33, -9, 5, -16, 22, 36, 37, 39, -11, 45, 28, 45, 19, -46, 10, 48, 44, 51, -27\}$ $(N = 20)$,

calculate the 95th percentile $(P = 95)$:

Step 1: Take the absolute value of each observation:
$\{7, 33, 9, 5, 16, 22, 36, 37, 39, 11, 45, 28, 45, 19, 46, 10, 48, 44, 51, 27\}$

Step 2: Sort the absolute values in ascending order:
$A = \{5, 7, 9, 10, 11, 16, 19, 22, 27, 28, 33, 36, 37, 39, 44, 45, 45, 46, 48, 51\}$

Step 3: Compute the percentile rank *n* for $P = 95$:

$$n = \left(\left(\left(\frac{P}{100}\right)*(N-1)\right)+1\right) = \left(\left(\left(\frac{95}{100}\right)*(20-1)\right)+1\right) = 19.05$$

The 95th percentile rank *(n)* of the sample data set is 19.05

Step 4: Compute the percentile value Q_p by interpolating between observations 19 and 20:

$$Q_p = \left(A[n_w]+\left(n_d *\left(A[n_w+1]-A[n_w]\right)\right)\right) = \left(48+\left(0.05*(51-48)\right)\right) = 48.15$$

The 95th percentile (Q_p) of the sample data set is 48.15.

Bibliography

Ager, Thomas. 2004. *An Analysis of Metric Accuracy Definitions and Methods of Computation.* Unpublished memo prepared for the National Geospatial-Intelligence Agency. InnoVision.

Aickin, M. 1990. Maximum likelihood estimation of agreement in the constant predictive probability model, and its relation to Cohen's kappa. *Biometrics.* Vol. 46, pp. 293–302.

American Society of Photogrammetry. 1960. *Manual of Photographic Interpretation.* ASP, Washington, DC.

American Society for Photogrammetry and Remote Sensing (ASPRS) Specifications and Standards Committee. 1990. ASPRS accuracy standards for large-scale maps. *Photogrammetric Engineering and Remote Sensing.* Vol. 56, No. 7, pp. 1068–1070.

Anderson, J. R., E. E. Hardy, J. T. Roach, and R. E. Witner. 1976. A land use and land cover classification system for use with remote sensor data. *USGS Professional Paper.* Vol. 964, 28 pp.

Aronoff, Stan. 1982. Classification accuracy: A user approach. *Photogrammetric Engineering and Remote Sensing.* Vol. 48, No. 8, pp. 1299–1307.

Aronoff, Stan. 1985. The minimum accuracy value as an index of classification accuracy. *Photogrammetric Engineering and Remote Sensing.* Vol. 51, No. 1, pp. 99–111.

ASPRS. 1989. ASPRS interim accuracy standards for large-scale maps. *Photogrammetric Engineering & Remote Sensing.* Vol. 54, No. 7, pp. 1038–1041.

ASPRS. 2004. *ASPRS Guidelines, Vertical Accuracy Reporting for Lidar Data.* American Society for Photogrammetry and Remote Sensing.

ASPRS. 2014. ASPRS positional accuracy standards for digital geospatial data. *Photogrammetric Engineering and Remote Sensing.* Vol. 81, No. 3, pp. A1–A26.

ASPRS and ASCE. 1994. *Glossary of the Mapping Sciences.* ASPRS, Bethesda, MD; ASCE, New York.

Biging, G. and R. Congalton. 1989. Advances in forest inventory using advanced digital imagery. Proceedings of Global Natural Resource Monitoring and Assessments: Preparing for the 21st Century. Venice, Italy. September, 1989. Vol. 3, pp. 1241–1249.

Biging, G., R. Congalton, and E. Murphy. 1991. A comparison of photointerpretation and ground measurements of forest structure. Proceedings of the Fifty-Sixth Annual Meeting of the American Society of Photogrammetry and Remote Sensing. Baltimore, MD. Vol. 3, pp. 6–15.

Bishop, Y., S. Fienberg, and P. Holland. 1975. *Discrete Multivariate Analysis: Theory and Practice.* MIT Press, Cambridge, MA. 575 pp.

Bitterlich, W. 1947. Die Winkelzahlmessung (Measurement of basal area per hectare by means of angle measurement). *Allg. Forst. Holzwirtsch. Ztg.* Vol. 58, pp. 94–96.

Blaschke, T. 2010. Object based image analysis for remote sensing. *ISPRS Journal of Photogrammetry and Remote Sensing.* Vol. 65, pp. 2–16.

Bolstad, Paul. 2005. *GIS Fundamentals.* 2nd edition. Eider Press, White Bear Lake, MN. 543 pp.

Bolstad, Paul. 2016. *GIS Fundamentals: A First Text on Geographic Information Systems.* 5th Edition. XanEdu Publishers, Ann Arbor, MI. 770 pp.

Brennan, R. and D. Prediger. 1981. Coefficient kappa: Some uses, misuses, and alternatives. *Educational and Psychological Measurement.* Vol. 41, pp. 687–699.

Brothers, G. L. and E. B. Fish. 1978. Image enhancement for vegetation pattern change analysis. *Photogrammetric Engineering and Remote Sensing.* Vol. 44, No. 5, pp. 607–616.

Campbell, James B. 1981. Spatial autocorrelation effects upon the accuracy of supervised classification of land cover. *Photogrammetric Engineering and Remote Sensing.* Vol. 47, No. 3, pp. 355–363.

Card, D. H. 1982. Using known map categorical marginal frequencies to improve estimates of thematic map accuracy. *Photogrammetric Engineering and Remote Sensing.* Vol. 48, No. 3, pp. 431–439.

Chrisman, N. 1982. Beyond accuracy assessment: Correction of misclassification. Proceedings of the 5th International Symposium on Computer-Assisted Cartography. Crystal City, VA. pp. 123–132.

Cliff, A. D. and J. K. Ord. 1973. *Spatial Autocorrelation.* Pion Limited, London, England. 178 pp.

Cochran, William G. 1977. *Sampling Techniques.* John Wiley & Sons, New York. 428 pp.

Cohen, Jacob. 1960. A coefficient of agreement for nominal scales. *Educational and Psychological Measurement.* Vol. 20, No. 1, pp. 37–40.

Cohen, Jacob. 1968. Weighted kappa: Nominal scale agreement with provision for scaled disagreement or partial credit. *Psychological Bulletin.* Vol. 70, No. 4, pp. 213–220.

Colwell, R. N. 1955. The PI picture in 1955. *Photogrammetric Engineering.* Vol. 21, No. 5, pp. 720–724.

Congalton, Russell G. 1981. The use of discrete multivariate analysis for the assessment of Landsat classification accuracy. MS Thesis, Virginia Polytechnic Institute and State University, Blacksburg, VA. 111 pp.

Congalton, R. G. 1984. A comparison of five sampling schemes used in assessing the accuracy of land cover/land use maps derived from remotely sensed data. PhD Dissertation, Virginia Polytechnic Institute and State University, Blacksburg, VA. 147 pp.

Congalton, R. G. 1988a. Using spatial autocorrelation analysis to explore errors in maps generated from remotely sensed data. *Photogrammetric Engineering and Remote Sensing.* Vol. 54, No. 5, pp. 587–592.

Congalton, R. G. 1988b. A comparison of sampling schemes used in generating error matrices for assessing the accuracy of maps generated from remotely sensed data. *Photogrammetric Engineering and Remote Sensing.* Vol. 54, No. 5, pp. 593–600.

Congalton, R. 1991. A review of assessing the accuracy of classifications of remotely sensed data. *Remote Sensing of Environment.* Vol. 37, pp. 35–46.

Congalton, R. 2009. Accuracy and error analysis of global and local maps: Lessons learned and future considerations. In: *Remote Sensing of Global Croplands for Food Security.* P. Thenkabail, J. Lyon, H. Turral, and C. Biradar (Editors). CRC/Taylor & Francis, Boca Raton, FL. pp. 441–458.

Congalton, R. 2015. Assessing positional and thematic accuracies of maps generated from remotely sensed data. In: *Remote Sensing Handbook; Vol. I: Data Characterization, Classification, and Accuracies.* P. Thenkabail (Editor). CRC/Taylor & Francis, Boca Raton, FL. pp. 583–601.

Congalton, R. and G. Biging. 1992. A pilot study evaluating ground reference data collection efforts for use in forest inventory. *Photogrammetric Engineering and Remote Sensing.* Vol. 58, No. 12, pp. 1669–1671.

Congalton, R. and M. Brennan. 1998. Change detection accuracy assessment: Pitfalls and considerations. Proceedings of the Sixty Fourth Annual Meeting of the American Society of Photogrammetry and Remote Sensing. Tampa, Florida. pp. 919–932 (CD-ROM).

Congalton, R. and M. Brennan. 1999. Error in remotely sensed data analysis: Evaluation and reduction. Proceedings of the Sixty Fifth Annual Meeting of the American Society of Photogrammetry and Remote Sensing. Portland, OR. pp. 729–732 (CD-ROM).

Congalton, R. and K. Green. 1993. A practical look at the sources of confusion in error matrix generation. *Photogrammetric Engineering and Remote Sensing.* Vol. 59, No. 5, pp. 641–644.

Congalton, R. and K. Green. 2009. *Assessing the Accuracy of Remotely Sensed Data: Principles and Practices.* 2nd edition. CRC/Taylor & Francis, Boca Raton, FL. 183 pp.

Congalton, R. G. and R. D. Macleod. 1994. Change detection accuracy assessment on the NOAA Chesapeake Bay pilot study. Proceedings of the International Symposium of Spatial Accuracy of Natural Resource Data Bases. Williamsburg, VA. pp. 78–87.

Congalton, R. G. and R. A. Mead. 1983. A quantitative method to test for consistency and correctness in photo-interpretation. *Photogrammetric Engineering and Remote Sensing.* Vol. 49, No. 1, pp. 69–74.

Congalton, R. and R. Mead. 1986. A review of three discrete multivariate analysis techniques used in assessing the accuracy of remotely sensed data from error matrices. *IEEE Transactions of Geoscience and Remote Sensing.* Vol. GE-24, No. 1, pp. 169–174.

Congalton, R. G., R. G. Oderwald, and R. A. Mead. 1983. Assessing Landsat classification accuracy using discrete multivariate statistical techniques. *Photogrammetric Engineering and Remote Sensing.* Vol. 49, No. 12, pp. 1671–1678.

Cowardin, L. M., V. Carter, F. Golet, and E. LaRoe. 1979. *A Classification of Wetlands and Deepwater Habitats of the United States.* Office of Biological Services. U.S. Fish and Wildlife Service. U.S. Department of Interior, Washington, DC. 103 pp.

CropScape—Cropland Data Layer, United States Department of Agriculture, National Agricultural Statistics Service. Available online: https://nassgeodata.gmu.edu/CropScape

Czaplewski, R. 1992. Misclassification bias in aerial estimates. *Photogrammetric Engineering and Remote Sensing.* Vol. 58, No. 2, pp. 189–192.

Czaplewski, R. and G. Catts. 1990. Calibrating area estimates for classification error using confusion matrices. Proceedings of the 56th Annual Meeting of the American Society for Photogrammetry and Remote Sensing. Denver, CO. Vol. 4, pp. 431–440.

Defense Mapping Agency). 1991. Error theory as applied to mapping, charting, and geodesy. Defense Mapping Agency Technical Report 8400.1. Fairfax, Virginia. 71 pages plus appendices.

Environmental Systems Research Institute, National Center for Geographic Information and Analysis, and The Nature Conservancy. 1994. Accuracy Assessment Procedures: NBS/NPS Vegetation Mapping Program. Report prepared for the National Biological Survey and National Park Service. Redlands, CA, Santa Barbara, CA, and Arlington, VA, United States.

Eyre, F. H. 1980. *Forest Cover Types of the United States and Canada.* Society of American Foresters, Washington, DC. 148 pp.

FEMA (Federal Emergency Management Agency). 2003. *Guidelines and Specifications for Flood Hazard Mapping Partners.*

FEMA, https://www.fema.gov/media-library/assets/documents/13948

FGDC (Federal Geographic Data Committee), Subcommittee for Base Cartographic Data. 1998. *Geospatial Positioning Accuracy Standards. Part 3: National Standard for Spatial Data Accuracy.* FGDC-STD-007.3-1998: Washington, DC, Federal Geographic Data Committee. 24 pp.

Ferris State University. 2007. http://www.ferris.edu/faculty/burtchr/sure340/notes/History.pdf

Fitzpatrick-Lins, K. 1981. Comparison of sampling procedures and data analysis for a land-use and land-cover map. *Photogrammetric Engineering and Remote Sensing.* Vol. 47, No. 3, pp. 343–351.

Fleiss, J., J. Cohen, and B. Everitt. 1969. Large sample standard errors of kappa and weighted kappa. *Psychological Bulletin.* Vol. 72, No. 5, pp. 323–327.

Foody, G. 1992. On the compensation for chance agreement in image classification accuracy assessment. *Photogrammetric Engineering and Remote Sensing.* Vol. 58, No. 10, pp. 1459–1460.

Foody, G. M. 2009. Sample size determination for image classification accuracy assessment and comparison. *International Journal of Remote Sensing*. Vol. 30, No. 20, pp. 5273–5291.

Freese, Frank. 1960. Testing accuracy. *Forest Science*. Vol. 6, No. 2, pp. 139–145.

Ginevan, M. E. 1979. Testing land-use map accuracy: Another look. *Photogrammetric Engineering and Remote Sensing*. Vol. 45, No. 10, pp. 1371–1377.

Gong, P. and J. Chen. 1992. Boundary uncertainties in digitized maps: Some possible determination methods. In: Proceedings of GIS/LIS'92. Annual Conference and Exposition. San Jose, CA. pp. 274–281.

Goodman, Leo. 1965. On simultaneous confidence intervals for multinomial proportions. *Technometrics*. Vol. 7, pp. 247–254.

Gopal, S. and C. Woodcock. 1994. Theory and methods for accuracy assessment of thematic maps using fuzzy sets. *Photogrammetric Engineering and Remote Sensing*. Vol. 60, No. 2, pp. 181–188.

Grassia, A. and R. Sundberg. 1982. Statistical precision in the calibration and use of sorting machines and other classifiers. *Technometrics*. Vol. 24, pp. 117–121.

Green, K. and R. Congalton. 2004. An error matrix approach to fuzzy accuracy assessment: The NIMA Geocover project. A peer-reviewed chapter. In: *Remote Sensing and GIS Accuracy Assessment*. R. S. Lunetta and J. G. Lyon (Editors). CRC Press, Boca Raton, FL. 304 pp.

Green, K., K. Schulz, C. Lopez, et al. 2015. Vegetation Mapping Inventory Project: Haleakalā National Park. National Park Service. Fort Collins, Colorado. Natural Resource Report NPS/PACN/NRR2015/986. https://irma.nps.gov/App/Reference/DownloadDigitalFile?code=525341&file=halerpt.pdf

Green, K., R. G. Congalton, and M. Tukman. 2017. *Imagery and GIS: Best Practices for Extracting Information from Imagery*. ESRI Press, Redlands, CA.

Greenwalt, Clyde and Melvin Schultz. 1962 and 1968. *Principles of Error Theory and Cartographic Applications*. United States Air Force. Aeronautical Chart and Information Center. ACIC Technical Report Number 96. St. Louis, MO. 60 pages plus appendices. This report is cited in the ASPRS standards as ACIC, 1962.

Griffin, J. and W. Critchfield. 1972. The distribution of forest trees in California. USDA Forest Service Research Paper PSW-82. Pacific Southwest Forest and Range Experiment Station, Berkeley, CA.

Hay, A. M. 1979. Sampling designs to test land-use map accuracy. *Photogrammetric Engineering and Remote Sensing*. Vol. 45, No. 4, pp. 529–533.

Hay, A. M. 1988. The derivation of global estimates from a confusion matrix. *International Journal of Remote Sensing*. Vol. 9, pp. 1395–1398.

Hill, T. B. 1993. Taking the " " out of "ground truth": Objective accuracy assessment. In: Proceedings of the 12th Pecora Conference. Sioux Falls, SD. pp. 389–396.

Hopkirk, P. 1992. *The Great Game. The Struggle for Empire in Central Asia*. Kodansha International. 565 pp.

Hord, R. M. and W. Brooner. 1976. Land-use map accuracy criteria. *Photogrammetric Engineering and Remote Sensing*. Vol. 42, No. 5, pp. 671–677.

Hudson, W. and C. Ramm. 1987. Correct formulation of the kappa coefficient of agreement. *Photogrammetric Engineering and Remote Sensing*. Vol. 53, No. 4, pp. 421–422.

Husch, B., Beers, T. W., and Kershaw, J. A., Jr. 2003. *Forest Mensuration*. 4th edition. John Wiley & Sons, Hoboken, NJ. 443 pp.

Jensen, John. 2016. *Introductory Digital Image Processing: A Remote Sensing Perspective*. 4th edition. Pearson Education, Glenview, IL. 623 pp.

Katz, A. H. 1952. Photogrammetry needs statistics. *Photogrammetric Engineering*. Vol. 18, No. 3, pp. 536–542.

Kearsley, M. J., K. Green, M. Tukman, M. Reid, M. Hall, T. Ayers and K. Christie. 2015. Grand Canyon National Park-Grand Canyon/Parashant National Monument vegetation classification and mapping project. Natural Resource Report. NPS/GRCA/NRR—2015/913. National Park Service. Fort Collins, CO. Published Report-2221240. https://irma.nps.gov/DataStore/DownloadFile/520521

Landis, J. and G. Koch. 1977. The measurement of observer agreement for categorical data. *Biometrics*. Vol. 33, pp. 159–174.

Lea, Chris and Anthony Curtis. 2010. Thematic accuracy assessment procedures. National Park Service Vegetation Inventory, Version 2.0. Natural Resources Report NPS/NRPC/NRR – 2010/204. 116 pp.

Liu, C., P. Frazier, and L. Kumar. 2007. Comparative assessment of the measures of thematic classification accuracy. *Remote Sensing of Environment*. Vol. 107, pp. 606–616.

Lopez, A., F. Javier, A. Gordo, and A. David. 2005. Sample Size and Confidence When Applying the NSSDA. XXII International Cartographic Conference (ICC2005). Hosted by The International Cartographic Association. Coruna, Spain. July 11–16, 2005.

Lowell, K. 1992. On the incorporation of uncertainty into spatial data systems. In: Proceedings of GIS/LIS'92. Annual Conference and Exposition. San Jose, CA. pp. 484–493.

Lunetta, R., R. Congalton, L. Fenstermaker, J. Jensen, K. McGwire, and L. Tinney. 1991. Remote sensing and geographic information system data integration: Error sources and research issues. *Photogrammetric Engineering and Remote Sensing*. Vol. 57, No. 6, pp. 677–687.

MacLean, M., M. Campbell, D. Maynard, M. Ducey, and R. Congalton. 2013. Requirements for labeling forest polygons in an object-based image analysis classification. *International Journal of Remote Sensing*. Vol. 34, No. 7, pp. 2531–2547.

MacLean, M. and R. Congalton. 2013. Applicability of multi-date land cover mapping using Landsat 5 TM imagery in the Northeastern US. *Photogrammetric Engineering and Remote Sensing*. Vol. 79, No. 4, pp. 359–368.

Macleod, R. and R. Congalton. 1998. A quantitative comparison of change detection algorithms for monitoring eelgrass from remotely sensed data. *Photogrammetric Engineering and Remote Sensing*. Vol. 64, No. 3, pp. 207–216.

Malila, W. 1985. Comparison of the information contents of Landsat TM and MSS data. *Photogrammetric Engineering and Remote Sensing*. Vol. 51, No. 9, pp. 1449–1457.

Massey, Richard, Temuulen T. Sankey, Kamini Yadav, Russell G. Congalton, and James Tilton. 2018. Integrating cloud-based workflows in continental-scale cropland extent classification. *Remote Sensing of Environment*. (In Review).

Maune, David (Editor). 2007. *Digital Elevation Model Technologies and Applications: The DEM Users Manual*. 2nd Edition. American Society of Photogrammetry and Remote Sensing, Bethesda, MD. 655 pp.

Mayer, K. and W. Laudenslayer (Editors). 1988. A guide to wildlife habitats in California. *California Department of Forestry and Fire Protection*. Sacramento, CA.

McGlone, J. C. (Editor). 2004. *Manual of Photogrammetry*. American Society for Photogrammetry and Remote Sensing, Bethesda, MD. 1151 pp.

McGuire, K. 1992. Analyst variability in labeling unsupervised classifications. *Photogrammetric Engineering and Remote Sensing*. Vol. 58, No. 12, pp. 1705–1709.

Mikhail, E. M. and G. Gracie. 1981. *Analysis and Adjustment of Survey Measurements*. Van Nostrand Reinhold. 340 pp.

MPLMIC. 1999. *Positional Accuracy Handbook. Using the National Standard for Spatial Data Accuracy to Measure and Report Geographic Data Quality*. Minnesota Planning land Management Information Center, St. Paul, MN. 29 pp.

National Vegetation Classification Standard. 2018. http://usnvc.org/data-standard/natural-vegetation-classification/. Site last visited June 23, 2018.

NDEP. 2004. *Guidelines for Digital Elevation Data*. Version 1.0. National Digital Elevation Program. May 10, 2004.

Pillsbury, N., M. DeLasaux, R. Pryor, and W. Bremer. 1991. Mapping and GIS database development for California's hardwood resources. California Department of Forestry and Fire Protection, Forest and Rangeland Resources Assessment Program (FRRAP). Sacramento, CA.

Pontius, R. and M. Millones. 2011. Death to Kappa: Birth of quantity disagreement and allocation disagreement for accuracy assessment. *International Journal of Remote Sensing*. Vol. 32, No. 15, pp. 4407–4429.

Prisley, S. and J. Smith. 1987. Using classification error matrices to improve the accuracy of weighted land-cover models. *Photogrammetric Engineering and Remote Sensing*. Vol. 53, No. 9, pp. 1259–1263.

Radoux, J., R. Bogaert, D. Fasbender, and P. Defourny, 2010. Thematic accuracy assessment of geographic object-based image classification. *International Journal of Geographical Information Science*. Vol. 25, No. 6, pp. 895–911.

Rhode, W. G. 1978. Digital image analysis techniques for natural resource inventories. National Computer Conference Proceedings. pp. 43–106.

Rosenfield, G. and K. Fitzpatrick-Lins. 1986. A coefficient of agreement as a measure of thematic classification accuracy. *Photogrammetric Engineering and Remote Sensing*. Vol. 52, No. 2, pp. 223–227.

Rosenfield, G. H., K. Fitzpatrick-Lins, and H. Ling. 1982. Sampling for thematic map accuracy testing. *Photogrammetric Engineering and Remote Sensing*. Vol. 48, No. 1, pp. 131–137.

Sammi, J. C. 1950. The application of statistics to photogrammetry. *Photogrammetric Engineering*. Vol. 16, No. 5, pp. 681–685.

Singh, A. 1986. Change detection in the tropical rain forest environment of northeastern India using Landsat. In: *Remote Sensing and Tropical Land Management*. M. J. Eden and J. T. Parry (Editors). John Wiley & Sons, London. pp. 237–254.

Singh, A. 1989. Digital change detection techniques using remotely sensed data. *International Journal of Remote Sensing*. Vol. 10, No. 6, pp. 989–1003.

Spurr, Stephen. 1948. *Aerial Photographs in Forestry*. Ronald Press, New York. 340 pp.

Spurr, Stephen. 1960. *Photogrammetry and Photo-Interpretation with a Section on Applications to Forestry*. Ronald Press, New York. 472 pp.

Stehman, S. 1992. Comparison of systematic and random sampling for estimating the accuracy of maps generated from remotely sensed data. *Photogrammetric Engineering and Remote Sensing*. Vol. 58, No. 9, pp. 1343–1350.

Story, M. and R. Congalton. 1986. Accuracy assessment: A user's perspective. *Photogrammetric Engineering and Remote Sensing*. Vol. 52, No. 3, pp. 397–399.

Teluguntla, P., P. Thenkabail, J. Xiong, et al. 2015. Global Food Security Support Analysis Data (GFSAD) at Nominal 1-km (GCAD) derived from remote sensing in support of food security in the twenty-first century: Current achievements and future possibilities. In: *Remote Sensing Handbook; Vol. II: Land Resources Monitoring, Modeling, and Mapping with Remote Sensing*. P. Thenkabail (Editor). CRC/Taylor & Francis, Boca Raton, FL. pp. 131–159.

Tenenbein, A. 1972. A double sampling scheme for estimating from misclassified multinomial data with applications to sampling inspection. *Technometrics*. Vol. 14, pp. 187–202.

Tortora, R. 1978. A note on sample size estimation for multinomial populations. *The American Statistician*. Vol. 32, No. 3, pp. 100–102.

U.S. Bureau of the Budget. 1941. 1947. *National Map Accuracy Standards*. Washington, DC.

United States National Vegetation Classification Database Ver. 2.02. Federal Geographic Data Committee, Vegetation Subcommittee. Washington DC. 2018. https://www.sciencebase.gov/catalog/item/5aa827a2e4b0b1c392ef337a

Van Genderen, J. L. and B. F. Lock. 1977. Testing land use map accuracy. *Photogrammetric Engineering and Remote Sensing*. Vol. 43, No. 9, pp. 1135–1137.

Van Genderen, J. L., B. F. Lock, and P. A. Vass. 1978. Remote sensing: Statistical testing of thematic map accuracy. Proceedings of the Twelfth International Symposium on Remote Sensing of Environment. ERIM. pp. 3–14.

Woodcock, C. 1996. On roles and goals for map accuracy assessment: A remote sensing perspective. Proc: Second International Symposium on Spatial Accuracy Assessment in Natural Resources and Environmental Sciences, USDA Forest Service Rocky Mountain Forest and Range Experiment Station, Gen. Tech. Rep. RM-GTR-277, Fort Collins, CO. pp. 535–540.

Woodcock, C. and S. Gopal. 1992. Accuracy assessment of the Stanislaus Forest vegetation map using fuzzy sets. In: *Remote Sensing and Natural Resource Management*. Proceedings of the 4th Forest Service Remote Sensing Conference, Orlando, FL. pp. 378–394.

Yadav, Kamini and Russell G. Congalton. 2018. Issues with large area thematic accuracy assessment for mapping cropland extent: A tale of three continents. *Remote Sensing*. Vol. 10, No. 1, p. 53. doi:10.3390/rs10010053

Young, H. E. 1955. The need for quantitative evaluation of the photo interpretation system. *Photogrammetric Engineering*. Vol. 21, No. 5, pp. 712–714.

Young, H. E. and E. G. Stoeckler. 1956. Quantitative evaluation of photo interpretation mapping. *Photogrammetric Engineering*. Vol. 22, No. 1, pp. 137–143.

Zadeh, L. A. 1965. Fuzzy sets. *Information and Control*. Vol. 8, pp. 338–353.

Zar, J. 1974. *Biostatistical Analysis*. Prentice-Hall. 620 pp.

Index